Studies in Organic Chemistry 8

SYNTHESIS OF ACETYLENES, ALLENES AND CUMULENES

A LABORATORY MANUAL

Studies in Organic Chemistry

Volume 1 Complex Hydrides, by A. Hajós
Volume 2 Proteoglycans — Biological and Chemical Aspects in Human Life, by J.F. Kennedy
Volume 3 New Trends in Heterocyclic Chemistry, edited by R.B. Mitra, N.R. Ayyangar, V.N. Gogte, R.M. Acheson and N. Cromwell
Volume 4 Inositol Phosphates: Their Chemistry, Biochemistry and Physiology, by D.J. Cosgrove
Volume 5 Comprehensive Carbanion Chemistry, edited by E. Buncel and T. Durst
Volume 6 New Synthetic Methodology and Biologically Active Substances, edited by Z.-I. Yoshida
Volume 7 Quinonediazides, by V.V. Ershov, G.A. Nikiforov and C.R.H.I. de Jonge

Studies in Organic Chemistry 8

SYNTHESIS OF ACETYLENES, ALLENES AND CUMULENES

A LABORATORY MANUAL

L. Brandsma and H.D. Verkruijsse

Laboratory of Organic Chemistry, State University Utrecht, Croesestraat 79, Utrecht 2503, The Netherlands

ELSEVIER SCIENTIFIC PUBLISHING COMPANY
Amsterdam — Oxford — New York 1981

ELSEVIER SCIENTIFIC PUBLISHING COMPANY
Molenwerf 1
P.O. Box 211, 1000 AE Amsterdam, The Netherlands

Distributors for the United States and Canada:

ELSEVIER/NORTH-HOLLAND INC.
52, Vanderbilt Avenue
New York, N.Y. 10017

Library of Congress Cataloging in Publication Data

Brandsma, L.
Synthesis of acetylenes, allenes, and cumulenes.

(Studies in organic chemistry ; 8)
Includes bibliographical references and index.
1. Acetylene compounds. 2. Allene. 3. Cumulenes.
4. Chemistry, Organic--Synthesis. I. Verkruijsse, H. D.
II. Title. III. Series: Studies in organic chemistry
(Elsevier) ; 8.
QD305.H8B84 547'.413 81-9807
ISBN 0-444-42009-6 AACR2

ISBN 0-444-42009-6 (Vol. 8)
ISBN 0-444-41737-0 (Series)

Printed in The Netherlands

CONTENTS

PREFACE

In 1971 we published "Preparative Acetylenic Chemistry", a collection of procedures for various acetylenic compounds. An important aim of this book was to make acetylenes more accessible, so that barriers for starting research in this field could be lowered.

The experimental work for the present book was started some years ago. It involved checking procedures from the literature and developing new ones, in some cases based on very concise instructions in preliminary papers. Most of the preparations in this book required more or less extensive research in order to find procedures for synthesizing the compounds on a reasonable scale. Important aspects of this research were monitoring the conversions, in order to avoid undesirably long reaction times, and searching for efficient methods of isolation. A number of methods from the literature which seemed attractive at first sight, are not exemplified in this book, either because they could not be reproduced at all by us or because they appeared to be impractical when performed on a larger scale (0.1-0.2 molar).

We have exemplified the principal methods for allenes, cumulenes and also some new methods for acetylenic compounds by procedures for relatively simple structures. For each of the compounds the most suitable and economic method was selected.

We hope that our book will stimulate research in the field of unsaturated compounds and will give rise to new ideas.

We are indebted to Prof.Dr. H.J.T. Bos for helping us over the barrier of starting this endless looking project, to Mr. S. Mommers, Mrs. A.J.M. van den Goorbergh and Mr. L. de Vries for re-checking about half of our procedures, and to Mr. R.G. Visser for reading the manuscript.

Utrecht, May 1981

L. Brandsma
H.D. Verkruijsse

Chapter I

REACTION VESSELS AND SOME SPECIAL PROCEDURES

The apparatus most often prescribed for the procedures in this book consists of a round-bottomed flask with three *vertical* necks, provided with a dropping funnel, a mechanical stirrer (glass rod with a blade) and a thermometer. It is not necessary to use a flask with more than three necks, since the inlet for nitrogen can be combined with the dropping funnel and the gas outlet with the thermometer (see Fig. 1). If, instead of dropwise addition, the reagent can be added in one portion, only a gas inlet is used. The thermometer may be replaced with a reflux condenser, having a gas outlet on the top, if no control of the temperature is necessary and the solvent is allowed to reflux (e.g. during Grignard preparations in diethyl ether).

For reactions in anhydrous liquid ammonia, performed at the boiling point (-33°C), no supply of nitrogen is necessary in most cases, since the flow of escaping gas prevents the reagents and products from being oxidized. A convenient apparatus is illustrated in Fig. 2. The liquid ammonia is introduced via a plastic tube, connected to a cylinder. This is turned upside down and, after turning on the tap, the liquid flows into the flask. As soon as the required volume has been obtained, the dropping funnel and a tube about 10 cm long with a curved gas outlet are placed on the flask (the stirrer being present already). The combination of the tube and the gas outlet is preferred to the use of a curved outlet directly connected with the flask, since during vigorous agitation of the reaction mixture some of the solvent may splash out through the outlet. If during vigorous reactions the flow of escaping ammonia is very strong, the outlet is omitted. In some cases it is necessary to hydrolyze the reaction mixture obtained in liquid ammonia, in a special way. After the conversion has been completed, the stirrer, dropping funnel and tube are removed and two rubber stoppers are placed tightly on the outer necks. In the middle neck is fitted a rubber stopper through which passes a glass tube with an internal diameter of about 5 mm reaching nearly to the bottom (distance 1-2 mm) of the flask. Prior to performing this operation the glass tube is connected to a plastic tube, the end of which lies on the bottom of a conical flask filled with finely crushed ice (see Fig. 3). The pressure of the ammonia which builds up causes the solution to flow within a short time through the tubes and to discharge under the ice. In order to effect efficient mixing, the flask with the ice is swirled continuously during this operation, which does not take longer than 2 min.

Some reactions in liquid ammonia can be, or have to be, carried out very rapidly. Since the ingress of moisture during the short reaction time is negligible, the most practical solution consists in using a relatively large, wide-necked flask with a mechanical stirrer (see Fig. 4).

An even less complicated reaction vessel may be used for reactions in liquid ammonia which produce only a small amount of "heat" over a relatively long period and which proceed under homogeneous conditions. The conversion can then be performed in a one-necked flask with a stopper + gas outlet or small hole.

It is useful to devote some comments to the following work-up procedure, which is often prescribed in the case of compounds, prepared in THF, that have a volatility comparable to that of THF (b.p. between 35 and 100°C at normal pressure). The problem is to remove the THF, the amount of which exceeds manyfold that of the product, and yet to isolate all of the volatile product which has been formed. The procedure to be followed consists in adding a high-boiling solvent (b.p. > 170°C at 760 mmHg) to the reaction mixture and subsequently extracting with water or, more efficiently, with dilute hydrochloric acid (if the compound is stable in acidic medium), until no more THF is present in the organic layer. As a rule this requires washing several times. The many aqueous layers obtained in this operation, may contain some of the product, but even this can be isolated by shaking the combined washings once or twice with the extraction solvent and subsequently removing the THF from these extracts by washing 5-6 times with water or dilute acid. It should be pointed out that the partition coefficient of the *product* between the extraction solvent and water (or the acid) must be large, but that of *THF* and the mentioned phases in the order indicated must be small. High-boiling ethers (e.g. diisoamyl ether) therefore seem unsuitable, and hydrocarbons (light petroleum fractions) are the solvents of choice. The further isolation procedure involves heating the dried extract *in vacuo* and condensing the vapour of the product in a strongly cooled receiver (see Fig. 5). The contents of the receiver always contain some extraction solvent and should therefore be redistilled, either at normal or under reduced pressure.

The usefulness of the procedure described above is demonstrated *inter alia* by the fact that even highly explosive hexatriyne, HC≡CC≡CC≡CH, can be isolated in almost 90% yield from the reaction of $ClCH_2C{\equiv}CC{\equiv}CCH_2Cl$ with KO-*tert*.-C_4H_9 in THF (see Chapter V, Exp. 17).

It requires a certain flexibility of mind to accept the proposal of using the same THF as *extraction* solvent in some cases. We discovered this possibility, when we tried to remove this solvent from carboxylic acids in a water-pump vacuum: it appeared difficult to remove the last traces of this solvent, even when heating at 70-80°C in a vacuum of 10-15 mmHg was applied. It seemed that there is some weak complexation. This led us to the idea of using THF for the extraction of carboxylic acids from the aqueous phase (after saturation with ammonium chloride) and it appeared that 6-7 extractions were sufficient, even in the cases of diacids such as HOOC-CH=C=CH-COOH. With *diethyl ether* continuous extraction appeared to be necessary in the latter case.

Finally, some remarks will be made concerning the monitoring of reactions. The chemist of to-day is in the fortunate situation of having advanced apparatus at his disposal, which give him direct and often complete information about what he has created. The need for following the reaction by smelling the mixture, by observing a change in colour or using some other sense organs has decreased. NMR apparatus is often used for monitoring reactions. Without having the intention of criticizing this, we would like to draw attention to some other, simple means for following processes. If a compound A is gradually transformed into B, there will in general be changes in the refractive index (n_D) and in the boiling point of the solution. If the dilution is not too strong and the mentioned changes are large enough, they can be used as criteria for the progress of the reaction. Outstanding examples in this book are the base-catalyzed isomerization of $H_2C{=}CH{-}CH_2{-}C{\equiv}CH$ into $H_2C{=}CH{-}CH{=}C{=}CH_2$ (large increase in n_D, due to the formation of a conjugated system), the copper(I) halide-catalyzed isomerization of $HC{\equiv}CCH_2Br$ into $BrCH{=}C{=}CH_2$ (decrease of 7-10°C in the boiling point of the solution), the copper(I) halide-promoted conversion of $HC{\equiv}CCH(CH_3)Cl$ into $ClCH{=}C{=}CHCH_3$ (increase in the boiling point by several degrees) and the formation of $(CH_3)_2C(OCH_3)_2$ and $(CH_3)_2C{=}C{=}CH{-}CH_2C({=}O)CH_3$ by heating a mixture of $H_2C{=}C(OCH_3){-}CH_3$ and $H_3C{-}C(CH_3)(OCH_3){-}O{-}C(CH_3)_2C{\equiv}CH$ in the presence of a small amount of *p*-toluenesulfonic acid. By monitoring in this way, reaction times of several hours reported in the literature can often be reduced dramatically.

GENERAL REMARKS AND ABBREVIATIONS

1. Most of the preparations in this book were carried out under nitrogen.
2. Mechanical stirring (glass rod with a curved flattened end) was carried out in most of the procedures. Magnetic stirring in general is inadvisable but may be performed during reactions which are weakly exothermic and which do not require strong heating or efficient cooling.
3. For general practical instructions concerning reactions in liquid ammonia one should consult "Preparative Acetylenic Chemistry", Elsevier, Amsterdam, 1971.
4. The compounds for which syntheses are described in this book have a purity of at least 95% (NMR, GLC), unless stated otherwise.
5. In most cases boiling points of small, intermediate fractions are given, instead of the more usual boiling ranges.

6. The original literature sources on which the various procedures are based are indicated by a reference number after the name of the compound to be prepared.
7. Frequently used abbreviations are as follows:
   DMSO = dimethyl sulfoxide
   THF = tetrahydrofuran
   HMPT = hexamethylphosphoric triamide (more correct than hexamethylphosphoramide HMPA).

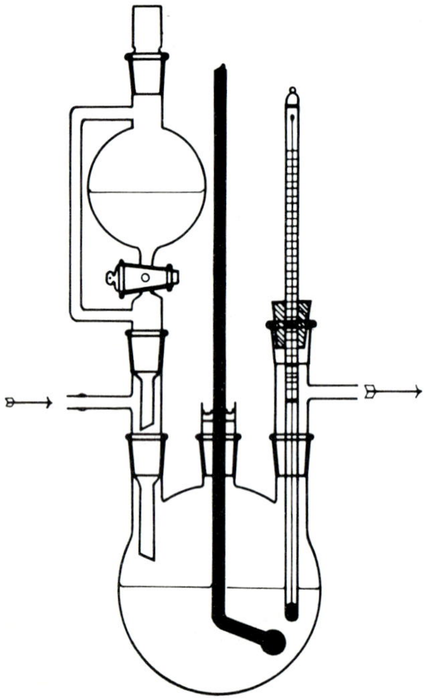

Fig. 1

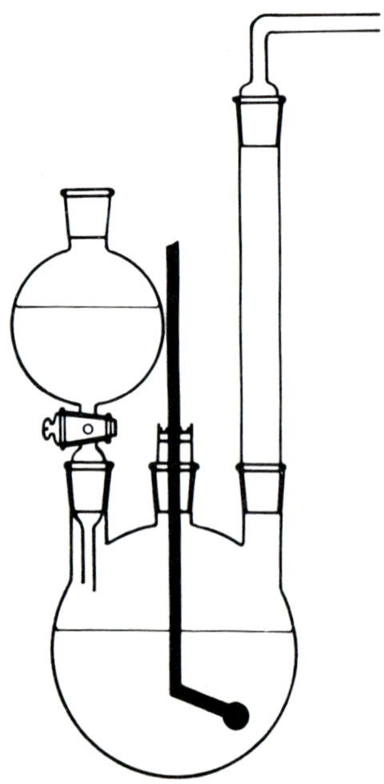

Fig. 2

Fig. 1. Addition of the reagent with temperature control and introduction of nitrogen.

Fig. 2. Reaction vessel suitable for conversions in liquid ammonia.

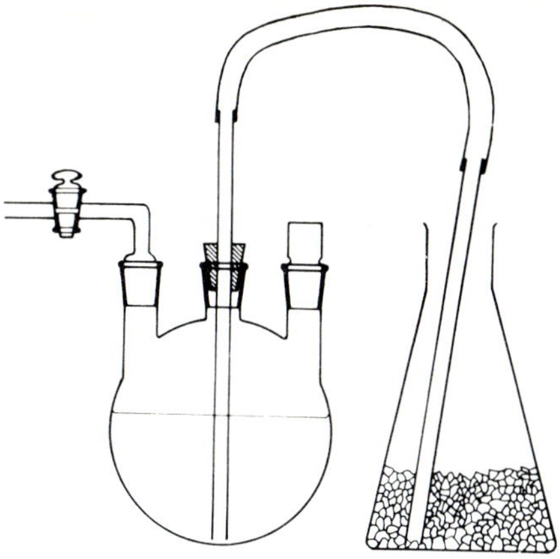

Fig. 3

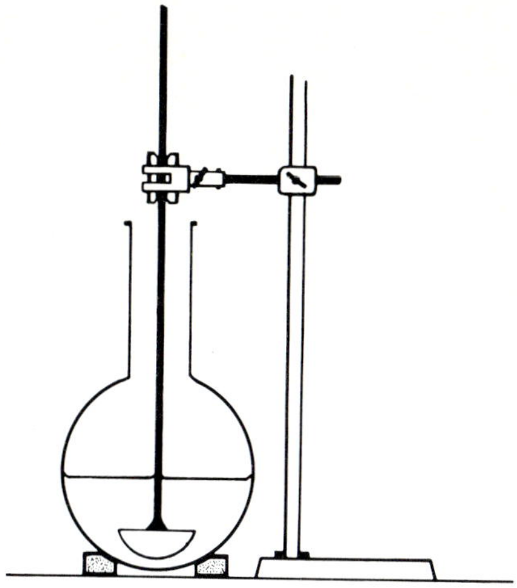

Fig. 4

Fig. 3. Transferring reaction mixtures in liquid ammonia under ice.

Fig. 4. Fast reactions in liquid ammonia.

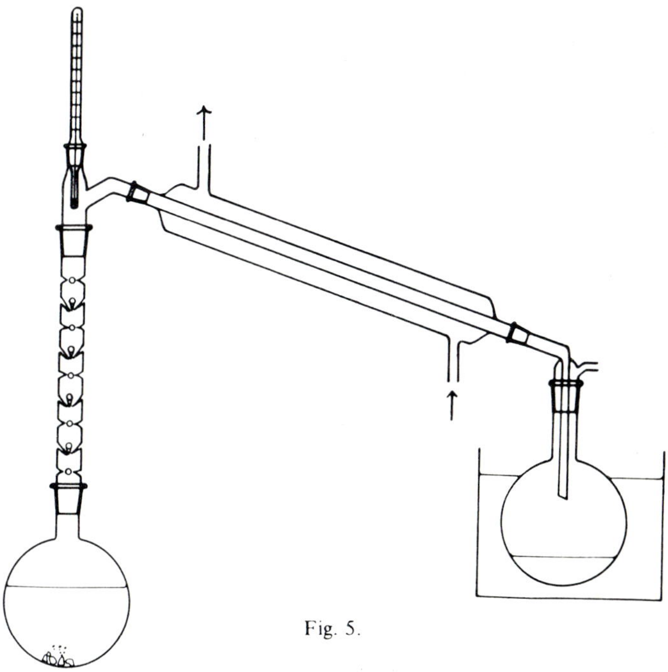

Fig. 5.

Fig. 5. Usual apparatus for the isolation of volatile and/or unstable compounds by evaporation or distillation *in vacuo* and condensation in a cooled receiver.

Chapter II

CONVERSION OF ACETYLENES AND CUMULENES INTO ANIONIC AND ORGANOMETALLIC INTERMEDIATES

1. ACIDITY OF ACETYLENES AND CUMULENES AND STABILITY OF THEIR ANIONS

One of the most versatile methods for the synthesis of derivatives with the acetylene or cumulene system consists in deprotonation of the starting acetylene or cumulene and subsequent reaction of the anionic or organometallic intermediate with an electrophilic reagent.

The conversion of acetylenes into acetylides, M-C≡C-R (M = Li, Na, K, MgBr), by means of alkyllithium or Grignard reagents in organic solvents or by alkali metal amides in liquid ammonia is well documented (for practical examples see ref. 1, for review articles consult *inter alia* refs. 2-5).

Allenic anions, $^{-}$C=C=C, in principle can be generated by deprotonation of allenes or their acetylenic isomers, C≡C-CH-. The (thermodynamic and kinetic) acidity of the corresponding protons is generally much lower than that of the ethynyl proton, unless strongly acidifying substituents (COOH, COOR, SO, SO_2, etc.) are present. Although no pK_a values of simple allenes, $H_2C=C=CHR$ (R = alkyl or aryl), are available as yet, some experimental data suggest thermodynamic acidities comparable to those of ammonia or aliphatic secondary amines. α-Lithioallenic ethers, e.g. $H_2C=C=C(Li)OCH_3$, undergo rapid ammonolysis in liquid ammonia, but are stable below -30°C for at least 1 h in the presence of dicyclohexylamine. Solutions of allenyl sulfide anions, $H_2C=C=\bar{C}$-S-alkyl, show a stability in liquid ammonia which is sufficient to allow alkylation with good yields. Alkylallenes are not lithiated by lithium dicyclohexylamide, they are clearly less acidic than the heterosubstituted representatives.

From monosubstituted allenes, $RCH=C=CH_2$, two different anions or organometallic derivatives can be derived: RCH=C=CHLi and $RC(Li)=C=CH_2$. Their relative stabilities depend strongly upon the nature of the group R. When R = alkyl the terminal lithioallenes LiCH=C=CHR are preferentially formed from allenes and alkyllithium, but in addition some (10-20%) $RC(Li)=C=CH_2$ is produced if R = primary alkyl. *tert.*-Butylallene undergoes only terminal lithiation, and so do phenylallene and some allenes with the system $H_2C=C=C-C=C-$[6, 8]. Whereas with *tert.*-butylallene steric hindrance may give rise to the regiospecificity, the terminal lithiation of phenylallene and 1,2,4-trienes gives a species in which the negative charge has better conjugative stabilization than in the nonterminally lithiated derivative. Interaction between allenic ethers or the corresponding sulfides, RCH=C=CHXR (X = O or S), and cumulenic ethers, H_2C=C=C=CHOR, and alkyllithium in organic solvents gives exclusively the α-lithiated derivatives. In the case of H_2C=C=CH-O-*tert.*-C_4H_9, lithiation with Li-N (cyclohexyl)$_2$ in THF gives also Li-CH=C=CH-O-*tert.*-C_4H_9, as may be concluded from the fact that subsequent alkylation results in the predominant formation of RCH=C=CH-O-*tert.*-C_4H_9[7]. Similar experiments conducted with *methoxy*allene give only H_2C=C=C(R)OCH_3[8].

Another type of lithiated allenic ether or sulfide is obtained from 2-alkynyl ethers and the analogous sulfides[9]:

$$\mathrm{RC{\equiv}C{-}CH_2{-}XR' + BuLi \longrightarrow RC{\equiv}C{-}\underset{}{\overset{Li}{\overset{|}{C}}}H{-}XR' \longleftrightarrow R{-}\overset{Li}{\overset{|}{C}}{=}C{=}CH{-}XR'}$$

These species are less stable than RCH=C=C(Li)-XR': in the presence of a sufficiently acidic proton donor, e.g. ammonia or dicyclohexylamine, they are transformed into the latter lithio derivatives or anions[8], probably via protonation and deprotonation. Such a process might also be involved during the reaction of 2-alkynoic acids with an excess of alkali metal amide in liquid ammonia[11]:

$$\mathrm{RCH_2C{\equiv}C{-}COOH + NaNH_2 \longrightarrow RCH{=}C{=}\bar{C}{-}COO^- \xrightarrow{NH_3} RCH{=}C{=}CH{-}COO^- \xrightarrow{^-NH_2}}$$

$$\mathrm{R{-}\underline{C}{=}C{=}CH{-}COO^- \longleftrightarrow R{-}C{\equiv}C{-}\underline{C}H{-}COO^-}$$

Note that in these cases the direction of the isomerization is the reverse of that observed with the 2-alkynyl ethers and sulfides.

Dilithiocompounds derived from acetylenes are becoming increasingly important in synthesis. They are formed by interaction between a lithium acetylide and butyllithium, using *n*-hexane, diethyl ether or THF as solvent. The abstraction of the propargylic hydrogen atom from the acetylide is a relatively slow process, but it can be accelerated considerably by using *NN*-tetramethylethylenediamine as a complexating reagent for the lithium[12-16]:

$$\mathrm{RR'CH{-}C{\equiv}CH + BuLi \xrightarrow[fast]{} RR'CH{-}C{\equiv}CLi + BuLi \xrightarrow[slow]{} RR'C(Li){-}C{\equiv}C{-}Li}$$

If one of the groups R is Ph or SR, the abstraction of the second proton is much faster. Allenes give the same dilithium compounds as their acetylenic isomers[8].

The *stability* of most acetylides, M-C≡CR, in organic solvents, even at room temperature or in liquid ammonia, allows a great variety of synthetic operations to be performed under different conditions (see *inter alia* refs. 1-5). Lithium acetylides with the structure Li-C≡C-C(R^1R^2)Hal are far less stable: they generally decompose at temperatures higher than -50°C to give carbenes, :C=C=C(R^1)(R^2). The stability decreases with increasing substitution and bulkiness of the halogen atom[8, 17]. The synthetic use is therefore limited.

The dilithio derivatives of acetylenes seem to be reasonably stable in organic solvents, even at 30-40°C[8].

The stability of the various cumulenic anions depends to a large extent upon the nature of the groups linked to the cumulenic system. Whereas solutions of lithiated allenic ethers and sulfides in diethyl ether or THF can be kept for a limited period at about 0°C, the lithiated hydrocarbons LiCH=C=CH-R are transformed into the isomeric lithium acetylides at temperatures above about -20°C, probably via HC≡C-C(Li)R^1R^2 [6, 18]. Lithiated 1,2,4-trienes, LiCH=C=C-C=C-, are reported to isomerize into mixtures of HC≡C-CH(Li)-C=C- and HC≡C-CH=C-C-Li at very low temperatures[19], so that clean derivatization reactions are hard to achieve.

2. ORGANOMETALLIC REAGENTS AND STRONG BASES

Alkyllithium solutions are extremely useful metallating reagents. The use of commercially available alkyllithium compounds, notably butyllithium, in inert solvents such as hexane, has some advantages over that of the reagents prepared from an alkyl halide and lithium in diethyl ether. Firstly the commercial reagents can be stored for very long periods, whereas the strength of alkyllithium in diethyl ether gradually decreases owing to reaction with the solvent. Methyllithium is more stable in diethyl ether, but at the same time is also much less reactive than butyllithium. Moreover, the preparation of the ethereal reagents is laborious. During the experimental work for this book we carried out some comparative experiments with butyllithium in hexane and ethyllithium in diethyl ether, using the same amounts of THF and HMPT as co-solvents. It appeared that subsequent alkylation of the lithiated compounds was much faster when butyllithium in hexane had been applied as the metallating agent. The experiments concerned were: O-lithiation of ethynylcyclohexanol, followed by reaction with methyl iodide, and lithiation of allene or methoxyallene, followed by alkylation with higher alkyl bromides. These results suggest that the LiBr, present in the ethyllithium solution, has a retarding effect upon the alkylation. We had the further impression that metallations of the weakly acidic allenic hydrocarbons with butyllithium-hexane-THF proceeded more smoothly than those with ethyllithium-diethyl ether-THF-LiBr.

In some cases, however, the use of ethyllithium-diethyl ether-LiBr is to be preferred, for example in the preparation of the cumulenic ether $H_2C=C=C=CH-OC_2H_5$ from $C_2H_5OCH_2C\equiv CCH_2OC_2H_5$ (see Chapter V, Exp. 7). If the elimination is carried out with commercial butyllithium in hexane and diethyl ether, some tar is produced, while the separation of the volatile and unstable cumulenic ether from the hexane is difficult. The elimination with ethyllithium-diethyl ether-LiBr is a clean reaction and gives excellent results.

In contrast to alkylations with most of the alkyl halides, the reactions of anions with ethylene oxide in organic solvents such as diethyl ether and THF seem to proceed more easily if lithium halide is present[8]. The salt probably coordinates with the oxygen atom of the epoxide, facilitating the ring opening by the anions.

Alkyllithium bases are generally less suitable for deprotonation of compounds with strongly electron-withdrawing groups such as C=O, COOR and C≡N. In these cases lithium dialkylamides, especially those with bulky groups (isopropyl, cyclohexyl), are the reagents of choice. They are very easily obtained from butyllithium and the dialkylamine in the desired solvent.

Alkali metal amides are weaker bases than lithium dialkylamides. Lithium and sodium amide are commercially available as solids, but freshly prepared suspensions in liquid ammonia are much more effective bases for most reactions. Although in many cases the soluble potassium amide gives the same results as obtained with $LiNH_2$ and $NaNH_2$, its use, involving prior handling of the more dangerous potassium metal, cannot always be avoided. For example, the "contra-thermodynamic" isomerization of 2-alkynes[1] into 1-alkynes via the acetylides gives good results only with potassium amide; the same holds for the isomerization of $H_2C=CH-C\equiv C-NR_2$ into $HC\equiv C-CH=CH-NR_2$ (see Chapter IV, Exp. 20).

Potassium *tert.*-butoxide is widely used in organic synthesis[20]. It is commercially available as the uncomplexed base: the product obtained by dissolving potassium in an excess of the alcohol and subsequently removing the excess by distillation *in vacuo* is the 1:1 complex KO-*tert.*-C_4H_9.HO-*tert.*-C_4H_9. In this book the base is used as a catalyst for isomerizations and as a dehydrohalogenating reagent. The uncomplexed product is soluble in DMSO, HMPT, THF and liquid ammonia. By addition of more or less *tert.*-C_4H_9OH the efficiency of the base-solvent pair can be decreased, which is necessary in some cases in order to stop a sequence of reactions at the desired intermediate stage (see, for example, the isomerization of ethynylcyclohexene, Chapter IV, Exp. 3).

In several procedures described in this book Grignard solutions in diethyl ether or THF are used. Though most chemists have prepared a Grignard solution at least once during their education, we judged it to be useful to give some extensive descriptions for the preparation of these reagents in this book. It is remarkable that, in spite of the known sensitivity of Grignard compounds towards oxygen, some practical textbooks do not prescribe the use of a nitrogen atmosphere during the preparation. The use of only a slight excess of magnesium in many of the procedures seems irrational. Towards the end of the preparation only a small surface area of magnesium is available for reaction with the halide, which is present in a very low concentration throughout the experiment. Such a situation requires relatively long additional periods of stirring and heating. By using a 100% excess of the metal (less than 100% for experiments on a large scale), one can save much time. Some Grignard reactions, especially those with tertiary alkyl chlorides and cyclohexyl chloride, are not easily started and it seemed desirable, therefore, to inform the user of this book about our experiences.

3. EXPERIMENTS

(1) *Ethyllithium in diethyl ether*

$$C_2H_5Br + 2\ Li \xrightarrow[-25^{\circ}C]{\text{diethyl ether}} C_2H_5Li + LiBr$$

Apparatus: 2-l three-necked round-bottomed flask with a dropping funnel, combined with a gas inlet, a mechanical stirrer and a thermometer, combined with a gas outlet.

In the flask were placed 800 ml (note 1) of dry diethyl ether. Twenty grams of lithium (note 2) were flattened (thickness about 1 mm) with a hammer (note 3) and cut into small pieces (about $10 \times 2 \times 1$ mm^3), which were introduced at the same time into the flask. The contents of the flask were cooled to -30°C, after the air in the flask had been replaced with nitrogen. From the dropping funnel, which contained 1.12 mol of ethyl bromide, were added 10-15 g of ethyl bromide. It took 5-10 min before the reaction started: this was visible by the appearance of turbidity of the diethyl ether and later by the appearance of a gloss on the pieces of lithium and a distinct increase in temperature. Care was taken that the temperature did not rise above -20°C (note 4). When the reaction had subsided, the addition of ethyl bromide was continued, now dropwise (note 5). The temperature was kept between -20 and -30°C (note 6). After the addition, which was carried out in 30-40 min, stirring was continued for about a further 1 h. The temperature was allowed to rise gradually to -10°C. When the gloss on the pieces of lithium had disappeared, the solution was poured into another flask through a small funnel, containing a loose plug of glass-wool (note 7). The strength of the solution can be determined by titration[160]. The "yield" was generally between 90 and 95%. The solution was stored at -20 to -30°C. As the strength diminishes gradually during storage, the solution should be used within 1-2 weeks.

Note 1. If less than this amount of diethyl ether is used, precipitation of salts may occur in the final stage of the preparation. This causes inactivation of the lithium.

Note 2. Lithium containing 0.5-1% of sodium should be used; the very pure metal reacts sluggishly and gives lower yields.

Note 3. The hammer and the surface on which the lithium is flattened should be free from rust: traces of iron salts may cause rapid decomposition of the alkyllithium.

Note 4. At higher temperatures the reaction between ethyl bromide and ethyllithium ("Würz coupling") becomes significant.

Note 5. Too fast a rate of addition may cause aggregation of the lithium. If this occurs, the addition should be interrupted and the rate of stirring diminished until the aggregate has disintegrated.

Note 6. Below -30°C salts precipitate from the solution and the reaction stops, probably because the pieces of lithium are covered with salt. If the temperature is increased, the salts dissolve and the gloss on the lithium re-appears.

Note 7. The lithium which remains on the glass-wool should be destroyed immediately with a mixture of ethanol and water.

(2) *Cyclohexylmagnesium chloride in diethyl ether*

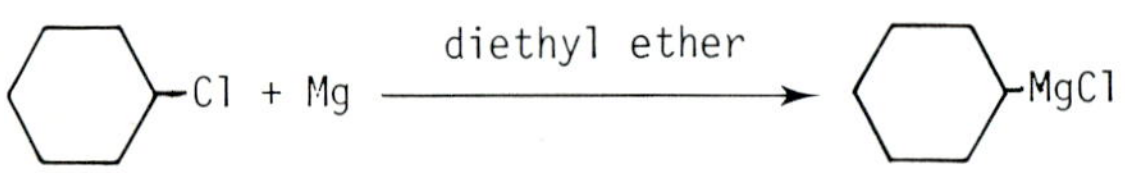

Apparatus: 1-l three-necked, round-bottomed flask with a dropping funnel, combined with a gas inlet, a mechanical stirrer and a reflux condenser.

The flask was charged with 48 g of magnesium and 200 ml of dry diethyl ether. 1,2-Dibromoethane (4 ml) was added. After a few minutes a vigorous reaction started and the diethyl ether began to reflux. When this reaction had subsided, 500 ml of dry diethyl ether were added. Stirring was started and 1.0 mol of chlorocyclohexane (note 1) was added from the dropping funnel at a rate such that the diethyl ether gently refluxed (note 2). After this addition, which was carried out in 1 h, the flask was heated under reflux for a further 1 h.

Note 1. The chlorocyclohexane should be free from cyclohexanol, which may lead to inactivation of the magnesium.

Note 2. It is advisable to add first about 15 g of the chlorocyclohexane in one portion. After 5-15 min the diethyl ether should again begin to reflux. If this does not occur, one should not add more, since the magnesium apparently has not been sufficiently activated.

Solutions of *tert.*-butylmagnesium chloride and cyclopentylmagnesium chloride in diethyl ether can be prepared in the same way. In these cases also the purity of the chlorides is of great importance for a successful and smooth conversion into the Grignard reagent.

(3) *n-Butylmagnesium chloride in diethyl ether*

$$n\text{-}C_4H_9Cl + Mg \xrightarrow{\text{diethyl ether}} C_4H_9MgCl$$

Apparatus and procedure: Closely similar to the preparation of *tert.*-C_4H_9MgCl, cyclohexyl-MgCl and cyclopentyl-MgCl (see Exp. 2). The yield (estimated from the results obtained from reactions with this reagent) is at least 90%. Here, too, it is essential to use *n*-butyl chloride which is free from butyl alcohol.

(4) *tert.-Butylmagnesium chloride in THF*

$$(CH_3)_3C\text{-}Cl + Mg \xrightarrow{\text{THF}} (CH_3)_3C\text{-}MgCl$$

Apparatus: 2-l flask, see Fig. 1.

The flask was charged with 60 g of magnesium turnings. After the air in the flask had been replaced completely with nitrogen, 250 ml of dry THF were added, and subsequently 12 g of 1,2-dibromoethane. Stirring was started and after a few minutes a vigorous reaction started, causing a rise in the temperature to 60°C or higher. The temperature was kept just below that of the b.p. of THF (66°C) by occasional cooling in a water-bath. When the evolution of ethene had stopped and the temperature had dropped to 50°C (without using a cooling bath!), 250 ml of THF were added. The temperature of the THF in the flask now became about 40°C. From the dropping funnel, containing 1.70 mol of *tert.*-C_4H_9Cl, 20 g were added.

The flask was immersed in a water-bath of the same temperature as that of the THF. This bath was removed as soon as the temperature of the THF became 2°C higher (note 1). It was kept just below 60°C by occasionally cooling the flask (for a few seconds) in a water-bath. After the temperature had dropped to 50°C, the remainder of the chloride was added during 1.5 h. It appeared to be possible to keep the reaction going at a temperature around 40°C. After the addition the flask was warmed for an additional 40 min at about 50°C. The mixture was then cooled to 30-35°C (note 2) and transferred, in a stream of nitrogen, into another flask (e.g. for storage) or, if it was to be used directly, into a dropping funnel. The excess of magnesium was rinsed with three 40-ml portions of THF and the rinsings were added to the main solution. The yield, estimated from further reactions performed with the Grignard compound, is only 60-65%. The compound can be stored for long periods at room temperature in well closed flasks. Before use, the flask should be warmed until all solid has disappeared.

Note 1. If there is no rise in temperature, a water-bath at 55 or 60°C should be applied. The reaction is slow compared with that with *n*-alkyl bromides.
Note 2. At lower temperatures part of the Grignard compound crystallizes out.

(5) *Alkyl- and arylmagnesium bromide in diethyl ether*

$$\text{RBr} + \text{Mg} \xrightarrow{\text{diethyl ether}} \text{RMgBr}$$

Apparatus: 1-l three-necked, round-bottomed flask with a dropping funnel combined with a gas inlet, a mechanical stirrer and a reflux condenser; in the case of CH_3MgBr, however, the apparatus in Fig. 1 is used.

The procedure is similar to that for cyclohexylmagnesium chloride. Although in most cases the reaction starts without previous activation of the magnesium, such activation makes it more likely that the reaction will start smoothly. As soon as refluxing, due to the activation of the magnesium with dibromoethane, has stopped, a few millilitres of RBr are added. If this results in refluxing of the diethyl ether, more ether can be added, after which the bromide is added dropwise. In the case of CH_3MgBr the following modified procedure is recommended. After the magnesium has been activated with dibromoethane and no more heat is evolved, the required amount of diethyl ether is added. The diethyl ether in the reaction flask is cooled to 15°C and about 5% of the solution of CH_3Br in diethyl ether (1:2, by weight) in the dropping funnel is added. If this results in an immediate rise in temperature, the remainder of the solution can be added dropwise with cooling to between 10 and 15°C.

The "yields" of the Grignard reagents are usually 90% or higher; too fast an addition of RBr, however, may result in the formation of substantial amounts of R-R, especially in the case of bromobenzene.

(6) n-Alkylmagnesium bromide and arylmagnesium bromide in THF

$$R\text{-}Br + Mg \xrightarrow{THF} R\text{-}MgBr$$

R = C_2H_5, higher alkyl or phenyl

Apparatus: 1-l flask, see Fig. 1.

In the flask (note 1) were placed 24 g of magnesium turnings and 150 ml of dry THF. A 3-ml volume of 1,2-dibromoethane (note 2) was added and stirring was started, during which ethene was evolved. The temperature of the THF was allowed to rise. When the exothermic reaction had ceased and the temperature had dropped to 40°C an additional 175 ml of THF was poured into the flask. From the dropping funnel, which contained 0.55 mol of the bromide, were added about 5 ml. This addition caused an immediate rise in temperature from 30 to 45°C or higher (note 3). When this reaction had subsided, the mixture was cooled to 30°C and the remainder of the halogen compound was added dropwise during 1.5 h, whilst keeping the temperature at about 30°C (note 4). After the addition the cooling bath was removed and stirring was continued for about 30 min. The solution obtained was decanted (under a stream of nitrogen) from the excess of magnesium and was transferred into another flask, or dropping funnel (if the Grignard solution had to be added to another compound), filled with nitrogen. The remaining magnesium was rinsed with two 20-ml portions of THF and the rinsings were added to the other solution. The solution obtained contained about 0.50 mol of R-MgBr (note 5).

General note: All reagents and glassware should be dry. The best way to obtain dry THF is to shake the commercial product with machine-powdered KOH (about 70 g/l) and distil the filtered liquid from a small amount of $LiAlH_4$.

Note 1. After the addition of the magnesium the air in the flask was replaced completely with nitrogen.

Note 2. Iodine can also be used.

Note 3. This means that the metal has been activated sufficiently. If the reaction does not start directly, one should wait (and observe the thermometer) for at least 30 min. After this period one can add a second portion of about 5 ml of R-Br if nothing has yet happened.

Note 4. At lower temperatures part of the reagent may crystallize out and the reaction becomes very slow or even stops.

Note 5. The solutions can be stored for several months at room temperature in well closed flasks. On storage at lower temperatures, and sometimes also at room temperature, part of the dissolved compound may crystallize out. If this occurs, the solution should be warmed until *all* solid material has dissolved, before part of it is used for some conversion.

(7) *Methylmagnesium bromide in THF*

$$CH_3Br + Mg \xrightarrow{\text{THF}} CH_3MgBr$$

Apparatus: 1-l flask, see Fig. 1.

Although the preparation of this reagent is similar to that for the other alkylmagnesium bromides (see Exp. 6), some modifications had to be introduced in view of the volatility of methyl bromide. After the magnesium had been activated by means of $BrCH_2CH_2Br$, the mixture was cooled to $5^{\circ}C$, more THF was added, and about one tenth of the solution of CH_3Br in THF (0.55 mol in 100 ml of THF) was added from the dropping funnel. After a few minutes the reaction started (note 1) and the temperature was allowed to rise to $30^{\circ}C$. When this reaction had ceased, the mixture was cooled again to 5 or $10^{\circ}C$ and the remainder of the solution was added dropwise from the dropping funnel during 1 h. The temperature of the mixture was held between 10 and $15^{\circ}C$. The conversion was terminated by stirring for a further 30 min (without using the cooling bath). For further details see Exp. 6.

Note 1. This requires careful and continuous observation.

(8) *Propargylmagnesium bromide in diethyl ether*

$$HC{\equiv}C\text{-}CH_2Br + Mg \xrightarrow{\text{diethyl ether}} HC{\equiv}C\text{-}CH_2MgBr$$

Apparatus: 1-l flask, see Fig. 1.

In the flask were placed 24 g of magnesium turnings and 100 ml of dry diethyl ether and 1.5 g of mercury(II) chloride were added successively. After stirring for 30 min at room temperature an additional 250 ml of diethyl ether were added and the mixture was cooled in a bath of ice and ice-water. After the temperature had reached $0^{o}C$, 5 ml of propargyl bromide were added from the dropping funnel, which contained 0.57 mol of this compound. As soon as the temperature began to rise (note 1), the bath was removed and replaced with a bath with dry-ice and acetone. In most cases the reaction started within 15 min (note 2). The temperature was lowered to $-5^{o}C$ as quickly as possible, the cooling bath was removed temporarily and a further 2 ml of propargyl bromide were added. When this resulted in a rapid rise in temperature, cooling at $-5^{o}C$ was continued and the remainder of the bromide was added dropwise in 40 min. The temperature of the mixture could be lowered gradually to about $-10^{o}C$ without causing a significant decrease in the reaction rate. After the addition cooling was applied occasionally and the temperature was allowed to rise to $5^{o}C$ in 30 min. The grey solution obtained was then decanted from the excess of magnesium and transferred into another reaction flask. The magnesium was rinsed with two 30-ml portions of diethyl ether and the rinsings combined with the main solution. The reagent should be used directly, but storage for 12-24 h at $-30^{o}C$ does not have an adverse influence on the result of further conversions. The "yield", estimated roughly from the results of reactions performed with the reagent, is at least 90%, so that the solution described above contains about 0.5 mol of $HC{\equiv}CCH_2MgBr$.

Note 1. In this way it is easy to see when the reaction starts.
Note 2. If this does not occur, a second portion of 1 g of $HgCl_2$ should be added.

(9) *Acetylenic Grignard and alkali metal derivatives*[1]

$$RC{\equiv}C\text{-}H + C_2H_5MgBr \longrightarrow RC{\equiv}C\text{-}MgBr$$

$$RC{\equiv}C\text{-}H + C_2H_5Li,\ C_4H_9Li \text{ or alkali metal amide} \longrightarrow RC{\equiv}C\text{-}Li,\ \text{-}Na$$

Acetylenic Grignard compounds or the corresponding organoalkali metal derivatives are important intermediates in many syntheses of acetylenic compounds. The various methods for their formation in organic solvents and in liquid ammonia have been discussed extensively[1] and we here give only a brief summary.

Deprotonation of terminal acetylenes by organolithium compounds in organic solvents or by alkali metal amides is an extremely fast reaction, even at very low temperatures in solvents of relatively low polarity (e.g. diethyl ether). As organic solvents THF and diethyl ether and as an "inorganic" solvent liquid ammonia are mostly used. The best way to prepare a solution or suspension of a lithium alkynylide in diethyl ether or THF consists in adding a solution of alkyllithium in diethyl ether or hexane dropwise to a solution of a slight (2-5%) excess of the acetylene in diethyl ether or THF. During this addition the temperature is kept below -10°C or, in the generation of the unstable carbenoid $LiC{\equiv}CCH_2Cl$, below -70°C. Inverse addition is generally inadvisable in view of the possibility of di-metallation or addition (especially in the case of conjugated diynes, $HC{\equiv}C{-}C{\equiv}CR$). Although alkali metal amides are much less basic than alkyllithium in a thermodynamic sense, the deprotonation of acetylenes by lithium, sodium or potassium amide in liquid ammonia at its boiling point (-33°C) or at lower temperatures proceeds almost instantaneously. Liquid acetylenes may be added in an undiluted state or (in the case of very volatile acetylenes) as a solution in an organic solvent or at temperatures below -33°C. The conversion of gaseous acetylenes ($HC{\equiv}CH$, $CH_3C{\equiv}CH$, $C_2H_5C{\equiv}CH$) into the alkali metal derivatives can be carried out by introducing the gas into a suspension or solution of alkali metal amide in the presence of a small amount of triphenylmethane. As long as alkali metal amide is present the solution is red owing to the triphenylmethyl anion. This colour disappears when all alkali amide has been converted: a slight excess of acetylene is generally required. We found that some acetylenes cannot be converted into their alkali metal "salts" in liquid ammonia: these are $HC{\equiv}C{-}Si(CH_3)_3$ [formation of $(CH_3)_3SiNH_2$], $HC{\equiv}C{-}CH_2{-}SR$ (partial isomerization into $H_2C{=}C{=}\bar{C}{-}SR$), $HC{\equiv}C{-}CH_2aryl$ (partial isomerization into $H_2C{=}C{=}CH$-aryl and formation of polymers), non-conjugated enynes $HC{\equiv}CCH_2CH{=}CHR$ and diynes $HC{\equiv}C{-}CH_2C{\equiv}C{-}R$ (partial or complete formation of conjugated acetylides) and $HC{\equiv}C{-}PR_2$ (disproportionation into $HC{\equiv}C^-$ and $R_2P{-}C{\equiv}C{-}PR_2$).

Grignardation of acetylenes is a much slower process than the conversion with the more polar organometallics mentioned above. There are some qualitative differences with respect to the rates with which the various acetylenes are metallated by RMgX. As expected, Grignardation proceeds more easily in THF than in diethyl ether. In diethyl ether at about 0°C the metallation is usually extremely slow, so that heating under reflux for 1.5-2 h is generally necessary. The conversion of very volatile or gaseous acetylenes in diethyl ether gives rise to some experimental problems: part of the acetylene is swept along with the ethane liberated from the reaction with C_2H_5MgBr. The best solution is to use a reflux condenser filled with dry-ice and acetone (-75°C). The gaseous acetylenes ($CH_3C{\equiv}CH$ or $C_2H_5C{\equiv}CH$), obtained by evaporation of a weighed amount of the liquified acetylene are slowly introduced into the ethereal solution of C_2H_5MgBr, which is heated at about 35°C. The gases evolved from the reaction mixture pass the "cold finger" and subsequently a cold trap connected to it. The cold trap usually contains some unconverted alkyne, which can be passed through the reaction mixture again.

Essentially the same procedure can be applied for the reaction of acetylenes with C_2H_5MgBr in THF (normal reflux condenser in the case of liquid acetylenes, cold finger for propyne, butyne and alkynes with a b.p. lower than 45^oC at normal pressure, temperature of the reaction mixture between 25 and 45^oC). For the preparation of HC≡CMgBr we recommend the method described in *Organic Synthesis*, Collect. Vol. *4*, 792: slow addition of a THF solution of C_2H_5MgBr to THF, through which acetylene is bubbled, taking care that the temperature of the reaction mixture remains between 25 and 30^oC. Some acetylenes appear to react markedly faster with C_2H_5MgBr in THF: this holds especially for $HC{\equiv}COC_2H_5$ and $HC{\equiv}CSi(CH_3)_3$ which are converted completely within 1 h at about 10^oC. The Grignardation of CH≡C-C≡CH with C_2H_5MgBr in THF proceeds almost instantaneously, even at -20^oC[8].

(*10*) *Butyllithium in pentane*

Apparatus: 2-l two-necked, round-bottomed flask, provided with a stopper and a 40-cm Vigreux column. This column was connected to a descending condenser and a receiver. A tube filled with KOH pellets was placed between the water pump and the receiver.

In the flask (previously filled with pure nitrogen) was placed a solution of 1.0 mol of (commercial) butyllithium in about 700 ml of hexane. Some boiling stones were added and the receiver was immersed in a bath of dry-ice and acetone (-75°). A tube filled with calcium chloride was placed between the water pump and the receiver. The apparatus was evacuated while the flask, containing the butyllithium solution, was gradually heated, as the solution became more concentrated. The last traces of hexane were removed from the butyllithium by heating the flask for 15-20 min at about 50^oC. The remaining viscous liquid was cooled to room temperature and pure nitrogen was admitted to the distillation apparatus. The stopper was quickly replaced with a dropping funnel, containing redistilled and dry pentane (b.p. $< 40^oC$; note 1). The desired amount of pentane was run in with swirling.

Note 1. Traces of dissolved oxygen from the air were replaced by passing a stream of dry and pure nitrogen through the pentane for a few minutes.

(*11*) *Lithium amide and sodium amide in liquid ammonia*[1]

$$M + NH_3 \xrightarrow{Fe(NO_3)_3} MNH_2\downarrow + \tfrac{1}{2}H_2\uparrow$$

$$M = Li, Na$$

Apparatus: 2-l flask, see Fig. 2 (note 1). Instead of mechanical stirring occasional swirling by hand may be applied: a one-necked flask provided with a stopper with a small hole can also be used.

Anhydrous liquid ammonia (note 2) (900 ml) was drawn from a cylinder and introduced into the flask. Iron(III) nitrate (100 mg) was added and, as soon as a uniformly brown solution had formed (after stirring for a few seconds), about 0.7 g of lithium (from the starting amount of 7 g) was cut into two or three pieces and immediately introduced into the flask. After 10-15 min the blue colour had disappeared completely and a white suspension of lithium amide had formed. The remainder of the 7 g (1 mol) of lithium was then cut up and introduced. In most cases the conversion was finished within about 30 min (note 3).

The conversion into sodium amide (1 mol) was carried out in a similar way. In contrast to the reaction with lithium amide, the sodium amide suspension immediately settles out after stopping the stirring and the supernatant ammonia has a grey or black colour, due to colloidal iron. In some cases it took a long time before all of the sodium had been converted (note 4). A further 0.1 g of iron(III) nitrate was then added to accelerate the reaction and some liquid ammonia was introduced to compensate for the losses due to evaporation.

Note 1. In view of the possibility that frothing may occur in the last stage, the use of a larger flask is advisable. With sodium this frothing does not occur.

Note 2. Liquid ammonia of good quality contains not more than 0.1% of water. Before starting the preparation, one should perform the following test. Add a very small piece of the metal; if after about 2 min (continuous swirling or stirring) the blue colour has disappeared, introduce a second piece. Repeat this procedure until the blue colour persists. The amount of metal required gives an indication of the water content.

Note 3. If the conversion takes longer, add some liquid ammonia to keep the volume of the suspension between 600 and 800 ml.

Note 4. The conversion of lithium and potassium into the alkali amides has never given problems.

(*12*) *Potassium amide in liquid ammonia*[1]

$$NH_3 + K \xrightarrow{Fe(NO_3)_3} KNH_2 + \tfrac{1}{2}H_2\uparrow$$

Apparatus: 1-l round-bottomed flask, provided with a rubber stopper, perforated by a small hole.

In the flask were placed 500 ml of anhydrous liquid ammonia (see Exp. 11), about 0.1 g of iron(III) nitrate was introduced and the flask was swirled by hand until the solution had become uniformly brown (a few seconds). Potassium (0.65 mol) previously cut into 1-2 g pieces under pentane, was introduced and the flask was swirled occasionally. In most cases the blue colour had disappeared completely within 30 min. The greyish solution was then filtered through a loose plug of glass-wool in a funnel (note 1). This operation was carried out in the following way. The funnel containing the glass-wool, was placed in the middle neck of the flask, used for the further reaction with the amide, and two very loose plugs of cotton-wool were placed on the other necks. The solution of potassium amide was cautiously poured on to the glass-wool, whilst keeping this in continuous contact with the funnel by means of a glass rod. When, instead of a three-necked flask, a flask with one neck was used for the further conversion, the funnel was placed in an iron ring at a height such that the ammonia vapour could escape freely. The glass-wool containing the very small (1 mm^3) pieces of unconverted potassium is placed in a beaker filled with water.

Note 1. The potassium always contains small pieces, which are surrounded by a crust of oxide and therefore are not converted into amide. If these are not filtered off, they can cause fire hazards and explosions during the aqueous work-up and extractions with organic solvents.

(*13*) *Allenyllithium*[28]

$$H_2C{=}C{=}CH_2 \xrightarrow[\text{hexane-THF}]{C_4H_9Li} H_2C{=}C{=}C(H)Li$$

Apparatus: 1-l three-necked, round-bottomed flask with a thermometer+gas outlet, a stirrer and a gas inlet. During the experiment a slow stream of nitrogen was passed through the flask. A Dewar flask containing liquid nitrogen (note 1) was used as a cooling bath.

The air in the flask was replaced with nitrogen and a solution of 0.30 mol of butyllithium in about 200 ml of hexane was transferred into the flask. The solution was cooled to $-60^{\circ}C$ and 180 ml of dry THF were added. The temperature of the mixture was lowered to $-95^{\circ}C$ and a mixture of 0.32 mol of allene (see Chapter V, Exp. 23) and 20 ml of dry THF, pre-cooled to $-50^{\circ}C$, was added in two equal portions (the gas inlet was removed temporarily) with an interval of 2 min. During this addition care was be taken that the temperature did not rise above $-75^{\circ}C$ and not drop below $-90^{\circ}C$. Therefore, very efficient cooling by occasionally immersing the flask in liquid nitrogen and vigorous agitation of the reaction mixture were necessary (note 1). After the addition the rate of stirring was diminished and the temperature of the mixture was allowed to rise to $-65^{\circ}C$ in 10 min. The resulting thin white suspension was then ready for further conversions (note 2).

Note 1. During the cooling with liquid nitrogen vigorous agitation of the reaction mixture is necessary in order to prevent the formation of a solid crust on the bottom of the reaction flask.

Note 2. Solutions of allenyllithium are very unstable and therefore should be used directly after their preparation. The temperature of the solutions should be kept below $-60^{\circ}C$.

The lithiation of allene can also be carried out with ethyllithium or butyllithium in diethyl ether (prepared from the alkyl bromides), using THF as a co-solvent. The salt suspension which is initially present when the solution of alkyllithium is cooled to $-60^{\circ}C$ or lower has disappeared almost completely when the reaction between allene and alkyllithium is finished.

(*14*) *Lithiated allenes*[28]

$$RR'C{=}C{=}CH_2 \xrightarrow[C_4H_9Li]{C_2H_5Li\ \text{or}} RR'C{=}C{=}C(H){-}Li$$

R = alkyl; R' = H or alkyl

Apparatus: 1-l flask, see Fig. 1.

To a solution of 0.20 mol of butyllithium in about 140 ml of hexane or 0.20 mol of ethyllithium in about 180 ml of diethyl ether (see Exp. 1) was added the desired amount of dry THF (usually between 110 and 160 ml) with cooling below

-20^oC. The mixture was cooled to -70^oC and the allene (0.22 mol) was added in 5-10 min while maintaining the temperature between -60 and -70^oC. After stirring for an additional 30 min at -60^oC the solution was ready for further conversions. In the metallation with ethyllithium the salts initially present had disappeared almost completely after this period. During the lithiation with commercial butyllithium the reaction mixture was continuously homogeneous. The solution of the lithiated allenes should be kept below -60^oC and used within a few hours.

Note 1. The lithiation of monoalkylallenes is not completely regiospecific. The ratio of α- to γ-lithiated allene varies from about 80:20 for methylallene to 93:7 for hexylallene.* *tert.*-Butylallene, however, is metallated exclusively on the terminal carbon atom.

* J.C. Clinet (Paris, France) made similar observations.

(*15*) α-*Lithiated allenic ethers*[29]

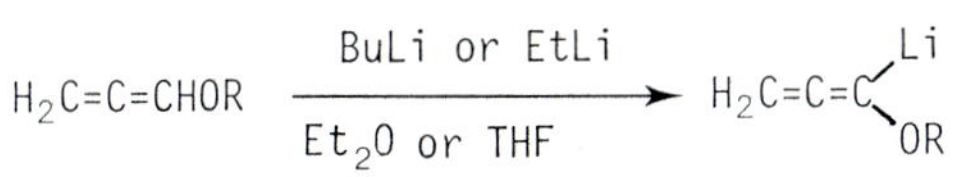

R = alkyl

Apparatus: 1-l flask, see Fig. 1

A solution of 0.40 mol of butyllithium in about 270 ml of hexane was cooled to -50^oC and 250 ml of dry THF or diethyl ether were added, while maintaining the temperature below -20^oC. The allenic ether (0.42 mol, freshly distilled) was subsequently added in 15 min at -30^oC. After an additional 10 min at this temperature the solution was ready for further conversions.

In some experiments the presence of hexane is undesirable in view of the volatility of the products. In these cases one can use butyllithium in pentane (prepared from butyllithium in hexane, by replacing the hexane with pentane; see Exp. 10) or ethyllithium in diethyl ether, prepared from ethyl bromide and lithium (see Exp. 1).

(*16*) *Lithiation of propargyl chloride*[1]

$$HC{\equiv}C\text{-}CH_2Cl + C_2H_5Li \xrightarrow{\text{diethyl ether}} LiC{\equiv}C\text{-}CH_2Cl$$

Apparatus: 1-l flask, see Fig. 1.

A mixture of 0.40 mol of propargyl chloride and 150 ml of dry diethyl ether was cooled at -90°C (liquid nitrogen bath) and a solution of 0.40 mol of ethyllithium (note 1) in about 350 ml of diethyl ether (see Exp. 1) was added with vigorous stirring and occasional cooling (note 2). The temperature of the reaction mixture was kept between -70 and -90°C. The formation of the lithium derivative proceeded almost instantaneously, so that the solution obtained could be used directly after the addition of the ethyllithium, which was carried out in 15-20 min. This lithium acetylide solution is very unstable and must be kept below -60°C.

Note 1. Butyllithium in hexane can also be used; in that case the propargyl chloride must be mixed with a volume of ether equal to or greater than that of the hexane solution of butyllithium.

Note 2. Continuous cooling would lead to formation of a solid crust on the bottom of the flask and too low a temperature of the mixture.

Substituted carbenoids such as $LiC{\equiv}C\text{-}C(CH_3)_2Cl$, are less stable than $LiC{\equiv}C\text{-}CH_2Cl$[17].

(*17*) *Dilithiated 1-alkynes in THF*

$$RCH_2C{\equiv}CH + 2\ BuLi \xrightarrow[-30^\circ \rightarrow +35^\circ C]{\text{THF-hexane}} R\text{-}CH(Li)\text{-}C{\equiv}C\text{-}Li$$

R = H, alkyl

Apparatus: 1-l flask, see Fig. 1.

In the flask were placed 350 ml of dry THF. The THF was cooled to -60°C and a solution of 0.50 mol (note 1) of butyllithium in about 350 ml of hexane was run in. After this addition the mixture was immediately cooled to -30°C and a cold solution of 0.20 mol of propyne or 1-butyne in 50 ml of dry THF, or 0.20 mol of propyne or 1-butyne in 50 ml of dry THF, or 0.20 mol of the other acetylenic compounds, was added in 10 min, while keeping the temperature below $-10^{o}C$. In the cases of propyne and butyne and other aliphatic alkynes the cooling bath was immediately removed thereafter and the mixture was heated to $25^{o}C$. At this temperature a slightly exothermic reaction took place and the temperature rose in about 30 min to $30\text{-}35^{o}C$. The mixture was kept for an additional period of 2 h between 30 and $35^{o}C$ (note 2) and was then ready for further conversions. $Li\text{-}CH_2\text{-}C{\equiv}CLi$ is a nearly white suspension, $LiC{\equiv}C\text{-}CH(CH_3)Li$ is a yellow suspension and $LiC{\equiv}C\text{-}CH(C_3H_7)Li$ is a brown solution.

Note 1. An excess of butyllithium is used, as some butyllithium is destroyed by the competing reaction with the THF.

Note 2. During this period the excess of butyllithium has completely reacted with the THF.

(*18*) *Preparation of the zinc-copper couple*[75b]

Finely powdered zinc (70 g, Merck, Darmstadt, G.F.R.) was transferred into a 500 ml conical flask. Dilute hydrochloric acid was prepared by mixing 50 ml of concentrated (ca 36%) acid with 550 ml of water. The zinc powder was swirled vigorously by hand with one third of the dilute hydrochloric acid for 40 s (note 1) then water (200 ml) was added in order to stop the evolution of hydrogen. The liquid was decanted from the zinc, which was treated subsequently with a second portion of 200 ml of dilute acid in the same way for 30 s. This treatment was carried out (after addition of water and decanting) for a third time with the remaining dilute acid. After decanting the liquid, the powder was shaken twice with 100 ml of distilled water. The flask was then provided with a mechanical stirrer and a third portion of 100 ml of distilled water was added. Stirring was started at a rate such that all zinc powder was homogeneously suspended. A solution of $CuSO_4$ (5 g) in 100 ml of distilled water was added to the stirred suspension during 10 s. Stirring was then stopped and the powder was allowed to precipitate. The supernatant liquid was removed by decantation, distilled water (100 ml) was added and the same procedure was repeated. After the third treatment with $CuSO_4$ the powder was washed successively three times wiht 75-ml portions of distilled water, three times with 75-ml portions of 96% ethanol and three times with 75-ml portions of 100% ethanol. The activated zinc powder obtained in this way was used directly: it was transferred, together with the third portion of 100% ethanol, into the reaction flask.

Note 1. From our first experiments with Zn-Cu couples, prepared from different batches of zinc powder, we found that the results (yield, and sometimes purity) of the allene preparations (see Chapter VI, Exp. 44-46) varied considerably from one batch to another. After many experiments we concluded that there is some connection between the results and the behaviour of the zinc during the treatment with hydrochloric acid and $CuSO_4$. A smooth reaction and good results were predicted and obtained whenever the evolution of hydrogen started immediately after addition of dilute acid, causing the powder to move slowly up and down, when swirling was stopped for a while. After washing with water the powder was still finely divided. The results were less satisfactory when, during the treatment with acid, the formation of porous spongy clusters of powder took place. During the evolution of hydrogen the powder remained on the bottom of the flask. Further, we had the impression that the treatment with $CuSO_4$ solution did not result in a satisfactory fixing of the copper on the zinc particles. Although we cannot give an explanation for the varying results, we believe that the structure of the powder (size of the particles?) has a considerable influence on the results of the allene syntheses. It also seems important to remove the water completely during the washing with ethanol. Traces of water can remain when aggregates of zinc are formed during the treatment with acid or by inefficient washing with ethanol.

Chapter III

DERIVATIZATION OF ACETYLENES AND CUMULENES

1. FUNCTIONALIZATION WITH VARIOUS ELECTROPHILIC REAGENTS

Cumulenic anions, $^{-}C{=}C{=}C$ and $^{-}C{=}C{=}C{=}C$, without strongly electron-withdrawing substituents are much stronger bases than acetylides[1], $^{-}C{\equiv}C-$ and are therefore also stronger nucleophiles. In view of the poor stability of the cumulenic anions at normal temperatures this is a fortunate circumstance: the usual functionalization reactions such as alkylation, trimethylsilylation and carboxylation in most cases proceed at a sufficient rate at low temperatures, provided that the polarity of the solvent system is satisfactory. The addition of a small amount of the very polar HMPT in reactions of cumulenic anions with alkyl halides[21] or epoxides[22] considerably enhances the rates of alkylation and hydroxyalkylation. The anions derived from allenic sulfides, $RCH{=}C{=}\bar{C}{-}SR'$, which are stable in liquid ammonia, appear to be exceptionally powerful nucleophiles[23].

The enormous difference in basicity between the two anionic centers in the dilithio compounds $LiC{\equiv}C{-}CH(Li)R$ may explain the regiospecific alkylation[12-15], silylation[16], hydroxyalkylation[16], sulfenylation[8], and carboxylation[8]:

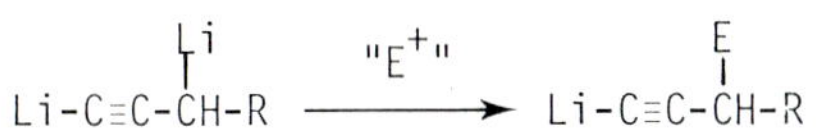

If two equivalents of the reagents are used, disubstitution to $EC{\equiv}C{-}CH(E)R$ occurs in most cases, but interestingly the reaction of $LiC{\equiv}CCH(Li)R$ with an excess of CO_2 gives mainly the allenic dicarboxylic acids[24]. These are probably the result of a rapid isomerization of the primary dilithium salt of the acetylenic diacid during the work-up or during the reaction of the dilithio compound with CO_2:

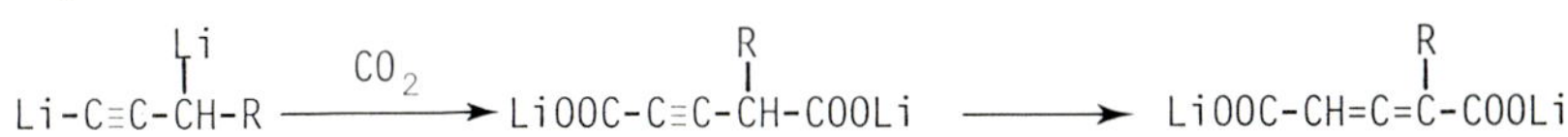

The reaction of cumulenic anions with electrophiles in principle may give two products:

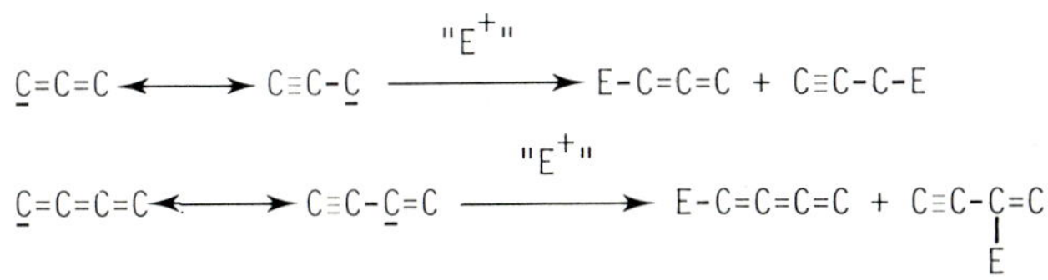

Although several experimental data concerning the functionalization of cumulenic anions are available (for a review see Ref. 10) (Table I), it remains difficult to predict the composition of the product of a particular reaction on a rationalistic basis. Especially the outcome of reactions with carbonyl compounds seems to be strongly dependent upon small variations in reaction parameters (solvent, counter ion, substituents in the anion and carbonyl compound[25-27]).Table I gives a selection of reactions with various electrophilic reagents, which afford reasonably clean products (see also the experimental section of this chapter).

2. EXPERIMENTS

(*1*) *1,2-Nonadiene*[22, 28]

$$H_2C{=}C{=}CH_2 \xrightarrow[\text{hexane-THF}]{C_4H_9Li} H_2C{=}C{=}CHLi \xrightarrow[\text{hexane-THF-HMPT}]{C_6H_{13}Br} H_2C{=}C{=}CH\text{-}C_6H_{13}$$

Apparatus: 1-l flask, see Fig. 1.

To a solution of 0.35 mol of allenyllithium in about 240 ml of hexane and 200 ml of THF, prepared according to Chapter II, Exp. 13, were added 35 g of dry HMPT at $-85^{\circ}C$. *n*-Hexyl bromide (0.30 mol) was then added dropwise in 15 min, whilst keeping the temperature of the reaction mixture close to $-70^{\circ}C$ (liquid nitrogen bath). After the addition the mixture was held at $-60^{\circ}C$ for an additional 1 h. The solution was then poured into 200 ml of 2 N HCl. The upper layer was washed eight times with 150-ml portions of 2 N HCl. The combined aqueous layers were extracted twice with 50-ml portions of redistilled pentane. These layers were shaken four times with 100-ml portions of 2 N HCl. The combined solutions (now practically free from THF) were dried over magnesium sulfate, after which the greater part of the solvents was distilled off at 760 mmHg through a 40-cm Vigreux column. Careful distillation of the residue through the same column gave 1,2-nonadiene, b.p. $45^{\circ}C/15$ mmHg, n_D^{22} 1.4418, in 88% yield.

In a similar way was prepared the allene $H_2C{=}C{=}CH(CH_2)_4CH{=}C{=}CH_2$ b.p. $75^{\circ}C/15$ mmHg, n_D^{20} 1.4883, in 70% yield, from 0.35 mol of allenyllithium and 0.15 mol of 1,4-dibromobutane.

The alkylations proceeded much more slowly, when ethyl- or butyllithium in diethyl ether, prepared from the alkyl bromides, had been used for the metallation of allene, in spite of the presence of THF and HMPT as co-solvents.

TABLE I

Starting compound	*Lithio compound or anion*	*Electrophilic reagent*	*Product* *Cumulene*	*Acetylene*	*Ref.*
$H_2C=C=CH_2$	$H_2C=C=CHLi$	RHal, CO_2	+	-	28
$R^1R^2C=C=CH_2$ R^1 = alkyl, R^2 = H or alkyl	$R^1R^2C=C=CHLi$	RHal, CO_2, RSSR, epoxides, Me_3SiCl	+	-	8, 25, 28
$ROCH=C=CH_2$	$RO\text{-}\overset{Li}{C}=C=CH_2$	various electro- philes, incl. carbonyl comp.	+	-	29-34
RSCH=C=CHR' R = alkyl, R' = H or alkyl	$RS\text{-}\bar{C}=C=CHR'$	H_2O, RHal, epoxides, carbo- nyl comp., NH_3 as solvent	+	-	23, 27, 32
$RS\text{-}C{\equiv}CCH_2R'$	$RS\text{-}\bar{C}=C=CHR'$	same reagents, NH_3 as solvent	+	-	23, 27 32
$RO\text{-}CH=C=C=CR^1R^2$ R = alkyl, R^1 and R^2 = H or alkyl	$RO\text{-}\overset{Li}{C}=C=C=CR^1R^2$	RHal, epoxides, carbonyl comp., organic solvents	+	-	21, 35
$RO\text{-}CH_2C{\equiv}C\text{-}R'$ R = alkyl, R' = alkyl or aryl	$RO\text{-}CH=C=\overset{Li}{C}\text{-}R'$	RHal, Me_3SiCl, organic solvents	+	-	36, 37
$RS\text{-}CH_2C{\equiv}C\text{-}R'$ R = alkyl, R' = alkyl	$RS\text{-}CH=C=\overset{Li}{C}\text{-}R'$	RHal organic solvents	-	+	8
$(RS)_2CHC{\equiv}C\text{-}CH_3$	$(RS)_2C=C=\overset{Li}{C}\text{-}CH_3$	RHal, ketones	-	+	38
		RSSR, $RSSO_2R$, organic solvents	+	-	38

(2) *6-Chloro-1,2-hexadiene*[22, 28]

$$H_2C{=}C{=}CHLi + Br(CH_2)_3Cl \xrightarrow[\text{hexane-THF-HMPT}]{} H_2C{=}C{=}CH\text{-}(CH_2)_3Cl$$

Apparatus: 1-l flask, see Fig. 1.

To a solution of 0.35 mol of allenyllithium in 240 ml of hexane and 200 ml of THF (see Chapter II, Exp. 13) were added 25 g of dry HMPT at $-80^{o}C$. Subsequently 0.30 mol of 1-bromo-3-chloropropane were added in 10 min. The reaction was very exothermic, but could be kept under control by occasional cooling in a bath with liquid nitrogen. After an additional 10 min the cooling bath was removed and the temperature was allowed to rise to $-30^{o}C$. The solution was then poured into 500 ml of water. The organic layer and three ethereal extracts were dried over magnesium sulfate. The solvents were distilled off as thoroughly as possible at normal pressure, using a 40-cm Vigreux column. Careful distillation of the remaining liquid afforded the allenic chloride, b.p. $40^{o}C/15$ mmHg, n_D^{20} 1.4738, in 88% yield.

(3) *5,6-Heptadien-1-ol*

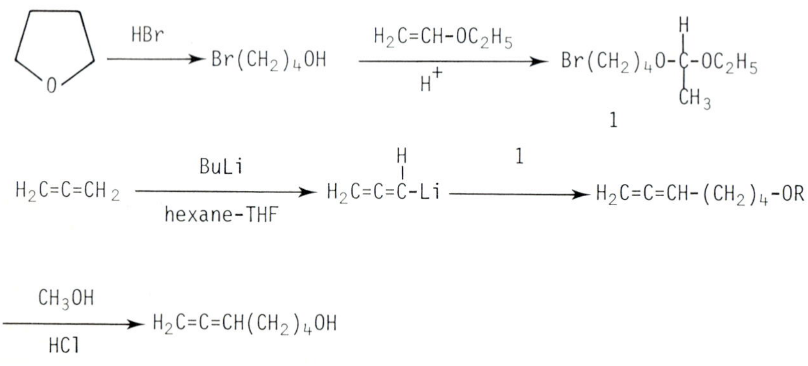

Apparatus: 250-ml round-bottomed flask for the cleavage of THF; 500 ml flask (see Fig. 1) for the coupling with ethyl vinyl ether; 1-l flask (see Fig. 1) for the reaction with allenyllithium; 500-ml round-bottomed flask for the deprotection.

Gaseous hydrogen bromide (note 1) was introduced in 100 ml of dry THF until the weight had increased by 33-34 g (0.40 mol). During the introduction of HBr the temperature of the solution was kept between 50 and $60^{o}C$ by occasional cooling. The conversion was terminated by warming the solution for an additional 30 min at 50°C. The solution of 4-bromo-1-butanol in THF obtained was then added in 20 min to 70 g (excess) of freshly distilled ethyl vinyl ether, in which 100-200 mg of anhydrous *p*-toluenesulfonic acid had been dissolved (after cooling below 0°C). During this addition the temperature of the mixture was kept between 0 and +5°C (bath with dry-ice and acetone). After stirring for an additional 30 min at 10°C, the solution was cooled below 0°C. A solution of 2 g of KOH in 4 ml of water was added with vigorous stirring. Anhydrous potassium carbonate (20 g) was then introduced, stirring being continued for an additional 5 min. The clear solution was poured off from the drying agent and, after rinsing the latter with diethyl ether, the solvents were removed by evaporation in a water-pump vacuum. The weight of the residue, the crude bromoether 1, corresponds to an overall yield of about 90%.

A solution of 0.35 mol of allenyllithium in about 240 ml of hexane and 240 ml of THF was prepared as described in Chapter II, Exp. 13. To the obtained solution (partly a suspension) were added 40 ml of dry HMPT at -80°C with vigorous stirring. Subsequently the crude bromoether was added over a period of 20 min, while keeping the temperature between -80 and -60°C by cooling in a bath of liquid nitrogen. The reaction was strongly exothermic. After the addition the cooling bath was removed and after 20 min the almost clear solution was poured into 500 ml of water. After shaking and separation of the layers the organic solution was washed five times with 200-ml portions of water in order to remove the HMPT (note 2). The combined aqueous layers were extracted twice with pentane or hexane and the extracts were subsequently washed twice with water. The combined organic solutions were concentrated in a water-pump vacuum (drying is not necessary). To the remaining liquid were added 150 ml of methanol and 3 ml of 36% hydrochloric acid. The mixture was warmed for 10 min at 60°C and subsequently the volatile compounds were removed in a water-pump vacuum (rotary evaporator). The residue was dissolved in 200 ml of diethyl ether and the ethereal solution was stirred for 20 min at 20°C with 25 g of machine-powdered KOH (note 3). After addition of water, extraction with diethyl ether and drying over magnesium sulfate, the solvent was removed by evaporation in a water-pump vacuum and the remaining liquid was distilled through a 25-cm Vigreux column to give the allenic alcohol, b.p. 88°/12 mmHg, n_D^{20} 1.4760, in an overall yield of 82%.

Note 1. If no cylinder is available, the HBr can be conveniently prepared by careful addition of the required amount of water to a vigorously stirred 1:1 ($^{v}/v$) mixture of phosphorus tribromide and carbon tetrachloride, using a very efficient reflux condenser.

Note 2. It is essential to remove the HMPT completely, as it cannot be separated by distillation from the allenic alcohol after the removal of the protecting group.

Note 3. By this treatment some bromobutanol, formed from unconverted $Br(CH_2)_4OCH(CH_3)OC_2H_5$, is removed by cyclization to tetrahydrofuran.

(4) *Tributylallenyltin*[56]

$$HC{\equiv}C\text{-}CH_2Br \xrightarrow[\text{diethyl ether}]{Mg} HC{\equiv}C\text{-}CH_2MgBr \xrightarrow[\text{diethyl ether}]{(C_4H_9)_3SnCl} H_2C{=}C{=}CH\text{-}Sn(C_4H_9)_3$$

Apparatus: 1-l flask, see Fig. 1.

To a solution of propargylmagnesium bromide, prepared according to Chapter II, Exp. 8 from 0.70 mol of propargyl bromide, was added 0.55 mol of tributyltin chloride in 10 min at -30°C. The cooling bath was then removed and the temperature was allowed to rise. Not much heat was evolved. The mixture was kept at room temperature for 12 h (overnight). Working up was carried out by pouring the mixture into a solution of 100 g of ammonium chloride in 500 ml of water. After vigorous shaking the upper layer was separated and dried over magnesium sulfate. Removal of the diethyl ether by evaporation in a water-pump vacuum gave an almost quantitative yield of a 95:5 mixture of $H_2C{=}C{=}CHSn(C_4H_9)_3$ and $HC{\equiv}CCH_2Sn(C_4H_9)_3$, n_D^{19} 1.4992.

Ph_3SnCl and propargylmagnesium bromide give a 90:10 mixture of the acetylenic and allenic tin derivatives, while in the case of Me_3SnCl this ratio is 30:70[56].

(5) *2,3-Butadienoic acid*

$$H_2C{=}C{=}\overset{\displaystyle H}{\overset{|}{C}}\text{-}Li \xrightarrow[\text{hexane-THF}]{CO_2} H_2C{=}C{=}\overset{\displaystyle H}{\overset{|}{C}}\text{-}COOLi \xrightarrow[H_2O]{H^+} H_2C{=}C{=}CH\text{-}COOH$$

Apparatus: 2-l flask, see Fig. 1 (without dropping funnel).

In the flask was placed a solution of 60 g of carbon dioxide in 250 ml of dry THF, made by introducing the gas from a cylinder into THF, cooled at about -90°C (liquid nitrogen bath). The gas inlet was removed and through the neck of the reaction flask was poured a cold solution (-70°C) of 0.40 mol of allenyllithium in 280 ml of hexane and 280 ml of THF (prepared as described in Chapter II,

Exp. 13). During this addition, which was carried out in 5 min, the mixture was stirred vigorously and its temperature was kept between -80 and -90°C by cooling in a bath of liquid nitrogen (care should be taken that no solid crust is formed on the bottom of the flask, in which case the addition of the allenyllithium solution should be interrupted). The cooling bath was removed directly after the addition and stirring was continued for an additional period of 10 min. The white suspension was then poured into 200 ml of a saturated solution of ammonium chloride. The mixture was acidified to pH 1 by addition of the required amount of 3 N HCl. After separation of the layers the aqueous layer was extracted six times with small portions of THF. The combined solutions were stirred for 2 h with about 100 g of magnesium sulfate. The drying agent was then filtered off on a sintered-glass funnel and rinsed with THF. The solution was concentrated in a water-pump vacuum. Subsequently traces of THF and water were removed by warming the remaining viscous residue for 1 h at 30°C in a vacuum of 0.1-0.5 mmHg. Crystallization of the remaining solid material from a 1:5 mixture of dichloromethane and pentane gave the pure allenic acid, m.p. 62°C, in 80% yield.

(6) *Allenic carboxylic acids*[8]

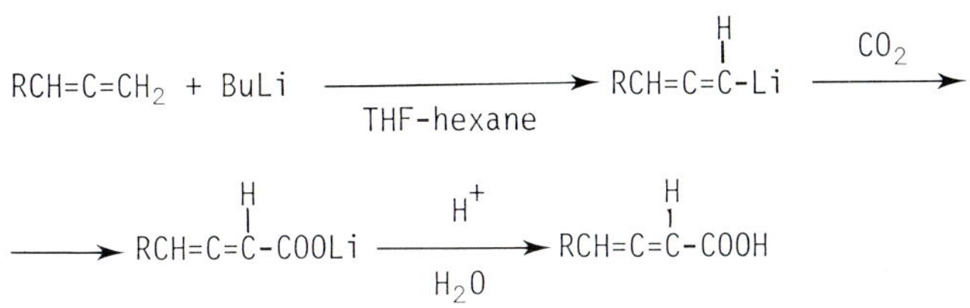

R = CH_3, n-C_4H_9, t-C_4H_9

Apparatus: 1-l flask, see Fig. 1.

A solution of 0.20 mol of butyllithium in about 140 ml of hexane was cooled to -50°C and 140 ml of dry THF were added. The mixture was cooled to about -80°C (liquid nitrogen bath) and 0.23 mol of the allenic hydrocarbon (see Chapter VI, Exp. 1, 2, 44) was added in 5 min (methylallene was added as a 1:1 solution in THF). The solutions were kept for 1 h at -65°C. Into another 1-l flask (see also Fig. 1, but without a dropping funnel), cooled at -90°C by immersion in liquid nitrogen, was poured a solution of dry carbon dioxide (from a cylinder) in 130 ml of dry THF. This solution was obtained by introducing about 40 g of carbon dioxide (note 1) into the THF at -90°C. The gas inlet was removed from the second flask and the solution of the lithiated allene (still cooled below -60°C) was poured through the open neck during 2 min. Care was taken that this solution was not poured along the walls of the flask but directly into the vigorously stirred carbon dioxide solution. The reactions were exothermic, so that cooling with liquid nitrogen was necessary to keep the temperature of the reaction mixtures between -70 and -95°C. After the addition the cooling bath was removed and the gas inlet tube was placed back on the flask. Stirring was continued for an additional 10 min. The white suspension was then poured cautiously into 300 ml of water in a 2-l conical flask. The flask was rinsed with water. After shaking, the upper layer was separated off and washed with two 40-ml portions of water. The washings were combined with the other aqueous layer. The organic layer was discarded. The combined aqueous layers were shaken twice with diethyl ether or (preferably) pentane, after which they were saturated with ammonium chloride. A mixture of 20 ml of 36% HCl and 50 ml of water was added, followed by six extractions with small portions of THF (when R = C_4H_9 or *tert.*-C_4H_9 diethyl ether can also be used). The combined extracts were dried by stirring them for 2 h with 50 g of magnesium sulfate. The drying agent was filtered off on a sintered-glass funnel and rinsed well with THF. The almost colourless filtrate was concentrated in a water-pump vacuum, after which the last traces of THF and water were removed by warming the residue for 1 h at 40°C (at 30°C when R = methyl) in a high vacuum (0.5-0.1 mmHg). The acid with R = *tert.*-C_4H_9, in 72% yield, solidified after cooling. The undistilled product appeared to be have a purity > 96%. Crystallization from pentane gave an m.p. of 43°C. The acids with R = CH_3 and C_4H_9 were obtained as liquids by the mentioned procedure. Distillation of the products through a very short column gave the acids in yields of about 45%. The main impurities (*ca.* 15%) were the acids $H_2C{=}C{=}C(R)COOH$ and $RC{\equiv}CCH_2COOH$, probably formed from the other lithiated allenes, $RC(Li){=}C{=}CH_2$. Considerable amounts of viscous residues remained after the distillation (note 2).

Note 1. The large excess was used to prevent any metallation of RCH=C=CHCOOLi by the lithiated allenes during the addition of the latter derivatives. For the same reason inverse addition was applied: introduction of carbon dioxide into the solution of RCH=C=CHLi has been shown to give bad results.

Note 2. The actual yields (based on MNR spectroscopy) are between 75 and 85%. The residues are probably dimers, formed during the strong heating during the distillations.

(7) *2-Methyl-2,3-butadienoic acid*[8]

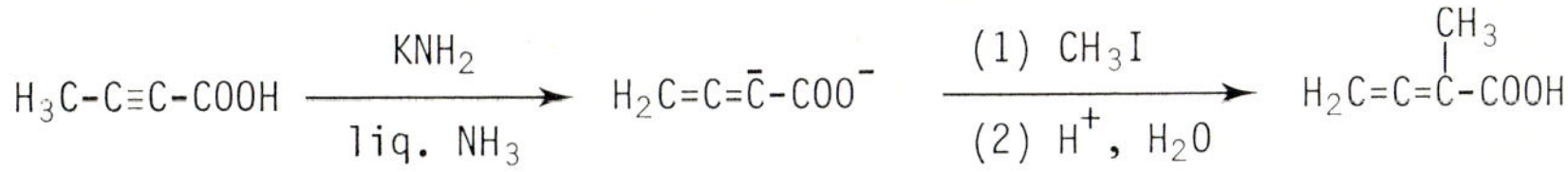

Apparatus: 3-l flask, see Fig. 2.

To a solution of 0.6 mol of potassium amide in 2 l of liquid ammonia (prepared as described in Chapter II, Exp. 12, and subsequently freed from small pieces of potassium by filtration through glass-wool) was added in about 5 min a solution of 0.20 mol of 2-butynoic acid (see Exp. 35) in 150 ml of liquid ammonia. During this addition, which was carried out by pouring the solution into the solution of potassium amide, both outer necks of the reaction flask were open. The ammoniacal solution of the acid was prepared by addition of the required amount of anhydrous liquid ammonia to the acid in a 500-ml round-bottomed flask. After the addition of the dissolved acid to the potassium amide the reaction flask was provided with the dropping funnel and gas outlet, as indicated in Fig. 2. The thick, greyish suspension was stirred for 90 min, then a mixture of 0.25 mol of methyl iodide and 100 ml of diethyl ether was added in 10 min. A considerable part of the suspended material passed into solution. Ten minutes after the addition of the methyl iodide the dropping funnel and gas outlet were removed and 100 g of powdered ammonium chloride were introduced in small portions with vigorous stirring. The ammonia was evaporated by warming the flask in a water--bath at 45°C. In order to effect complete removal of the ammonia, the flask was evacuated by means of the water pump as soon as the stream of escaping ammonia vapour had become faint. During the evacuation, which took about 1 h, the flask was immersed in a water-bath at 40°C. The remaining solid material was dissolved in 300 ml of ice-water, then the solution was acidified to pH 1 with 3 N hydrochloric acid. The mixture was then extracted ten times with diethyl ether and the ethereal extracts were dried (without previous washing) over magnesium sulfate and subsequently concentrated in a water-pump vacuum. Distillation of the remaining liquid in a high vacuum (0.1-0.5 mmHg) gave 18 g of crude product, which was dissolved in 50 ml of pentane. After standing for 12 h at -25 to -35°C the crystalline material was sucked off on a sintered-glass funnel. After drying in a water-pump vacuum the m.p. was 47-48°C. From the mother liquor a second batch of reasonably pure product was obtained, making the yield 45%.

(8) *6,6-Dimethyl-3,4-heptadien-1-ol*[22]

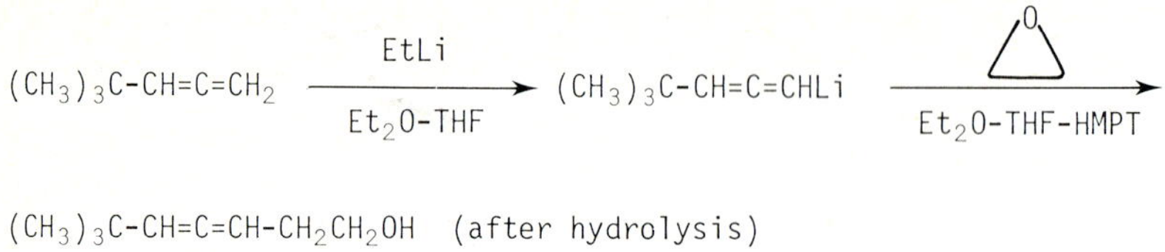

Apparatus: 1-l flask, see Fig. 1.

To a solution of 0.20 mol of ethyllithium in about 180 ml of diethyl ether (see Chapter II, Exp. 1) were added 280 ml of dry THF with cooling below -30°C. The mixture was cooled to -70°C and 0.21 mol of *tert.*-butylallene (see Chapter VI, Exp. 2) was added in 10 min while keeping the temperature between -60 and 65°C. The salt gradually dissolved. After an additional 30 min (at -60°C) the solution was cooled to -75°C and 19 ml of dry, pure HMPT and 0.4 mol (large excess) of ethylene oxide (cooled below 0°C) were added successively in 1-2 min. The temperature of the mixture was held at -60°C for 2 h, and was then allowed to rise gradually in 2 h to 0°C. Ice-water (200 ml) was added (with stirring) and, after separation of the layers, five extractions with diethyl ether were carried out. The combined organic solutions were washed five times with saturated sodium chloride solution and subsequently dried over magnesium sulfate. After concentration of the extract in a water-pump vacuum the residue was distilled through a 25-cm Vigreux column. The allenic alcohol, b.p. 87°C/18 mmHg, n_D^{21} 1.4664, was obtained in 76% yield.

The same procedure with allene gave a mixture of comparable amounts of $H_2C{=}C{=}CH{-}CH_2CH_2OH$ and $CH_3C{\equiv}C{-}CH_2CH_2OH$.

(9) *4,4-Dimethyl-1-trimethylsilyl-1,2-pentadiene*[22]

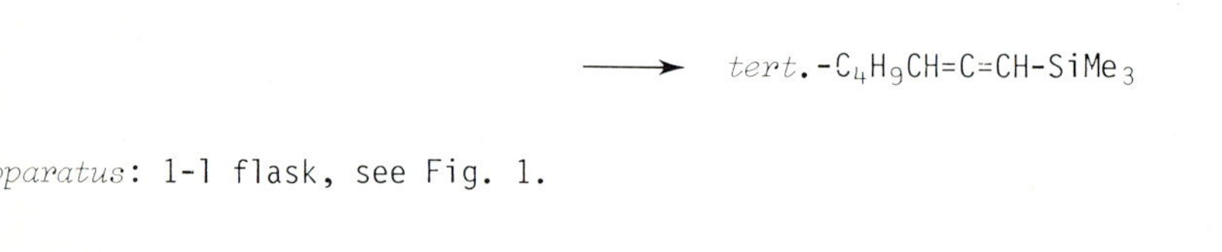

Apparatus: 1-l flask, see Fig. 1.

To a solution of 0.22 mol of butyllithium in about 150 ml of hexane were added 120 ml of dry THF with cooling below -10°C. The mixture was cooled to -70°C and 0.22 mol of *tert.*-butylallene (see Chapter VI, Exp. 2) was added in 10 min, while maintaining the temperature at about -70°C. After 30 min the solution of the lithiated allene was cooled to -100°C in a bath of liquid nitrogen. The cooling bath was then removed and 0.20 mol of trimethylchlorosilane was added with vigorous stirring in a few seconds. Fifteen minutes later the reaction mixture was poured into 200 ml of saturated ammonium chloride solution. After vigorous shaking the layers were separated and the aqueous layer was extracted with two 40-ml portions of diethyl ether. The combined solutions were dried over magnesium sulfate. After the greater part of the solvents had been distilled off at normal pressure through a 40-cm Vigreux column, the remaining liquid was distilled. The trimethylsilylated allene, b.p. 50°C/15 mmHg, $n_D^{22.5}$ 1.4406, was obtained in 75% yield.

Trimethylsilylation of allenyllithium afforded predominantly $HC{\equiv}CCH_2SiMe_3$, while in the cases of the homologues of allene (R = CH_3 or primary alkyl) 10-20% contamination by $RC{\equiv}CCH_2SiMe_3$, probably formed by trimethylsilylation of $RC(Li){=}C{=}CH_2$, was present.

(10) 3-Methoxy-1,2-heptadiene[29]

$$H_2C{=}C{=}CHOCH_3 \xrightarrow[(2)\ C_4H_9Br\ +\ HMPT]{(1)\ EtLi/Et_2O} H_2C{=}C{=}C(C_4H_9)(OCH_3)$$

Apparatus: 1-l flask, see Fig. 1.

To a solution of 0.30 mol of ethyllithium (note 1) in about 270 ml of diethyl ether (see Chapter II, Exp. 1) was added 0.30 mol of methoxyallene at -20°C (see Chapter IV, Exp. 4) at a rate such that the temperature could be kept between -15 and -20°C. Fifteen minutes later a mixture of 0.27 mol of *n*-butyl bromide and 100 ml of pure, dry HMPT was added in 5 min with efficient cooling, so that the temperature of the reaction mixture remained below 0°C. The cooling bath was then removed and the temperature was allowed to rise. After 4 h the brown reaction mixture was poured into 200 ml of ice-water. The aqueous layer was extracted twice with diethyl ether. The combined solutions were washed with concentrated ammonium chloride solution (which had been made slightly alkaline by addition of a few millilitres of aqueous ammonia, note 2) and dried over potassium carbonate. After addition of a small amount (2-5 ml) of diethylamine the greater part of the diethyl ether was distilled off at normal pressure through a 40-cm Vigreux column. The residue was carefully distilled through a 40-cm Widmer column, giving the alkylated allenic ether, b.p. 40°C/20 mmHg, n_D^{18} 1.4462, in 72% yield.

Note 1. If the lithiation of the allenic ether is performed with butyllithium in hexane and THF as a co-solvent, subsequent alkylation (in the presence of a small amount of HMPT) is much faster. The separation of the volatile product from the hexane and THF is difficult, however.

Note 2. The product can isomerize in the presence of traces of acids, giving $H_2C{=}CH{-}C(OCH_3){=}CH{-}C_3H_7$. All glassware of the distillation apparatus was rinsed with a dilute solution of diethylamine in diethyl ether.

(*11*) *3-Methoxy-1,2-nonadiene*[29]

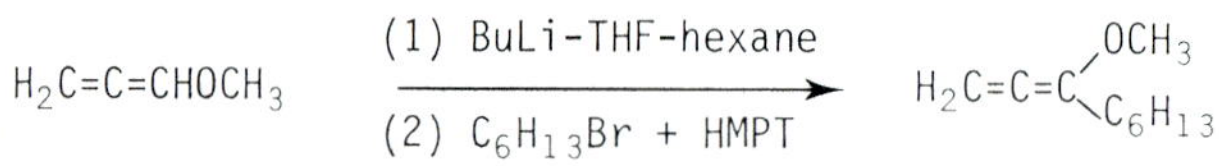

Apparatus: 1-l flask, Fig. 1.

A solution of 0.30 mol of the lithiated allenic ether, prepared according to Chapter II, Exp. 15, was cooled to -30°C and 0.28 mol of hexyl bromide was added in 5 min, while maintaining the temperature at -30°C. Hardly any reaction was observable. The temperature was lowered to -35°C and 40 ml of dry HMPT was added dropwise in 10 min. Efficient cooling was necessary to keep the temperature below -15°C. Ten minutes after the addition of the HMPT the cooling bath was removed and the temperature was allowed to rise to 15-20°C. The solution was then poured into 200 ml of ice-water, to which 2 ml of aqueous ammonia had been added (note 1). After shaking the layers were separated. The aqueous layer was extracted three times with diethyl ether. The combined solutions were washed three times with a saturated aqueous solution of NaCl, to which 2 ml of aqueous ammonia had been added and were subsequently dried over potassium carbonate. Evaporation of the solvents in a water-pump vacuum, followed by distillation (note 1) of the residue through a 40-cm Vigreux column, afforded the desired allenic ether, b.p. 72°C/15 mmHg, n_D^{20} 1.4492, in 85% yield.

Note 1. This addition is to ensure that no traces of acids are present which cause isomerization to $H_2C{=}CH{-}C(OCH_3){=}CH{-}C_5H_{11}$. All glassware of the distillation apparatus was therefore rinsed with a dilute solution of diethylamine in acetone, prior to distillation.

(*12*) *3-Methoxy-3,4-pentadien-1-ol*[30]

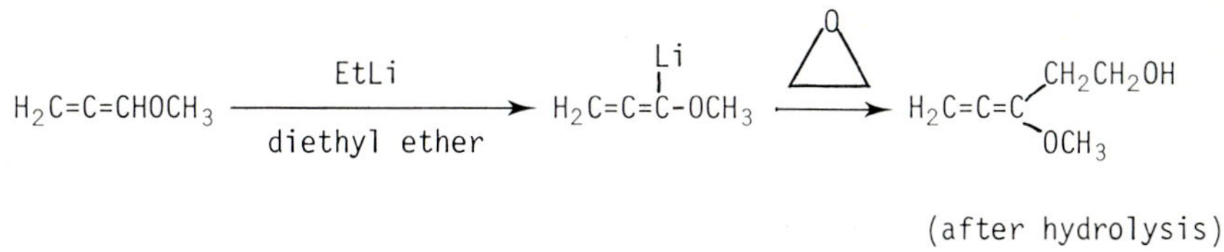

Apparatus: 1-l flask, see Fig. 1.

A solution of α-lithiomethoxyallene was prepared from methoxyallene and 0.20 mol of ethyllithium (note 1) in about 200 ml of diethyl ether (see Chapter II, Exp. 15). The solution was cooled to $-50^{\circ}C$ and 0.20 mol of ethylene oxide was added immediately. The cooling bath was removed temporarily and the temperature was allowed to rise to $-15^{\circ}C$ and was kept at this level for 2.5 h. The mixture was then poured into 200 ml of saturated ammonium chloride solution, to which a few millilitres of aqueous ammonia had been added (note 2). After shaking the layers were separated. The aqueous layer was extracted six times with small portions of diethyl ether. The combined ethereal solutions were dried over sodium sulfate and subsequently concentrated in a water-pump vacuum. Distillation of the remaining liquid through a short column gave the desired alcohol, b.p. $\sim 55^{\circ}C$/1 mmHg, n_D^{20} 1.4768, in 76% yield. Repeated distillation of a small portion in a water-pump vacuum gave b.p. $76^{\circ}C$/12 mmHg, n_D^{20} 1.4768.

Note 1. A solution of 1-lithiomethoxyallene, prepared from methoxyallene and BuLi-hexane-THF, did not react with ethylene oxide below $20^{\circ}C$. The reaction started at about $30^{\circ}C$, but the reaction mixture became very dark. An impure product was obtained in a considerably lower yield.

Note 2. The ammonia was added to ensure the absence of traces of acids, which may cause cyclization of the product. All glassware, used for the work-up and distillation must be rinsed with gaseous ammonia or with a dilute solution of triethylamine in acetone or diethyl ether.

(*13*) *3-Methoxy-4-ethoxy-1,2-butadiene*[33]

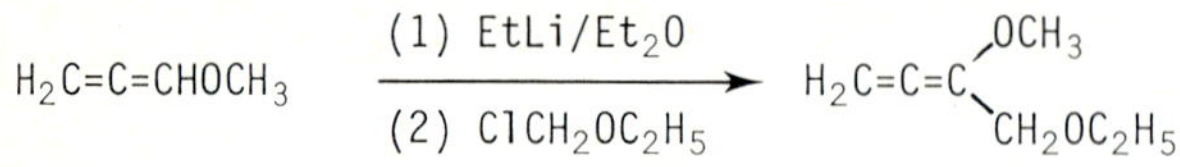

Apparatus: 1-l flask, see Fig. 1 (note 1).

To a solution of 0.33 mol of ethyllithium in about 270 ml of diethyl ether (see Chapter II, Exp. 1) was added 0.33 mol of methoxyallene (see Chapter IV, Exp. 4) in 10 min with cooling at $-30^{o}C$. After an additional 15 min 0.30 mol of α-chloromethyl ethyl ether (note 2) was introduced in 10 min, while keeping the temperature between -20 and $-30^{o}C$. A white precipitate of lithium chloride was formed. The cooling bath was then removed and the temperature was allowed to rise to $+10^{o}C$. The mixture was hydrolyzed by shaking it with 200 ml of a solution of 30 g of ammonium chloride, to which 5 ml of aqueous ammonia had been added. The aqueous layer was extracted with diethyl ether. The combined ethereal solutions were dried over potassium carbonate, after which the greater part of the diethyl ether was distilled off at normal pressure through a 40-cm Vigreux column (bath temperature $< 90^{o}C$). Careful distillation of the remaining liquid afforded the bis-ether, b.p. $47\text{-}49^{o}C$/18 mmHg, n_D^{20} 1.4469, in 78% yield.

Note 1. Compare also Exp. 10 for notes, and precautions during the work-up.

Note 2. Prepared by introducing gaseous HCl at $-10^{o}C$ into a mixture of 30 g of paraformaldehyde (corresponding to 1 mol of formaldehyde) and 1 mol of ethanol, until copious fumes escaped from the mixture. This was cooled (without stirring) to $-70^{o}C$ and the upper layer was decanted from the solid (frozen hydrochloric acid) and mixed with 50 g of *NN*-diethyl(or dimethyl)aniline. Subsequent distillation in a partial water-pump vacuum afforded the desired chloroether (b.p. about $40^{o}C$/40-50 mmHg).

(*14*) *3-Methoxy-4-ethoxy-1,2-pentadiene*[33]

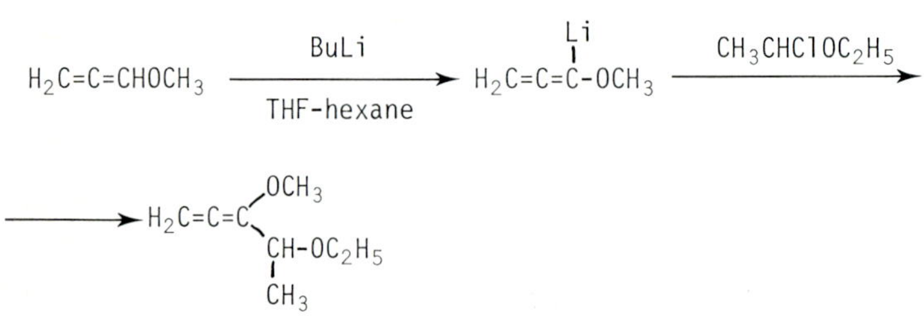

Apparatus: 1-l flask, see Fig. 1.

A solution of 0.33 mol (note 1) of lithiated methoxyallene in about 240 ml of hexane and 250 ml of THF (see Chapter II, Exp. 15) was cooled to $-30^{\circ}C$. A mixture of 0.30 mol of α-chloroethyl ethyl ether and ethyl vinyl ether (note 2) was added at this temperature in 15 min. After an additional 30 min the cooling bath was removed and the temperature was allowed to rise to $10^{\circ}C$. The white suspension was then poured into 200 ml of a solution of 30 g of ammonium chloride to which 5 ml of aqueous ammonia had been added. After vigorous shaking the upper layer was separated off and the aqueous phase was extracted three times with diethyl ether. The extracts (washing with water was not carried out) were dried over potassium carbonate. The greater part of the solvents was distilled off at normal pressure through a 40-cm Vigreux column (bath temperature $< 100^{\circ}C$) (note 3). After cooling to room temperature the remaining liquid was carefully distilled through a 40-cm Widmer column, giving the bis-ether, b.p. $47^{\circ}C$/15 mmHg, n_D^{20} 1.4438, in 80% yield.

Note 1. This excess was used to be absolutely sure that all α-chloroether would have reacted. Traces of this compound, if still present in the reaction mixture, will hydrolyse during the aqueous work up. The acid that is liberated can cause hydrolysis of the product to $H_2C{=}CH{-}C({=}O)CH(CH_3)OC_2H_5$.

Note 2. Prepared by introducing 0.30 mol of dry gaseous HCl (weight increase) into 45 ml of freshly distilled ethyl vinyl ether (excess) at $-30^{\circ}C$. This mixture should be used directly.

Note 3. All glassware of the distillation apparatus must be rinsed before use with a dilute solution of triethylamine or another volatile amine in diethyl ether. Traces of acid on the glass walls may cause isomerization to $H_2C{=}CH{-}C(OCH_3){=}C(CH_3)(OC_2H_5)$.

(15) *1-Trimethylsilyl-1-methoxyallene*

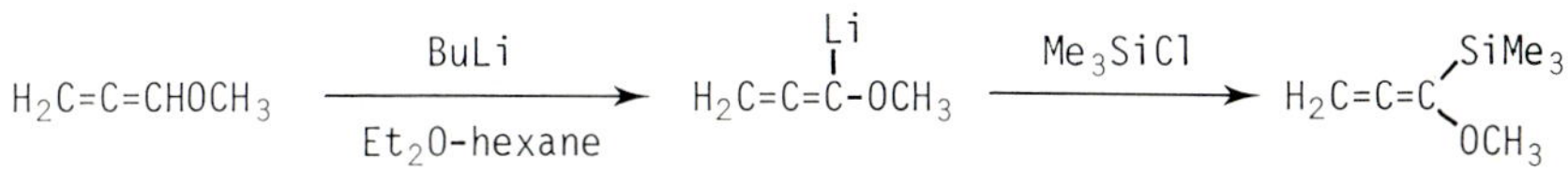

Apparatus: 250-ml flask, see Fig. 1.

To a solution of 0.12 mol of butyllithium in about 80 ml of hexane were added 100 ml of dry diethyl ether with cooling below $-10^{\circ}C$. Subsequently 0.14 mol of methoxyallene (see Chapter IV, Exp. 4) was introduced in 5 min at $-30^{\circ}C$. Fifteen minutes later 0.10 mol of trimethylchlorosilane was added dropwise in 2-3 min. The cooling bath was removed and the temperature was allowed to rise to 10-15°C. A white precipitate of lithium chloride was formed gradually. The mixture was poured into 100 ml of a solution of 20 g of ammonium chloride. After shaking and separation of the layers three extractions with diethyl ether were carried out. The combined organic solutions were dried over magnesium sulfate. After the greater part of the solvents had been distilled off at normal pressure through a 40-cm Vigreux column (bath temperature not higher than 90°C) the remaining liquid was carefully distilled in a partial vacuum through the same column. The silylated allenic ether, b.p. ~ 70°C/100 mmHg, n_D^{20} 1.4442, was obtained in 65% yield (note 1).

Note 1. This relatively low yield is probably due to the difficult separation of hexane and the product. Better results can presumably be obtained if the lithiation of methoxyallene is performed with ethyllithium in diethyl ether (see Chapter II, Exp. 1).

In a similar way $H_2C{=}C{=}C(OCH_3)(SnBu_3)$, n_D^{20} 1.4955 (undistilled) was prepared in almost quantitative yield from 0.12 mol of butyllithium in 75 ml of hexane and 75 ml of diethyl ether, 0.14 mol of methoxyallene and 0.10 mol of tributyltin chloride. The product contained 8-10% of an impurity, possibly $Bu_3Sn{-}CH_2C{\equiv}C{-}OCH_3$.

(16) *1-Methoxy-1-methylthioallene*[30]

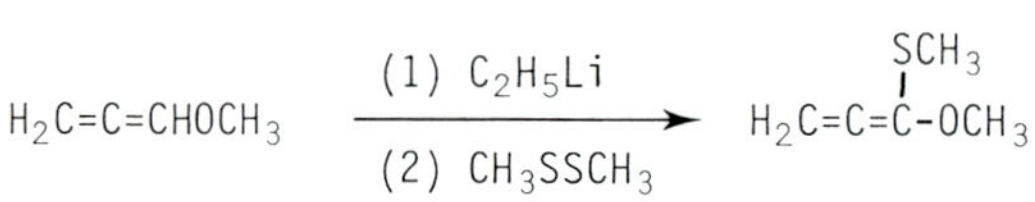

Apparatus: 1-l flask, see Fig. 1.

To a solution of 0.20 mol of ethyllithium (see Chapter II, Exp. 1) in about 200 ml of diethyl ether was added 0.22 mol of methoxyallene (see Chapter IV, Exp. 4) with cooling to about -20°C. The addition was carried out in about 10 min. Ten minutes later the solution was cooled to -60°C and a mixture of 0.20 mol of dimethyl disulfide and 50 ml of diethyl ether was added in 15 min with efficient cooling, so that the temperature could be kept below -40°C. Five minutes after the addition the mixture was poured into ice-water and three extractions with diethyl ether were carried out. The combined solutions were dried over magnesium sulfate and concentrated in a water-pump vacuum. Distillation of the residue through a 30-cm Vigreux column gave the allenic derivative, b.p. 50°C/15 mmHg, n_D^{20} 1.5215, in 76% yield (note 1).

Note 1. The product is very sensitive to oxygen. All operations during the work-up must therefore be carried out under nitrogen.

Similar results are probably obtained when the metallation of the allenic ether is carried out with butyllithium in hexane-THF or diethyl ether.

(*17*) *Carbinols derived from methoxyallene*[30]

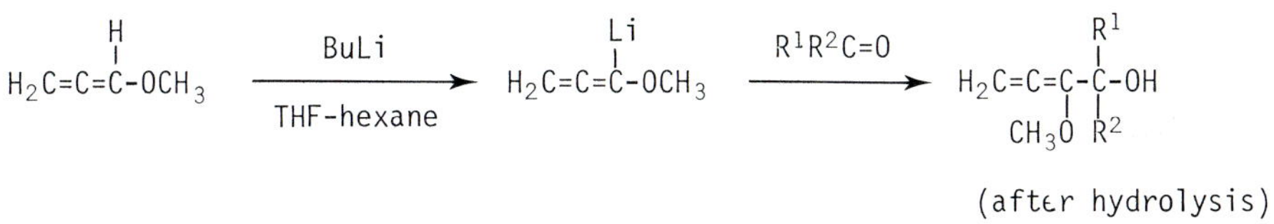

R^1 = H or alkyl
R^2 = alkyl

Apparatus: 500-ml flask, see Fig. 1.

A solution of 0.10 mol of lithiated methoxyallene in about 70 ml of hexane and 50 ml of THF (see Chapter II, Exp. 15) was cooled to -40°C. Dry, pure acetone (0.12 mol) was added dropwise during 10 min, while keeping the temperature at about -30°. Five minutes after the addition 100 ml of saturated NH_4Cl solution, to which 5 ml of aqueous ammonia had been added (note 1), were run in with vigorous stirring. The product was extracted three times with diethyl ether. The combined organic solutions were dried over potassium carbonate and subsequently concentrated in a water-pump vacuum. Distillation through a 30-cm Vigreux column afforded the desired carbinol, b.p. 52°C/12 mmHg, n_D^{20} 1.4640, in 79% yield.

The following carbinols were prepared in an analogous way (note 2) in yields varying from 80 to 88%:

$R^1 = R^2 = C_2H_5$, b.p. 68°C/12 mmHg, n_D^{20} 1.4594;
$R^1 = H$, $R^2 = CH_3$, b.p. 58°C/12 mmHg, n_D^{20} 1.4720;
$R^1 = H$, $R^2 = C_2H_5$, b.p. 66°C/12 mmHg, n_D^{20} 1.4688;
$R^1R^2C = (CH_2)_5$, b.p. 100°C/12 mmHg, n_D^{20} 1.4955.

Note 1. The carbinols are sensitive toward acids. All glassware used for the work-up and distillation must be rinsed with a dilute solution of di- or triethylamine in diethyl ether or acetone.

Note 2. The aldehydes were distilled before use.

(*18*) *1-Iodo-1-methoxyallene*[57]

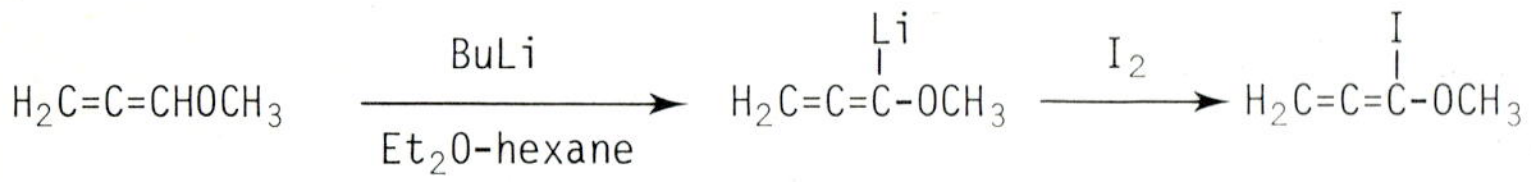

Apparatus: 500-ml flask, see Fig. 1.

A solution of 0.13 mol of 1-lithiomethoxyallene in about 80 ml of hexane and 90 ml of diethyl ether (see Chapter II, Exp. 15) was cooled to $-40^{o}C$. From the dropping funnel was added a solution of 0.10 mol of iodine in 150 ml of dry diethyl ether in 15-25 min, while keeping the temperature between -30 and $-40^{o}C$ (note 1). Five minutes after the addition the light-yellow solution was poured into 200 ml of ice-water. The aqueous layer was extracted twice with diethyl ether (note 2). The combined organic solutions were dried over magnesium sulfate and subsequently concentrated in a water-pump vacuum, keeping the bath temperature below $25^{o}C$. After complete removal of the hexane a light-brown residue remained, consisting of fairly pure 1-iodo-1-methoxyallene, n_D^{20} 1.573 in *ca.* 80% yield. Attempted distillation in a water-pump vacuum often gave rise to vigorous decomposition.

Note 1. When an excess of iodine is used a very impure product is obtained which may decompose explosively after removal of the solvent.

Note 2. All operations during the aqueous work-up must be carried out without any delay: the product is probably water-sensitive.

(*19*) *3-Methyl-1-tert.-butoxy-1,2-pentadiene*[9]

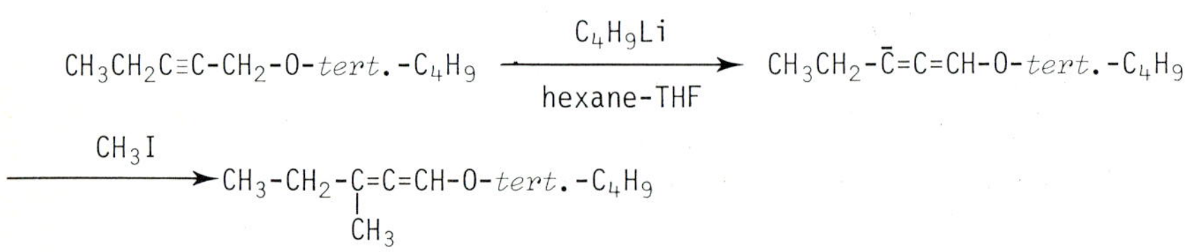

Apparatus: 250-ml flask, see Fig. 1.

To a mixture of 50 ml of dry THF and 0.050 mol of 1-*tert.*-butoxy-2-pentyne (prepared by ethylation of $HC{\equiv}CCH_2O$-*tert.*-C_4H_9 in liquid ammonia[1] was added 0.055 mol of butyllithium in about 35 ml of hexane in 10 min at -30°C. After stirring for 20 min at -25°C the solution was cooled to -50°C and 0.06 mol of methyl iodide was added in one portion, followed 10 min later by 50 ml of water. The aqueous layer was separated and extracted twice with diethyl ether. The solutions were dried over magnesium sulfate and concentrated in a water-pump vacuum. Distillation of the residue gave the allenic ether, b.p. 60°C/15 mmHg, n_D^{20} 1.4523, in 94% yield.

Trimethylsilylation with trimethylchlorosilane affords the corresponding allene derivative, hydroxyalkylation with aldehydes and ketones gives mixtures of comparable amounts of acetylenic and allenic carbinols.

Note 1. To a suspension of 0.40 mol of lithium amide in 400 ml of liquid NH_3 (see Chapter II, Exp. 11) was added 0.30 mol of $HC{\equiv}CCH_2O$-*tert.*-C_4H_9[1, 76]. Subsequently 0.45 mol of C_2H_5Br was introduced in 30 min. After an additional 1 h the NH_3 was removed by placing the flask in a water-bath at 40°C. Addition of water, extraction with diethyl ether and distillation gave $C_2H_5C{\equiv}CCH_2O$-*tert.*-C_4H_9 in more than 85% yield.

(20) *4-Ethoxy-1,2,3-pentatriene*[21]

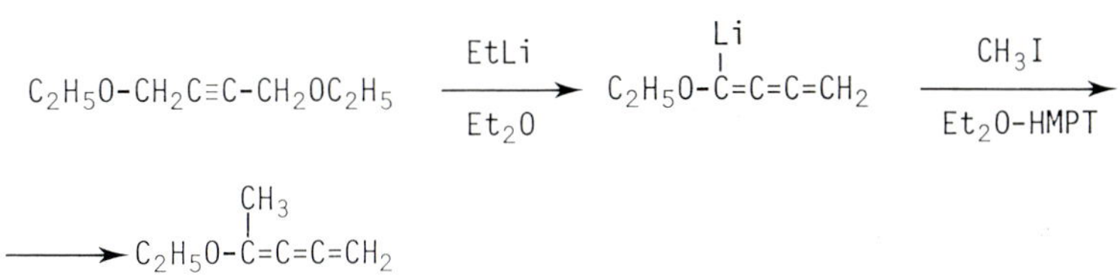

Apparatus: 1-l flask, see Chapter I, Fig. 1.

A solution of 0.60 mol of ethyllithium (note 1) in about 400 ml of diethyl ether (see Chapter II, Exp. 1) was added in 30 min to a mixture of 0.25 mol of 1,4-diethoxy-2-butyne (see Chapter VIII-6, Exp. 8) and 100 ml of dry diethyl ether. The temperature of the reaction mixture was kept between -40 and -45°C. Fifteen minutes after the addition had been completed, 0.5 mol of methyl iodide was added at -40°C, then 100 ml of dry HMPT (for the purification see ref. 1) were added dropwise in 15 min while keeping the temperature at about -40°C. Thirty minutes after this addition the cooling bath was removed, the temperature was allowed to rise and stirring was continued for 3 h. The mixture was then poured into ice-water (note 2). After shaking the ethereal layer was separated off and the aqueous layer was extracted twice with small portions of diethyl ether. The combined ethereal solutions were washed with a saturated solution of ammonium chloride and dried over anhydrous potassium carbonate. The diethyl ether was evaporated in a water-pump vacuum on a water-bath at 15°C and the remaining liquid was distilled through a 30-cm Vigreux column. During this distillation the receiver was cooled in ice. The cumulenic ether, b.p. 42°C/17 mmHg, n_D^{20} 1.5130, was obtained in 78% yield. The compound can be stored for a limited period (a few days) under nitrogen in a refrigerator at -30°C.

Note 1. If commercial BuLi in hexane is used with diethyl ether or THF as co-solvent, a dark brown reaction mixture is formed, from which the desired product can be isolated in lower yields.

Note 2. In view of the sensitivity of cumulenic ethers to oxygen the work up was carried out under nitrogen with water that had been saturated with nitrogen.

The reaction of lithiated cumulenic ethers with ethylene oxide, trimethylchlorosilane and carbonyl compounds shows the same regiospecificity as does the alkylation.

(21) *Alkylation and hydroxyalkylation of allenic sulfides*[23]

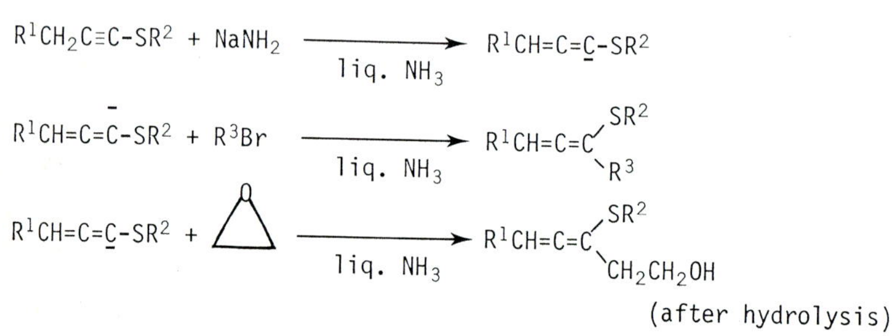

R^1 = H or alkyl; R^2 = CH_3, C_2H_5; R^3 = primary or *sec.*-alkyl (note 1).

Apparatus: 1-l flask, see Fig. 2.

To a suspension of 0.43 mol of sodium amide in 350 ml of liquid ammonia (see ref. 1 and Chapter II, Exp. 11) was added 0.40 mol of the acetylenic sulfide[1, 63] in 3 min. The dropping funnel was rinsed with diethyl ether and 0.45 mol of the alkyl halide (note 2) or a solution of 0.6 mol of ethylene oxide in 50 ml of diethyl ether was placed in it; 1, 2 or 4 min after the addition of the propynyl sulfides, butynyl sulfides or higher homologues, respectively, the alkyl halides or ethylene oxide were introduced in 10 min. The reactions were very vigorous, and almost directly after completion of the addition of the alkylating reagents or ethylene oxide the brown colour of the solution (indicating the presence of the anions) disappeared. The greater part of the ammonia was evaporated by placing the flask in a water-bath at 40-45°C: in the case of the reaction with ethylene oxide 25 g of solid ammonium chloride were cautiously added prior to carrying out this operation. To the residue were successively added 150 ml of diethyl ether and 200 ml of water. After stirring the layers were separated and the aqueous layer was extracted three times with diethyl ether. The combined ethereal solutions were dried over potassium carbonate and subsequently concentrated in a water-pump vacuum (note 3). The remaining liquid was distilled through a short Vigreux column in the cases of higher alkyl halides or ethylene oxide. When R^3 was methyl or ethyl, however, careful distillation through a 30-40-cm Widmer column was carried out in order to separate some starting compound or unalkylated allenic sulfide, $R^1CH=C=CH-SR^2$, from the alkylation product. In all cases yields were higher than 70%. Some typical examples are as follows:

$C_3H_7CH=C=C(SCH_3)CH_2CH_2OH$, b.p. 100°C/0.4 mmHg, n_D^{20} 1.5250, in 88% yield from $C_4H_9C\equiv C-SCH_3$

$H_2C=C=C(SCH_3)CH_2CH_2OH$, b.p. ~ 50°C/0.2 mmHg, n_D^{20} 1.5465, in 70% yield from $CH_3C\equiv C-SCH_3$

$CH_3CH=C=C(SC_2H_5)C_4H_9$, b.p. 70°C/0.3 mmHg, n_D^{20} 1.4968, in 84% yield from $C_2H_5C\equiv C-SC_2H_5$

$H_2C=C=C(SC_2H_5)C_2H_5$, b.p. 48°C/12 mmHg, n_D^{20} 1.5043, in 72% yield from $CH_3C\equiv C-SC_2H_5$

$CH_3CH=C=C(SC_2H_5)C_2H_5$, b.p. 62°C/8 mmHg, n_D^{20} 1.5019, in 74% yield from $C_2H_5C\equiv C-SC_2H_5$

$H_2C=C=C(SC_2H_5)$-*iso*-C_3H_7, b.p. 52°C/9 mmHg, n_D^{20} 1.4980, in 70% yield from $CH_3C\equiv C-SC_2H_5$

Note 1. Cyclohexyl and *tert.*-butyl halides give very low yields, due to dehydrohalogenation.

Note 2. In most cases *bromides* can be used, but in methylation methyl iodide will be more convenient.

Note 3. All glassware used for the work-up and distillation must be rinsed with a dilute solution of triethylamine in diethyl ether or acetone in order to be sure that traces of acids on the glass walls have been neutralized. Allenic sulfides with the structure C=C=C(SR)-CH- isomerize under the influence of acids to give conjugated dienes, C=C-C(SR)=C.

(22) *4-Phenyl-1-butyne*[16]

$$H_3C\text{-}C{\equiv}CH \xrightarrow[\text{THF-hexane}]{\text{2 BuLi}} LiCH_2\text{-}C{\equiv}C\text{-}Li \xrightarrow{Ph\text{-}CH_2Cl} PhCH_2CH_2C{\equiv}C\text{-}Li$$

$$\xrightarrow{H_2O} Ph\text{-}CH_2CH_2C{\equiv}CH$$

Apparatus: 1-l flask, see Fig. 1.

A suspension of dilithiopropyne in THF and hexane was prepared from 0.25 mol of propyne (see Chapter II, Exp. 17). The suspension was cooled to -10°C and 0.22 mol of benzyl chloride was added dropwise in 30 min, while maintaining the temperature of the mixture at about -10°C. The cooling bath was then removed temporarily and the temperature was allowed to rise to +5°C. After stirring for 1 h at this temperature the mixture was poured into a solution of 30 g of ammonium chloride in 300 ml of water. The upper layer, formed after shaking, was separated off and the aqueous phase was extracted twice with diethyl ether. The combined solutions were dried over magnesium sulfate and then concentrated in a water-pump vacuum. The residue was carefully distilled through a 30-cm Widmer column, giving the acetylene, b.p. 72°C/15 mmHg, n_D^{20} 1.5212, in 64% yield.

(23) *1-Chloro-5-hexyne*[12]

$$HC{\equiv}C\text{-}CH_3 \xrightarrow[\text{hexane-THF}]{\text{2 BuLi}} LiC{\equiv}C\text{-}CH_2Li \xrightarrow[\text{(2) } H_2O]{\text{(1) } Br(CH_2)_3Cl} HC{\equiv}C(CH_2)_4Cl$$

Apparatus: 1-l flask, see Fig. 1.

A suspension of $LiC{\equiv}C\text{-}CH_2Li$, prepared from 0.20 mol of propyne according to Chapter II, Exp. 17, was cooled to -40°C and 0.17 mol of 1-bromo-3-chloropropane was added in 5 min. The cooling bath was then removed. Above -20°C the temperature began to rise more rapidly and after 10-15 min the suspension had disappeared almost completely. After 1 hour the solution was warmed to 40°C and held at that temperature for another 1 h. It was then cooled to -20°C and 200 ml of ice-water were added with vigorous stirring. After separation of the layers the aqueous layer was extracted three times with diethyl ether. The combined organic solutions were dried over magnesium sulfate, after which the solvents were distilled off at 760 mmHg through a 40-cm Vigreux column. Distillation of the residue gave

1-chloro-5-hexyne, b.p. 60°C/45 mmHg, n_D^{20} 1.4465, in 73% yield.

The higher homologues of propyne, e.g. 1-decyne, can be obtained in a similar way. Starting from 1-butyne and its homologues, alkylation with alkyl halides leads to 1-alkynes with a branched substituent.

(24) *3-Butynoic acid*[8]

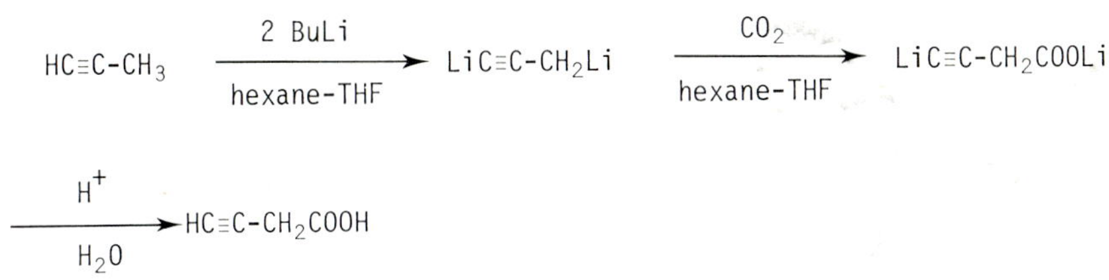

Apparatus: 2-l round-bottomed flask, see Fig. 1.

The air in the flask was completely replaced with nitrogen. After the flask had been cooled sufficiently by immersion in a bath of liquid nitrogen, a cold (-85 to -95°C) solution of 0.23 mol of carbon dioxide in 150 ml of dry THF (prepared by introducing the carbon dioxide into THF, cooled below -80°C) was poured in it. The inlet for nitrogen was removed, so that this neck of the flask was open. A pre-cooled (-50°C) suspension of dilithiopropyne (prepared from 0.20 mol of propyne and butyllithium, see Chapter II, Exp. 17) in THF-hexane was poured in 5 min through the neck. During this addition the mixture was agitated vigorously and its temperature was kept between -100 and -85°C. Stirring at about -90°C was continued for another 3 min, after which the suspension was poured with vigorous swirling into a mixture of 50 ml of 36% HCl and 250 ml of water. The aqueous layer was saturated with ammonium chloride whereupon the layers were separated. The aqueous layer was extracted five times with small portions of THF. The combined solutions were dried by stirring for 2 h with 50 g of magnesium sulfate. After suction filtration and rinsing of the drying agent with THF the solvents were removed in a water-pump vacuum by means of a rotary evaporator. High vacuum distillation gave a crystalline product, containing small amounts of impurities, *inter alia* some 2-butynoic acid. Crystallization from a 3:1 mixture of pentane and diethyl ether at low temperature gave the pure acid, m.p. 77°C, in 38-45% yields.

Homologues of propyne can be converted in a similar way into HC≡C-CH(R)COOH, but as these compounds are liquids, purification by crystallization is not possible.

(25) *2,3-Pentadienedioic acid*[24]

$$CH_3C{\equiv}CH \xrightarrow[\text{THF-hexane}]{\text{2 BuLi}} LiCH_2C{\equiv}C\text{-}Li \xrightarrow{CO_2} LiOOC\text{-}CH_2C{\equiv}C\text{-}COOLi \longrightarrow$$

$$\longrightarrow LiOOC\text{-}CH{=}C{=}CH\text{-}COOLi \xrightarrow[H_2O]{H^+} HOOC\text{-}CH{=}C{=}CH\text{-}COOH$$

Apparatus: 1-l flask (see Fig. 1) for the dilithiation; 2-l flask with a gas inlet, a mechanical stirrer and a thermometer, combined with a gas outlet.

A suspension of dilithiopropyne was made from 0.20 mol of propyne and 0.45 mol of butyllithium, as described in Chapter II, Exp. 17. The suspension was cooled to -50°C. The 2-l reaction flask was filled with dry nitrogen and cooled in a liquid nitrogen bath. A solution of 40-45 g of dry carbon dioxide (from a cylinder) in 250 ml of dry THF (made by introducing the required amount of carbon dioxide at -90°C) was poured into the flask and the gas inlet tube was replaced with a funnel. The solution was stirred vigorously, and the cooled suspension was poured through the funnel in 3 min. During this operation the flask was cooled by occasionally keeping it in the bath of liquid nitrogen so that the temperature of the reaction mixture remained between -60 and -90°C. After the addition the gas inlet tube was placed on the flask and a slow stream (∿ 300 ml/min) of dry carbon dioxide was passed through the suspension. This was brought to a temperature of +20°C by warming in a water-bath at 20-25°C. The suspension was then poured into 200 ml of ice water. After vigorous swirling a solution of 6 g of KOH in 20 ml of water was added (note 1). After standing for 1 h at room temperature the aqueous layer was saturated with ammonium chloride and 3 N HCl was added to render the pH of the solution 1. After separation of the layers the aqueous layer was extracted eight times with small portions of THF (note 2). The combined brown solutions were dried for 12 h over 100 g of magnesium sulfate. The drying agent was filtered off on a sintered-glass funnel and rinsed with some THF. The solution was transferred into a 500-ml wide-necked flask and concentrated in a water-pump vacuum. In order to remove the last traces of THF and some water the flask was connected directly (without a column) to a condenser and a receiver, cooled at -190°. The flask was evacuated by means of a mercury pump ($p < 0.1$ mmHg) and heated for 1.5 h at 85°C. The greater part of the monoacids (2-butynoic and 3-butynoic acid) passed over during this operation and remained as a solid crust in the condenser. The brown solid residue, weighing 14.2 g, seemed to be fairly pure HOOCCH=C=CHCOOH, as judged from the NMR spectrum. Crystallization from 30 g of dry acetonitrile at -25°C gave an NMR-pure, but still brown, crystalline product in 32% yield. The mother liquor contained appreciable amounts of the acid, but it appeared to be very difficult to effect crystallization.

Note 1. The primary product of the carboxylation is probably LiOOC-CH_2C≡C-COOLi. We presume that the dilithiopropyne causes partial isomerization into the allenic derivative. The treatment with KOH is to complete this transformation. When the latter treatment was not carried out, considerable amounts of the acetylenic diacids were sometimes found in the crude product.

Note 2. With diethyl ether continuous extraction has to be performed.

The low yield of the diacid may be explained by incomplete carboxylation, due to the slight solubility of the dilithio compound.

(26) *2-Methyl-2,3-pentadienedioic acid*[24]

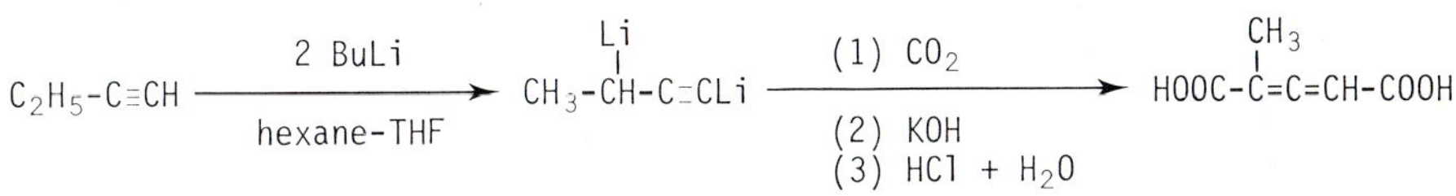

Apparatus: 2-l flask, see Fig. 3, but without dropping funnel.

The yellow suspension obtained from 0.20 mol of 1-butyne and butyllithium in a THF-hexane mixture (see Chapter II, Exp. 17) was cooled to -50°C and poured in 3 min through an open funnel, placed on the reaction flask, which contained a solution of 40 g of carbon dioxide in 150 ml of dry THF, cooled below -80°C (note 1). During the addition of the suspension the reaction mixture was stirred vigorously and cooled in a bath of liquid nitrogen: the temperature was held between -100 and -80°C. Directly after the addition the bath was removed and stirring was continued until the temperature had risen to +20°C. During this period carbon dioxide (0.2 l/min) was passed through the solution. The mixture was then cautiously poured into 150 ml of water. After vigorous shaking the aqueous layer was separated off. The organic layer was extracted twice with 40 ml portions of water and the aqueous layers were added to the first one, and subsequently a solution of 6 g of KOH in 20 ml of water (note 2) was added. After standing for 1 h at room temperature the aqueous solution was saturated with ammonium chloride and a mixture of 60 g of 35% hydrochloric acid and 50 ml of ice-water was cautiously added with swirling (some frothing due to evolution of carbon dioxide may occur at the beginning), so that the pH of the aqueous layer became 1. Subsequently seven extractions with 30-ml portions of THF were carried out. The extracts were dried over magnesium sulfate and then concentrated in a water-pump vacuum, using a bath temperature of 80°C. The syrupy residue was heated under reflux for 10 min with 50 ml of dichloromethane (note 3). Subsequently 150 ml of pentane were added to the warm mixture and, after cooling to room temperature, the solvent was decanted. The residue was heated for the first 15 min in a water-pump vacuum, and subsequently for 1.5 h at 80°C in a vacuum of 0.5 mmHg or lower. Dry acetonitrile(50 ml) was then added. After heating for 10 min under reflux the yellow-brown solution was cooled to about 50°C, then decanted from the brown solid material (salts). The latter was rinsed twice with 5-ml portions of acetonitrile, which were added to the main solution. The acetonitrile solution was placed in a refrigerator. The crystalline material was sucked off on a small sintered-glass funnel, rinsed with cold (-30°C) diethyl ether and dried in a water-pump vacuum. The crystalline product, still yellow, appeared to be the pure allenic diacid, yield 38%. From the mother liquor a second crop of crystalline material could be obtained (after concentration in a water-pump vacuum and cooling below -20°C), bringing the yield to 42%. Recrystallization of a small part from dry diethyl ether gave an m.p. of 163°C (uncorrected).

Note 1. Carbon dioxide from a cylinder was introduced into THF, cooled below -80°C, until the weight increase amounted to 40 g.

Note 2. If this treatment with KOH is omitted, the product may contain considerable amounts of $HOOC\text{-}CH(CH_3)C{\equiv}C\text{-}COOH$.

Note 3. In this way the monoacids, mainly $HOOC\text{-}CH(CH_3)C{\equiv}CH$, are removed: they dissolve in the CH_2Cl_2-pentane mixture.

(27) *3-Trimethylsilyl-1-heptyne*[16]

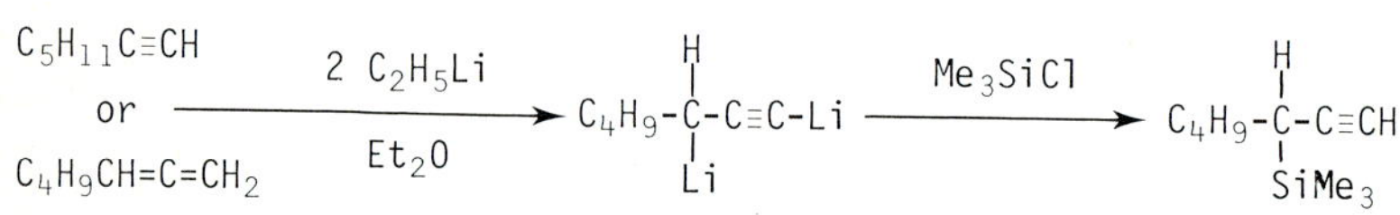

Apparatus: 1-l flask, see Fig. 1

To a solution of 0.50 mol of ethyllithium in about 450 ml of diethyl ether (see Chapter II, Exp. 1) was added 0.20 mol of 1-heptyne[1] or butylallene (see Chapter VI, Exp. 1) with cooling below $0^{o}C$. After the addition the cooling bath was removed and the thermometer-gas outlet combination was replaced with a reflux condenser. The solution was heated under reflux for 6 h. The thermometer-gas outlet was again placed on the flask and the yellow suspension was cooled to $-50^{o}C$. Trimethylchlorosilane (0.20 mol) was added dropwise in 10 min, while keeping the temperature between -40 and $-35^{o}C$. After having kept the mixture for an additional 30 min at $-30^{o}C$, it was poured into 200 ml of ice-water. The aqueous layer was extracted three times with small portions of diethyl ether. The combined ether solutions were dried over $MgSO_4$ and the greater part of the ether was distilled off at normal pressure through a 40-cm Vigreux column. The residue was distilled through the same column, giving the trimethylsilylated acetylene, b.p. $55^{o}C/17$ mmHg, n_D^{20} 1.4350, in 58-64% yield.

(28) *3-Ethynyl-1-hexanol*[16]

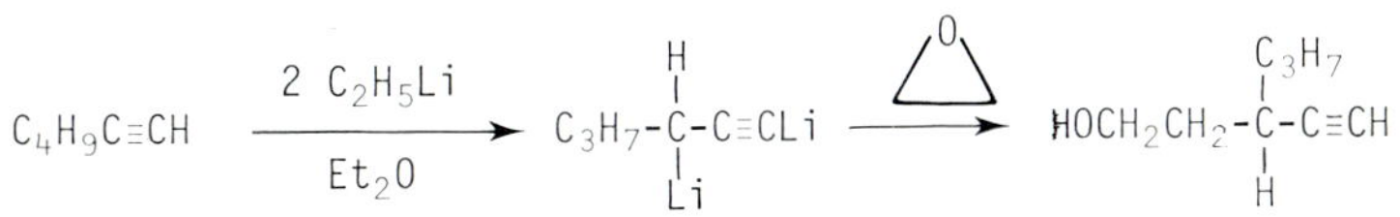

Apparatus: 1-l flask, see Fig. 1.

A suspension of dilithiohexyne in diethyl ether was made from 0.20 mol of 1-hexyne[1] and 0.5 mol ethyllithium in 400 ml of diethyl ether in the same way as described for 1-heptyne (see this chapter, Exp. 27). The suspension was cooled to $-40^{o}C$ and at this temperature a solution of 0.20 mol of ethylene oxide in 50 ml of diethyl ether was added in 15 min, the brown colour changing into yellow. Subsequently the temperature was allowed to rise gradually during 1 h to $+5^{o}C$. The solution was then poured into 200 ml of saturated ammonium chloride solution. After vigorous shaking the upper layer was separated off and the aqueous layer was extracted three times with diethyl ether. The combined ethereal solutions were dried over magnesium sulfate and then concentrated in a water-pump vacuum. The residue was distilled through a 30-cm Vigreux column, giving the desired alcohol, b.p. $80^{o}C/15$ mmHg, n_D^{20} 1.4472, in 69% yield.

(29) *3-Phenyl-3-trimethylsilyl-1-propyne*[16]

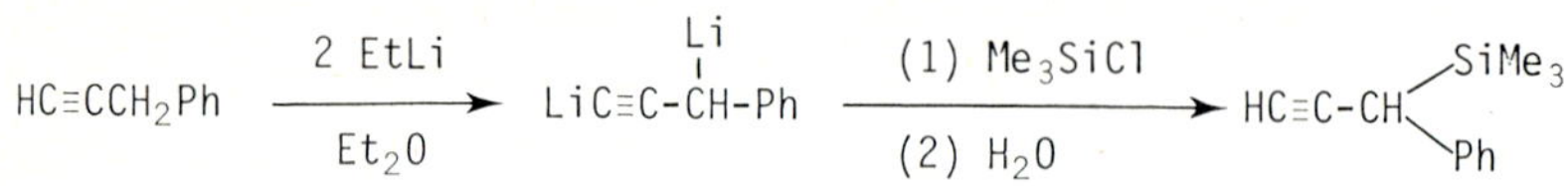

Apparatus: 1-l flask, see Fig. 1.

To a solution of 0.26 mol of ethyllithium (see Chapter II, Exp. 1) in 200 ml of diethyl ether was added 0.12 mol of benzylacetylene (see Chapter VI, Exp. 38) with cooling between -20 and -40°C. After the addition the brown solution was kept for 2 h at 15°C. It was then cooled to -45°C and 0.12 mol of freshly distilled trimethylchlorosilane was added in 10 min, while keeping the temperature of the reaction mixture at about -40°C. The cooling bath was then removed and after an additional 30 min the mixture was poured into ice-water. The upper layer, formed after shaking, was separated. The aqueous layer was extracted with diethyl ether and, after drying the combined solutions over magnesium sulfate, the diethyl ether was evaporated in a water-pump vacuum. Distillation of the residue through a 30-cm Vigreux column afforded the desired compound, b.p. 102°C/18 mmHg, n_D^{20} 1.5172, in 86% yield.

In a similar way HC≡C-CH(CH_3)Ph, b.p. 62°C/12 mmHg, n_D^{20} 1.5168, was prepared in 82% yield by addition of 0.12 mol of methyl iodide to the solution of LiC≡C-CH(Li)Ph[15].

(30) *3-Trimethylsilyl-3-methylthio-1-propyne*[16]

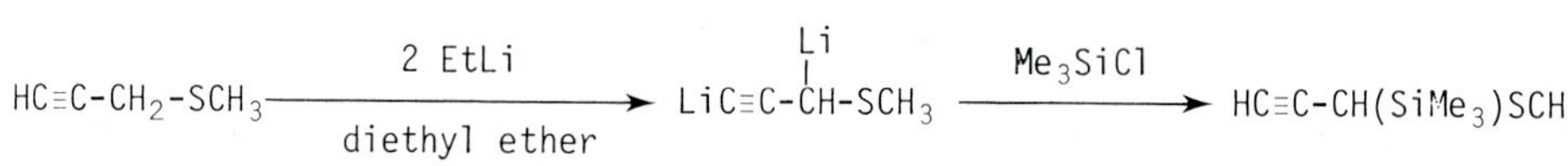

Apparatus: 1-l flask, see Fig. 1.

To a solution of 0.21 mol of ethyllithium in about 200 ml of diethyl ether (see Chapter II, Exp. 1) was added 0.10 mol of methyl propargyl sulfide (note 1) in 10 min with cooling at -20°C. After stirring for an additional 2 h at -15 to -20°C the brown solution was cooled to -60°C and 0.10 mol of trimethylchlorosilane was added in 20 min, while maintaining the temperature between -60 and -65°C. The cooling bath was then removed and the temperature was allowed to rise to -10°C. The mixture was poured into 300 ml of ice-water and after separation of the layers the aqueous phase was extracted three times with diethyl ether. The combined solutions were dried over magnesium sulfate and subsequently concentrated in a water-pump vacuum. Distillation of the residue through a 30-cm Vigreux column gave the silylated propargyl sulfide, b.p. 65°C/18 mmHg, n_D^{20} 1.4832, in 80% yield.

Alkylation of the dilithiated sulfide can be performed by adding at -10°C 0.12 mol of an alkyl bromide, allowing the temperature to rise to 20-30°C and hydrolyzing the reaction mixture after stirring for an additional 45 min at about 25°C.

The dilithiation can also be carried out with butyllithium in a 1:1 mixture of hexane and THF at -20°C (reaction time about 45 min). Subsequent alkylation is much faster than in diethyl ether.

Note 1. To a solution of 0.55 mol of sodium hydroxide in 160 ml of methanol and 40 ml of water was added a cold solution of 0.55 mol of methanethiol in 50 ml of methanol with cooling below 10°C. Subsequently 0.50 mol of propargyl chloride was added with swirling in 15 min, while keeping the temperature between 20 and 40°C. After an additional 30 min 1 l of water was added and the product was extracted six times with small amounts of pentane. The extracts were washed with water, dried and the solvent was carefully distilled off through a 40-cm Widmer column. Methyl propargyl sulfide, b.p. 65°C/100 mmHg, n_D^{20} 1.4920 was obtained in 80% or higher yield.

(*31*) *Trimethylsilylacetylene*

$$\mathrm{HC{\equiv}CH} \xrightarrow[\mathrm{THF},\ 25^{\circ}C]{C_2H_5MgBr} \mathrm{HC{\equiv}C\text{-}MgBr} \xrightarrow{Me_3SiCl} \mathrm{HC{\equiv}C\text{-}SiMe_3}$$

Apparatus: 3-l flask, see Fig. 1.

A solution of about 1.1 mol of ethynylmagnesium bromide (note 1) was cooled to -50°C and 1.0 mol of trimethylchlorosilane was added in 15 min with cooling between -5 and +5°C. After the addition stirring at 0-5°C was continued for 2 h. The mixture was allowed to stand at room temperature overnight (6 h will probably be sufficient). The mixture was then poured into 1 l of 1 N HCl. High-boiling light petroleum (b.p. > 180°C) (250 ml) was added and the mixture was shaken vigorously. After separation of the layers the organic solution was washed twelve times with 250-ml portions of 1 N HCl. The collected aqueous layers were extracted once with 50 ml of petroleum. The upper layer was washed six times with 1 N HCl and then combined with the other organic layer. After drying over magnesium sulfate the light petroleum solution was transferred into a 1-l round-bottomed flask, which was equipped for a vacuum distillation: 40-cm Vigreux column, condenser and single receiver, cooled at -70^{0}C (see Fig. 5). The apparatus was evacuated (10-15 mmHg) with a water pump and the flask was gradually heated. This operation was stopped when light petroleum began to reflux in the top of the column. Trimethylsilylacetylene, b.p. 52^{0}C/760 mmHg, n_D^{20} 1.3935, was obtained in 81-89% yield after redistillation through a 40-cm Vigreux column.

Note 1. The required solution of C_2H_5MgBr was prepared from 1.25 mol of C_2H_5Br, 45 g of Mg turnings and 600 ml of THF (see Chapter II, Exp. 6). This solution was added dropwise at 25^{0}C to 500 ml of THF, through which acetylene was bubbled, in accordance with the procedure described in *Organic Synthesis*, Collect. Vol. IV, p. 792.

(32) *Bis(trimethylsilyl)acetylene*[58]

$$HC{\equiv}CH \xrightarrow[\text{THF}]{C_2H_5MgBr} BrMgC{\equiv}CMgBr \xrightarrow{2\ Me_3SiCl} Me_3SiC{\equiv}CSiMe_3$$

Apparatus: 3-l flask, see Fig. 1.

A solution of ethylmagnesium bromide, prepared from 35 g of magnesium, 146 g of ethyl bromide and 1400 ml of THF, was warmed to 50^{0}C and acetylene (note 1) was introduced. The flow was adjusted such that the temperature of the mixture remained between 50 and 55^{0}C (note 2) and foaming did not become excessive. The introduction was stopped when the temperature had dropped to 40^{0}C. The jelly-like reaction mixture was then heated for an additional 1 h at 50^{0}C. Trimethylchlorosilane (130 g) was added in 20 min, while keeping the temperature of the reaction mixture between 55 and 60^{0}C. The gel disappeared gradually. After the addition, warming at 60^{0}C was continued for 2 h. The mixture was then cooled below 30^{0}C and subsequently poured into 1 l of a solution of 100 g of ammonium chloride. Pentane (100 ml) was added and, after shaking, the upper layer was separated off. The aqueous layer was extracted twice with 150-ml portions of pentane. The combined organic solutions were washed ten times with 150-ml portions of saturated ammonium chloride solution and subsequently dried over magnesium sulfate. After filtering through a sintered-glass funnel and rinsing the drying agent with pentane, the solvents were distilled off at normal pressure through a 40-cm Vigreux column. The product passed over between 128 and 138°C, the greater part between 134 and 136°C/760 mmHg. The yield was 87%. After a short time the product solidified.

Note 1. The acetylene was freed from acetone by means of two traps, cooled at -75°C.

Note 2. Because ethane escaped from the solution, the net heat effect was not very strong.

Although the original procedure also gives excellent yields, our procedure seems more economic because the use of the expensive butyllithium is avoided.

(*33*) *1-Ethoxy-2-trimethylsilylacetylene*

$$HC{\equiv}COC_2H_5 \xrightarrow[\text{THF-hexane}]{\text{BuLi}} LiC{\equiv}C\text{-}OC_2H_5 \xrightarrow{Me_3SiCl} Me_3SiC{\equiv}C\text{-}OC_2H_5$$

Apparatus: 1-l flask, see Fig. 1.

In the flask were mixed a solution of 0.20 mol of butyllithium in hexane and 120 ml of dry THF below -10°C. The mixture was cooled to -50°C and 0.22 mol of ethoxyacetylene (see Chapter V, Exp. 29) was added in 10-15 min, while keeping the temperature below -20°C. Subsequently 0.20 mol of trimethylchlorosilane was added in 10 min at -20°C. The cooling bath was then removed. After 3 h 2.5 ml of triethylamine (note 1) were added, then the mixture was poured into 200 ml of saturated NaCl solution. After shaking, the upper layer was separated off and the aqueous layer was extracted twice with diethyl ether. The combined solutions were dried over magnesium sulfate. Concentration of the dried solutions in a water-pump vacuum (bath temperature below 35°C) and distillation of the remaining liquid through a 40-cm Vigreux column (bath temperature not higher than 80°C, note 2) gave the trimethylsilyl derivative, b.p. 50°C/20 mmHg, n_D^{20} 1.4255, in 78% yield.

Note 1. The amine was added to ensure that after addition of water no traces of acid were liberated (from unconverted Me_3SiCl); the acid would catalyze the addition of water to the triple bond.

Note 2. 1-Alkynyl ethers may split off alkenes with the formation of ketenes (e.g. $Me_3SiCH{=}C{=}O$) upon strong heating (see Chapter XI in ref. 3a).

(*34*) *3-Trimethylsilyl-2-propyn-1-ol*

$$HC{\equiv}C\text{-}CH_2OH \xrightarrow[\text{diethyl ether}]{2\ C_2H_5Li} LiC{\equiv}C\text{-}CH_2OLi \xrightarrow{2\ Me_3SiCl} Me_3SiC{\equiv}C\text{-}CH_2OSiMe_3$$

$$\xrightarrow[H^+]{H_2O} Me_3Si\text{-}C{\equiv}C\text{-}CH_2OH$$

Apparatus: 2-l flask, see Fig. 1.

To a solution of 1.00 mol of ethyllithium in 800-900 ml of diethyl ether (see Chapter II, Exp. 1) was added, with cooling between -20 and -10°C, 0.50 mol of dry propargyl alcohol, dissolved in 100 ml of diethyl ether. Subsequently 1.1 mol of trimethylchlorosilane was introduced over a period of 25 min with cooling between -15 and +5°C. After stirring for an additional 2 h at about 30°C the suspension was poured into a solution of 30 g of acetic acid in 150 ml of water. After stirring for 1 h at room temperature the layers were separated and the aqueous layer was extracted four times with diethyl ether. The combined ethereal solutions were washed with sodium hydrogen carbonate solution in order to neutralize acetic acid, and were then dried over magnesium sulfate. The diethyl ether was removed by evaporation in a water-pump vacuum and the residue distilled through a 40-cm Vigreux column, giving the desired alcohol, b.p. 71°C/15 mmHg, n_D^{20} 1.4500, in 86% yield (note 1).

Note 1. If $Me_3SiC{\equiv}C\text{-}CH_2OSiMe_3$ appears to be present, the product should be shaken for a few minutes with 1 N HCl at room temperature.

Butyllithium in a mixture of hexane and diethyl ether or THF can presumably also be used for the dilithiation of propargyl alcohol.

(35) *2-Butynoic acid*[1]

$$CH_3C\equiv CH \xrightarrow[\text{THF-hexane}]{\text{BuLi}} CH_3C\equiv C\text{-}Li \xrightarrow{CO_2} CH_3C\equiv C\text{-}COOLi \xrightarrow[H_2O]{H^+} CH_3C\equiv C\text{-}COOH$$

Apparatus: 2-l round-bottomed flask, with a gas inlet tube, a mechanical stirrer and a thermometer, combined with a gas outlet.

In the flask was placed 0.50 mol of butyllithium in about 350 ml of hexane. The solution was cooled below -30°C while 400 ml of dry THF were run in. To the mixture obtained was added in a few portions a solution of 0.55 mol of propyne in 50 ml of THF, cooled at -30°C (introduction of nitrogen was interrupted during this short period). During this addition, which took about 5 min, the temperature of the mixture was held below -10°C by efficient cooling. The flow of nitrogen was then stopped and dry carbon dioxide was passed through the vigorously stirred suspension, while maintaining its temperature between 0 and -10°C. The rate of introduction of carbon dioxide was about 0.5 l/min, but it can be increased if cooling is sufficiently efficient. When no more heat was evolved, the cooling bath was removed. Introduction of carbon dioxide was continued for an additional 10 min. The suspension was then cautiously poured into a 3-l conical flask containing 200 ml of a solution of 60 g of ammonium chloride. The carboxylic acid was liberated from its salt by cautious addition of a mixture of 50 ml of 36% HCl and 50 ml of water. The upper layer was separated off and the aqueous phase was extracted five times with small portions of THF. After drying the combined solutions over magnesium sulfate the solvent was removed in a water-pump vacuum. The residue was distilled through a 30-cm Vigreux column, connected to an air condenser. After a preliminary aqueous fraction of the carboxylic acid the main fraction passed over at 100°C/15 mmHg. The compound solidified in the receiver and (partly) in the condenser. The yield was almost quantitative.

(36) *Methyl-1-propynyl sulfoxide*[59]

$$CH_3C\equiv CH \xrightarrow[\text{THF}]{\text{BuLi}} CH_3C\equiv CLi \xrightarrow{CH_3\overset{O}{\overset{\|}{S}}\text{-}Cl} CH_3C\equiv C\text{-}\overset{O}{\overset{\|}{S}}\text{-}CH_3$$

Apparatus: 1-l flask, see Fig. 1.

A solution of 0.30 mol of *n*-butyllithium in 210 ml of hexane was cooled to about -50^{0}C, 200 ml of dry THF were added and the temperature was immediately lowered to -60^{0}C. A solution of 0.32 mol of propyne in 50 ml of dry THF, pre--cooled below -30^{0}C, was added with efficient cooling, care being taken that the temperature of the mixture did not rise above 0^{0}C. Subsequently the mixture was cooled to -65^{0}C and a mixture of 0.30 mol of methanesulfinyl chloride (prepared according to the method described in ref. 73) and 50 ml of dry diethyl ether was added, while keeping the temperature below -60^{0}C. A white suspension was formed. Ten minutes after the addition the reaction mixture was poured into 100 ml of water, the upper layer was separated off and the aqueous layer was extracted ten times with 30-ml portions of chloroform (note 1). The chloroform extracts were dried (without previous washing) over magnesium sulfate. The THF--hexane-diethyl ether layer was also dried over magnesium sulfate. Concentration of the combined organic solutions in a water-pump vacuum gave a residue consisting of the almost pure sulfoxide, yield 96%. Distillation in a high vacuum gave the pure compound, b.p. 60-70^{0}C/0.5 mmHg, n_D^{18} 1.5122, in 78% yield.

Note 1. The sulfoxide has a very good solubility in water. Extraction with diethyl ether is inefficient.

Under similar conditions (diethyl ether or THF as solvent) the reaction of lithium alkynylides with methanesulfonyl chloride (CH_3SO_2Cl) gave the corresponding alkynyl sulfones in low to moderate yields.

(37) *Di(1-alkynyl) sulfides*[60]

$$RC{\equiv}CH \xrightarrow[Et_2O\text{-hexane}]{BuLi} RC{\equiv}CLi \xrightarrow{SCl_2} RC{\equiv}C\text{-}S\text{-}C{\equiv}CR$$

$R = CH_3$, C_2H_5, *tert.*-C_4H_9, Me_3Si

Apparatus: 2-l flask, see Fig. 1.

In the flask was placed a solution of 0.44 mol of butyllithium in about 300 ml of hexane. To this solution were added, with cooling below -20^{0}C, 800, 600 and 400 ml of dry diethyl ether (note 1) in the case of $R = CH_3$, C_2H_5 and *tert.*-C_4H_9 or Me_3Si, respectively. Subsequently 0.46 mol of the alkyne [in the case of $R = CH_3$, C_2H_5 a cooled (-30^{0}C) solution in 50 ml of diethyl ether] was added in about 10 min, while keeping the temperature below -20^{0}C. The suspension (in the case of R = CH_3 or C_2H_5) or solution of the lithium alkynylide was cooled in a liquid nitrogen bath and a cold (-30°C) solution of 0.20 mol of freshly distilled sulfur dichloride (note 2) (b.p. 58-62°C/760 mmHg) in 40 ml of diethyl ether was added in 15 min with vigorous stirring and cooling between -85 and -95°C. After the addition the cooling bath was removed and the temperature was allowed to rise to -50°C. The reaction mixture was then poured into 500 ml of water and,after vigorous shaking, the upper layer was separated off. The aqueous layer was extracted twice with small amounts of diethyl ether. The organic solutions were dried over magnesium sulfate (in the case of R = CH_3 and C_2H_5 some tarry material precipitated) and subsequently concentrated in a water-pump vacuum. The remaining liquids were distilled by means of the oil pump (note 3), using a short distillation column. The compounds with R = CH_3 and C_2H_5 were collected in a cooled receiver (0°C). The following results were obtained:

R = CH_3 , b.p. ∿ 30°C/0.2 mmHg, n_D^{20} 1.5335, yield 65-73%;
R = C_2H_5 , b.p. ∿ 40°C/0.2 mmHg, n_D^{20} 1.5193, yield 68-75%;
R = *t*-C_4H_9, b.p. ∿ 50°C/0.2 mmHg, n_D^{20} 1.4841, yield 82-86%;
R = Me_3Si , b.p. ∿ 55°C/0.2 mmHg, n_D^{20} 1.4930, yield 78-84%.

Note 1. When THF was used instead of diethyl ether as a co-solvent, the yields of $(CH_3C{\equiv}C)_2S$ and $(C_2H_5C{\equiv}C)_2S$ were only 30-40%.

Note 2. Sulfur dichloride was prepared by introducing the calculated amount of chlorine into a mixture of dry powdered sulfur and iron powder[74].

Note 3. During the distillation of the sulfide with R = CH_3 the bath temperature should be kept below 70°C. Too strong heating may result in explosive decomposition of the residue.

(38) *Di(1-alkynyl) sulfoxides*[60]

$$RC{\equiv}CH \xrightarrow[\text{or } C_2H_5Li/Et_2O]{BuLi/Et_2O\text{-hexane}} RC{\equiv}CLi \xrightarrow{SOCl_2} RC{\equiv}C\text{-}\overset{O}{\overset{\|}{S}}\text{-}C{\equiv}CR$$

R = CH_3, C_2H_5, *tert.*-C_4H_9

Apparatus: 1-l flask, see Fig. 1.

To a solution of 0.22 mol of butyllithium in 300 ml of diethyl ether and about 160 ml of hexane (obtained by adding the diethyl ether to the hexane solution of butyllithium below $0^{o}C$) was added with cooling between -20 and $-30^{o}C$ 0.24 mol of the alkyne [propyne and butyne were added as a cold ($-30^{o}C$) solution in 50 ml of diethyl ether]. The suspension (when R = CH_3 or C_2H_5) or solution (R = *tert.*-C_4H_9) of the alkynyllithium was cooled to $-90^{o}C$ and a cold ($-20^{o}C$) mixture of 0.10 mol of freshly distilled thionyl chloride and 40 ml of dry diethyl ether was added with vigorous stirring. During the addition which was carried out in about 10 min, the temperature was kept between -80 and $-90^{o}C$, care being taken that no solid crust was formed on the bottom of the flask (note 1). After stirring for an additional 10 min without cooling 200 ml of water were added. The upper layer was separated off and the aqueous layer was extracted three times with small portions of dichloromethane (when R = CH_3 or C_2H_5) or diethyl ether (when R = *tert.*-C_4H_9). The combined organic solutions were dried over magnesium sulfate and subsequently concentrated in a water-pump vacuum. The sulfoxide with R = C_2H_5 remained as a viscous oil, n_D^{20} 1.5220, yield 86%; the other sulfoxides were solids. Recrystallization at low temperatures from a 1:1 mixture of diethyl ether and pentane gave the pure sulfoxides: R = CH_3, m.p. $69.5^{o}C$; yield 70%; R = *tert.*-C_4H_9, m.p. $47^{o}C$, yield 88%.

Note 1. Partial contact of the flask with liquid nitrogen is sufficient to maintain the indicated temperature.

(39) *2-Butynenitrile and homologues*[61]

$$RCH_2C{\equiv}CH \xrightarrow[\text{diethyl ether}]{C_2H_5Li} RCH_2C{\equiv}CLi$$

$$RCH_2C{\equiv}CLi + ClC{\equiv}N \xrightarrow[\text{diethyl ether}]{} RCH_2C{\equiv}C\text{-}C{\equiv}N \qquad R = H, \text{alkyl}$$

Apparatus: 1-l flask, Chapter I, Fig. 1.

A solution of 0.40 mol of ethyllithium in about 350 ml of diethyl ether (see Chapter II, Exp. 1) was transferred into the flask, which previously had been filled with nitrogen. The solution was cooled to $-50^{o}C$ and a cold solution ($-30^{o}C$) of 0.43 mol of propyne in 50 ml of dry diethyl ether was added at a rate such that the temperature could be kept below $-20^{o}C$. A solution of 0.45 mol of cyanogen chloride in 100 ml of diethyl ether, cooled at about $0^{o}C$, was then added in

15 min, while the temperature of the reaction mixture (suspension) was maintained between -10 and -20°C. After this addition the cooling bath was removed and the temperature was allowed to rise to 15 or 20°C. After stirring for 3 h at this temperature the brown reaction mixture was cautiously poured into 500 ml of ice--water. After vigorous shaking (caution: some dissolved ethane may escape, causing an over-pressure in the separating funnel) the ethereal layer was separated off and the aqueous layer was extracted three times with small portions of diethyl ether. The combined solutions were dried over magnesium sulfate, after which the greater part of the diethyl ether was carefully distilled off through a 40-cm Widmer column. The remaining liquid was also distilled at normal pressure, using a 30-cm Vigreux column. 2-Butynenitrile, b.p. 109°C/760 mmHg, n_D^{20} 1.4340, was obtained in 84% yield. [In the last stage of the distillation a partial vacuum (*ca.* 50 mmHg) was applied in order to minimize the hold up.]

In essentially the same way were prepared the following compounds:

2-Pentynenitrile, $C_2H_5C{\equiv}C\text{-}C{\equiv}N$, b.p. 125°C/760 mmHg, n_D^{20} 1.4387, yield 86%,
2-Hexynenitrile, $C_3H_7C{\equiv}C\text{-}C{\equiv}N$, b.p. 40°C/15 mmHg, n_D^{20} 1.4444, yield 90%, and
2-Heptynenitrile, $C_4H_9C{\equiv}C\text{-}C{\equiv}N$, b.p. 56°C/15 mmHg, n_D^{20} 1.4501, yield 92%.

(*40*) *1,1,4-Triethoxy-2-butyne* (note 1)

$$HC{\equiv}C\text{-}CH(OC_2H_5)_2 \xrightarrow[(2)\ C_2H_5O\text{-}CH_2Cl]{(1)\ BuLi\text{-}hexane\text{-}THF} C_2H_5O\text{-}CH_2C{\equiv}C\text{-}CH(OC_2H_5)_2$$

Apparatus: 1-l flask, see Fig. 1.

To a solution of 0.40 mol of butyllithium in about 280 ml of hexane were added 280 ml of dry THF with cooling below -10°C. Subsequently 0.40 mol of 1,1-diethoxy--2-propyne (see Chapter V, Exp. 28) was introduced in 15 min at -30 to -10°C. To the solution obtained was then added in 15 min with cooling at about -15°C 0.40 mol of chloromethyl ethyl ether (note 2). After the addition stirring was continued for 1 h without cooling. The mixture was then shaken with concentrated ammonium chloride solution and the ethereal layer was separated off. The aqueous layer was extracted twice with diethyl ether. After drying the ethereal solutions over magnesium sulfate the diethyl ether was evaporated in a water-pump vacuum. Distillation of the residue through a 30-cm Vigreux column gave the desired compound, b.p. 105°C/15 mmHg, n_D^{20} 1.4336, in 88% yield.

Note 1. This compound can also be prepared by heating $C_2H_5O\text{-}CH_2C{\equiv}C\text{-}MgBr$ for 24 h with $HC(OC_2H_5)_3$ in diethyl ether under reflux. The reaction mixture was hydrolysed by cautious addition of an aqueous solution of ammonium chloride.

Note 2. For the preparation of $ClCH_2OC_2H_5$ see Exp. 13, note 2.

(*41*) *5-Hexen-3-yn-1-ol*

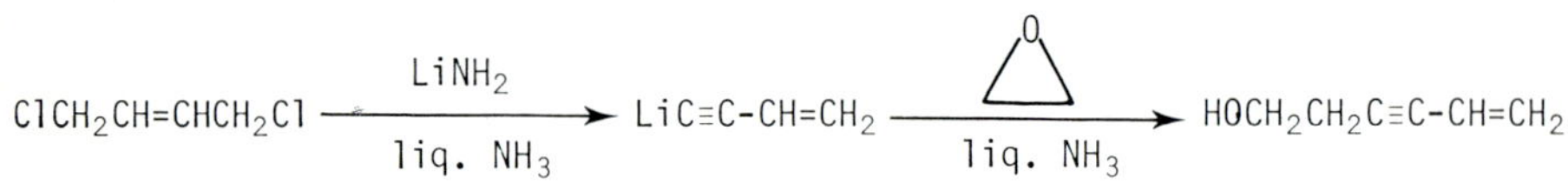

(after hydrolysis)

Apparatus: 4-l flask (see Fig. 2) for the reaction with lithium amide; 3-l silvered Dewar flask, provided with a rubber stopper and a gas outlet for the hydroxyalkylation (no stirring was applied).

To a suspension of 3.1 mol of lithium amide in 3 l of liquid ammonia (see Chapter II, Exp. 11) was added in 30 min with vigorous stirring 1.0 mol of *trans*-1,4-dichloro-2-butene (commercially available) or *cis*-1,4-dichloro-2-butene.* A thick brown reaction mixture was formed, which after an additional 15 min, was transferred into the Dewar flask. The remaining material on the walls of the first flask was rinsed into the Dewar flask with liquid ammonia. Ethylene oxide (55 g) was then added in five portions at intervals of 15 min. After the addition of each portion the mixture was swirled for a few seconds. The flask was allowed to stand for at least 15 h. Powdered ammonium chloride (50 g) was then added cautiously with manual swirling. The ammonia was allowed to evaporate (this took several days; if desired one can pour the reaction mixture into a normal round-bottomed flask and rinse the Dewar flask with ice-water). To the remaining brown solid residue were added 500 ml of water and, after dissolution of the salts, six extractions with diethyl ether were carried out. The combined solutions were dried (washing was not carried out) over magnesium sulfate and concentrated in a water-pump vacuum. Distillation of the residue through a 25-cm Vigreux column gave the enyne alcohol, b.p. 75 °C/15 mmHg, n_D^{20} 1.4956, in 66-74% yield.

*Prepared from the commercially available *cis*-diol and thionylchloride.

(42) *4-Chloro-2-butyn-1-ol*[62]

$$\mathrm{HC{\equiv}CCH_2Cl} \xrightarrow[\mathrm{Et_2O\text{-}hexane}]{\mathrm{C_4H_9Li}} \mathrm{LiC{\equiv}CCH_2Cl} \xrightarrow{\mathrm{(CH_2O)_n}} \mathrm{HO\text{-}CH_2C{\equiv}C\text{-}CH_2Cl}$$

(after hydrolysis)

Apparatus: 2-l flask, see Fig. 1.

To a mixture of 0.55 mol of propargyl chloride and 325 ml of dry diethyl ether was added with cooling between -70 and -90°C (liquid nitrogen bath) a solution of 0.50 mol of butyllithium in about 350 ml of hexane during 20 min. After an additional 5 min 17.4 g of dry powdered paraformaldehyde, corresponding to 0.60 mol of the monomer, were added in a few seconds at -70°C. The temperature was allowed to rise in 10 min to -40°C and was kept at that level for 2.5 h. The cooling bath was then removed and stirring was continued for 3 h at room temperature. The reaction mixture was hydrolysed by addition of 200 ml of ice-water. After separation of the layers the aqueous layer was extracted eight times with small portions of diethyl ether. The combined organic extracts were dried (without previous washing) over magnesium sulfate and then concentrated in a water-pump vacuum. Distillation of the residue through a 25-cm Vigreux column gave 4-chloro-2-butyn-1-ol, b.p. 95°C/20 mmHg, n_D^{20} 1.4963, in 51% yield.

(43) *1-Methylthio-3-chloro-1-propyne*[63]

$$\mathrm{HC{\equiv}C\text{-}CH_2Cl} + \mathrm{C_2H_5Li} \xrightarrow[\text{ether}]{\text{diethyl}} \mathrm{LiC{\equiv}C\text{-}CH_2Cl}$$

$$\mathrm{LiC{\equiv}C\text{-}CH_2Cl} + \mathrm{CH_3S\overset{\overset{\displaystyle O}{\|}}{\underset{\underset{\displaystyle O}{\|}}{S}}\text{-}CH_3} \xrightarrow[\text{ether}]{\text{diethyl}} \mathrm{CH_3S\text{-}C{\equiv}C\text{-}CH_2Cl}$$

Apparatus: 1-l flask, see Fig. 1.

To a solution of 0.20 mol of $\mathrm{LiC{\equiv}C\text{-}CH_2Cl}$ in about 300 ml of diethyl ether (see Chapter II, Exp. 16) was added in 15 min with vigorous stirring and cooling between -75 and -90°C (liquid nitrogen bath, note 1) a mixture of 0.20 mol of methyl methanethiosulfonate[68] and 100 ml of diethyl ether (note 2). A white precipitate was formed immediately. Subsequently the temperature was allowed to rise to -50°C and the mixture was then poured into ice-water. The ethereal layer was separated off as soon as possible and was immediately dried over magnesium sulfate (note 3). The aqueous layer was extracted, without any delay, twice with pentane and these extracts were added to the main solution. After concentration in a water-pump vacuum the residue was distilled through a 30-cm Vigreux column and 1-methylthio-3-chloro-1-propyne, b.p. 65°C/15 mmHg, n_D^{20} 1.5500, was obtained in 85% yield.

Note 1. Occasional immersion of the flask in liquid nitrogen is sufficient to keep the mixture at the required temperature. Care should be taken that no solid crust is formed on the bottom of the flask, due to too rigorous cooling.

Note 2. The sulfenylation with the disulfide proceeds too slowly at -80°C, but methyl thiocyanate could be used as an alternative to the thiosulfonate.

Note 3. $CH_3S\text{-}C{\equiv}C\text{-}CH_2Cl$, being the ethynyl analogue of CH_3SCH_2Cl, may be water-sensitive.

(*44*) *1-Chloro-3-trimethylsilyl-2-propyne*

$$HC{\equiv}C\text{-}CH_2Cl \xrightarrow[(2)\ Me_3SiCl]{(1)\ C_2H_5Li/Et_2O} Me_3Si\text{-}C{\equiv}C\text{-}CH_2Cl$$

Apparatus: 1-l flask, see Fig. 1.

To a solution of 0.50 mol of $LiC{\equiv}C\text{-}CH_2Cl$, prepared from $HC{\equiv}C\text{-}CH_2Cl$ and C_2H_5Li in diethyl ether (see Chapter II, Exps. 1 and 16) was added at -70°C in one portion 0.50 mol of trimethylchlorosilane followed by 35 ml of dry HMPT. The temperature was allowed to rise gradually in 2.5 h to +10°C. At about -40°C salt began to separate from the solution. The work-up was carried out by pouring the mixture into 200 ml of ice-water and extracting the aqueous layer twice with diethyl ether. The ethereal solutions were dried over magnesium sulfate, after which the greater part of the diethyl ether was distilled off at normal pressure through a 40-cm Vigreux column. The product was distilled in a water-pump vacuum: b.p. 45°C/15 mmHg, n_D^{20} 1.4546, with a yield of 85%.

Similar experiments with propargyl bromide failed.

(45) 1-Chloro-2-phenylacetylene[64]

$$\text{Ph-C}\equiv\text{CH} \xrightarrow[\text{Et}_2\text{O-hexane}]{\text{BuLi}} \text{PhC}\equiv\text{CLi} \xrightarrow{\text{Cl}_2} \text{PhC}\equiv\text{C-Cl} + \text{LiCl}$$

Apparatus: 1-l flask, see Fig. 1.

To a solution of 0.40 mol of butyllithium (note 1) in about 280 ml of hexane were added 300 ml of dry diethyl ether with cooling below -10°C. Subsequently 0.40 mol of phenylacetylene was added at a rate such that the temperature could be held below -20°C. The clear solution was then cooled to -60°C and a cold (-40°C) solution of 0.40 mol of chlorine in 100 ml of dichloromethane (note 2) was added in 10 min with vigorous stirring. Cooling with liquid nitrogen was performed so that the temperature of the reactiom mixture could easily be kept between -60 and -75°C during the chlorination. Two minutes after the addition of chlorine 300 ml of water were added with vigorous stirring. The upper layer was separated off and dried over magnesium sulfate. Concentration in a water-pump vacuum, followed by distillation through a 40-cm Vigreux column, gave the chloroacetylene, b.p. 66°C/17 mmHg, n_D^{20} 1.5760, in 90% yield.

Note 1. Butyl- or ethyllithium in diethyl ether, prepared from the alkyl bromide, contains LiBr, which may react with chlorine to form bromine, so that RC≡C-Br will also be formed.

Note 2. Introduction of gaseous chlorine may lead to an explosion, owing to reaction with the diethyl ether.

(46) 1-Propynyl thiocyanate[65]

$$CH_3C\equiv CH + \text{BuLi} \xrightarrow[\text{hexane-diethyl ether}]{} CH_3C\equiv CLi \xrightarrow{S_8} CH_3C\equiv C\text{-SLi}$$

$$CH_3C\equiv C\text{-SLi} + \text{Cl-C}\equiv\text{N} \xrightarrow[\text{hexane-diethyl ether}]{} CH_3C\equiv C\text{-S-C}\equiv\text{N}$$

Apparatus: 1-l flask, see Fig. 1.

To a cooled (-30°C) solution of 0.30 mol of butyllithium in about 200 ml of hexane were added 500 ml of dry diethyl ether. The mixture was cooled below -30°C and a solution of 0.35 mol of propyne in 50 ml of diethyl ether, cooled at -30°C, was added in 10 min while keeping the temperature between -10 and -20°C. To the white suspension was subsequently added 0.30 mol of finely powdered sulfur. The cooling bath was removed and the temperature was allowed to rise (additional cooling may be necessary if too fast a rise is observed). The solid disappeared gradually and a brown solution was formed. After stirring for an additional 45 min at 15-20°C the mixture was transferred (in portions) into the dropping funnel of another apparatus. The solution was added in 25 min with cooling between -30 and -40°C to a solution of 0.35 mol of cyanogen chloride in 100 ml of dry diethyl ether. After the addition the cooling bath was removed and the temperature was allowed to rise to 10°C. The reaction mixture was poured into 300 ml of ice--water and, after vigorous shaking, the layers were separated and the aqueous layer was extracted twice with 50-ml portions of diethyl ether. After drying over magnesium sulfate the solution was concentrated in a water-pump vacuum. The residue was distilled at low pressure (< 0.5 mmHg) using a single receiver cooled at -30°C. The bath temperature was kept below 100°C. The viscous residue was discarded. The contents of the receiver were redistilled through a 30-cm Vigreux column, affording the thiocyanate, b.p. 40°C/12 mmHg, n_D^{20} 1.5041, in 45% yield.

In a similar way was prepared $(CH_3)_3C\text{-}C{\equiv}C\text{-}S\text{-}C{\equiv}N$, b.p. 60°C/12 mmHg, n_D^{20} 1.4748, in 64% yield.

(*47*) *1-Trimethylsilylthio-1-propyne*[66]

$$CH_3C{\equiv}C\text{-}H \xrightarrow[Et_2O\text{-hexane}]{BuLi} CH_3C{\equiv}C\text{-}Li \xrightarrow{S_8} CH_3C{\equiv}C\text{-}S\text{-}Li$$

$$CH_3C{\equiv}C\text{-}S\text{-}Li + Me_3SiCl \longrightarrow CH_3C{\equiv}C\text{-}S\text{-}SiMe_3$$

Apparatus: 1-l flask, see Fig. 1.

To a solution of 0.20 mol of butyllithium in about 140 ml of hexane were added 250 ml of dry diethyl ether below -10°C. Subsequently a solution of 0.25 mol of propyne in 25 ml of ether, cooled below -25°C, was added in 10 min, keeping the temperature of the reaction mixture below -20°C. Powdered sulfur (0.20 at) was added in one portion and the cooling bath was removed. The mixture was stirred for 2 h and the suspension disappeared gradually and a brown solution was formed. This solution was transferred into a 500-ml dropping funnel. In another flask (1 l) were placed 0.20 mol of trimethylchlorosilane and 100 ml of dry diethyl ether. The solution of the alkynethiolate was added at -40°C during 15 min. The cooling bath was then removed and the temperature was allowed to rise to about 20°C. The reaction mixture was transferred into a 1-l round-bottomed flask and was concentrated in a water-pump vacuum (a tube filled with anhydrous calcium chloride was placed between the flask and the water pump). The flask was then equipped for distillation under a pressure of 1 mmHg (or lower). The reaction products were collected in a receiver cooled at -70°C. Careful redistillation of the contents of the receiver (note 1) through a 40-cm Widmer column gave 1-trimethylsilylthio-1-propyne, b.p. 50°C/12 mmHg, n_D^{20} 1.4852, in 48-57% yield.

Note 1. The product is very water-sensitive and the distillation apparatus must therefore be thoroughly dry.

(*48*) *Methyl 1-propynyl disulfide*[67]

$$CH_3C{\equiv}C\text{-}H \xrightarrow[Et_2O\text{-hexane}]{BuLi} CH_3C{\equiv}C\text{-}Li \xrightarrow{S_8} CH_3C{\equiv}C\text{-}S\text{-}Li$$

$$CH_3C{\equiv}C\text{-}S\text{-}Li + CH_3SSO_2CH_3 \longrightarrow CH_3C{\equiv}C\text{-}SS\text{-}CH_3$$

Apparatus: 500-ml flask, see Fig. 1.

A solution of lithium propynethiolate was prepared on a 0.10 molar scale as described in Exps. 46, 47. The brown solution was transferred into the dropping funnel and was added during 40 min to a mixture of 0.10 mol of methyl methanethiosulfonate[68] and 50 ml of dry diethyl ether. During this addition the temperature was maintained between -40 and -45°C. After removing the cooling bath the temperature was allowed to rise to -10°C. The reaction mixture was poured into ice-water and, after shaking, the layers were separated. The aqueous layer was extracted twice with diethyl ether. After drying the combined solutions, the diethyl ether and hexane were removed completely in a water-pump vacuum, keeping the bath temperature below 25°C (note 1). The flask was then equipped for a distillation in a high vacuum (pressure lower than 0.2 mmHg). When the pressure had dropped below 0.5 mmHg, the (single) receiver was immersed in a bath at -50°C. The distillation flask was completely immersed in a bath at 25-30°C. The contents of the receiver consisted of pure acetylenic disulfide, n_D^{20} 1.5726, in a yield of 54-68%.

Note 1. The acetylenic disulfide is very unstable in the presence of alkaline reagents and oxygen, and may polymerize during heating at slightly elevated temperatures. Therefore, clean glassware must be used for the work-up and distillation. Bath temperatures higher than 30-35°C must be avoided. After removal of the solvents and after distillation in an oil-pump vacuum nitrogen must be used for the devacuation. If the MNR spectrum and n_D^{20} of the crude product indicate that the purity is satisfactory, distillation should not be carried out.

(49) *1,4-Pentadiyne*[69]

$$HC{\equiv}CH \xrightarrow[THF]{C_2H_5MgBr} HC{\equiv}C\text{-}MgBr$$

$$HC{\equiv}C\text{-}MgBr + HC{\equiv}CCH_2OTs \xrightarrow[CuBr]{THF} HC{\equiv}C\text{-}CH_2\text{-}C{\equiv}CH$$

Apparatus: 2-l flask, see Chapter I, Fig. 1.

A solution of ethynylmagnesium bromide in 1.2 l of THF was prepared according to the instructions given in *Organic Synthesis*, Collect. Vol. *4*, 792, starting from 0.80 mol of ethyl bromide (for the preparation of C_2H_5MgBr, see Chapter II, Exp. 6). The solution was cooled to 0°C , and 3 g of CuBr were added. After 15 min 0.50 mol of propargyl tosylate (see Ref. 1 *and* Chapter VIII-3, Exp. 3) was added to the suspension in 10 min. During this addition the temperature was kept between 0 and 5°C. The temperature was then allowed to rise gradually over a period of 1 h to about 15°C (with occasional cooling). Most of the precipitated salt originally present disappeared. To the brown solution were added 300 ml of high-boiling light petroleum, (b.p. $\geq$ 180°C/76 cmHg), then the mixture was poured into 1 l of 2 N hydrochloric acid. After vigorous shaking the upper layer was separated off and washed fifteen times with 200-ml portions of 2 N HCl in order to remove the THF. The combined washings and the original aqueous layer were combined and extracted twice with 50-ml portions of the light petroleum. The two light petroleum layers were washed ten times with 100-ml portions of 2 N HCl and then combined with the main light petroleum solution. After drying over a small amount of magnesium sulfate the extract was transferred into a 2-l round-bottomed flask. Some boiling stones were added and the flask was connected to a 40-cm Vigreux column, descending condenser and receiver (see Chapter I, Fig. 5). The receiver, immersed in a bath at -75°C, was connected, via a tube filled with KOH pellets, with the water pump. The contents of the flask were gradually heated in a water-bath. This procedure was stopped by admitting nitrogen to the system when a few millilitres of light petroleum had passed over. This procedure was repeated once more with the contents of the receiver, now heating the flask with the "distillate" below 55°C, so that the light petroleum did not distil. The second "distillate" was redistilled in a slow stream of nitrogen through a 30-cm Vigreux column, affording 1,4-pentadiyne, b.p. 62°C/760 mmHg, n_D^{23} 1.4283, in 63% yield. The NMR spectrum showed that a small amount (4-5%) of $HC{\equiv}CCH{=}C{=}CH_2$ was present in the product.

(*50*) *1,4-Hexadiyne, 1,4-nonadiyne and 5-trimethylsilyl-1,4-pentadiyne*[69]

$$RC{\equiv}CH + C_2H_5MgBr \xrightarrow[THF]{} RC{\equiv}C\text{-}MgBr$$

$$HC{\equiv}C\text{-}CH_2\text{-}O\text{-}Ts + RC{\equiv}C\text{-}MgBr \xrightarrow[THF]{CuBr} HC{\equiv}C\text{-}CH_2\text{-}C{\equiv}C\text{-}R$$

$$R = CH_3,\ n\text{-}C_4H_9,\ Me_3Si$$

Apparatus: 1-l flask, see Fig. 1.

To a solution of ethylmagnesium bromide in 350 ml of THF, prepared from 0.5 mol of ethyl bromide (see Chapter II, Exp. 6) was added in 10 min at 10°C 0.47 mol of 1-hexyne (Exp. 62) and at 0°C 0.47 mol of trimethylsilylacetylene (Exp. 31) or a solution of 0.60 mol of propyne in 70 ml of THF (cooled below -20°C). With trimethylsilylacetylene an exothermic reaction started almost immediately, so that efficient cooling in a bath of dry-ice and acetone was necessary in order to keep the temperature between 10 and 15°C. When the exothermic reaction had subsided, the mixture was warmed to 20°C and was kept at that temperature for 1 h. With 1-hexyne the cooling bath was removed directly after the addition and the temperature was allowed to rise to 40-45°C and was maintained at that level for 1 h. After the addition of the propyne the thermometer-gas outlet combination was replaced with a "cold finger" filled with dry-ice and acetone. The top of this reflux condenser was connected via a plastic tube with a cold trap (-75°C) containing 50 ml of dry THF. The cooling bath was removed and the conversion of propyne started gradually, as was visible from the intensity of the reflux from the condenser. The slow stream of nitrogen through the apparatus prevented the THF from being sucked back from the cold trap. When refluxing had ceased, the contents of the trap (a solution of ethane and propyne in THF) were cautiously poured into the dropping funnel and subsequently it was added in a few minutes to the reaction mixture. The condenser was again connected with the empty trap and stirring was continued for another 1 h. The conversion of propyne took (from the moment of addition) about 3 h and for hexyne and trimethylsilylacetylene 1-2 h. The solutions of the acetylenic Grignard compounds were cooled to $10^{o}C$, 3 g of CuBr were added and, after stirring for 10 min at about $10^{o}C$, the temperature was lowered to $0^{o}C$. Propargyl tosylate (see Chapter VIII-3, Exp. 3, for a general procedure for tosylates) (0.47 mol) was added in 10 min at $0^{o}C$ (a cooling bath of dry-ice and acetone is required). The reaction mixture was kept for 30 min between 0 and $5^{o}C$, and subsequently for 1 h between 10 and $15^{o}C$. When R = n-C_4H_9 or Me_3Si the reaction mixtures were hydrolysed by cautious addition of 200 ml of a solution containing 50 g of NH_4Cl and 7 g of NaCN or KCN. After vigorous stirring the upper layers were separated and the aqueous layers were extracted three times with diethyl ether. The combined solutions were dried over magnesium sulfate. Most of the solvent was distilled off at normal pressure through a 40-cm Vigreux column (bath temperature not higher than $125^{o}C$, note 1). The residue was distilled through the same column. $Me_3SiC{\equiv}C\text{-}CH_2\text{-}C{\equiv}CH$, b.p. $42^{o}C$/15 mmHg, n_D^{18} 1.4497, was obtained in 72% yield, $C_4H_9C{\equiv}C\text{-}CH_2\text{-}C{\equiv}CH$, b.p. $56^{o}C$/18 mmHg, n_D^{20} 1.4525, in 78% yield.

The reaction mixture, containing $CH_3C{\equiv}C\text{-}CH_2\text{-}C{\equiv}CH$, was poured into 500 ml of 2 N HCl. High-boiling light petroleum (b.p. $\geq 170^{o}C$/760 mmHg) (250 ml) was added and the upper layer, formed after shaking, was separated off and washed twelve times with 250-ml portions of 2 N HCl in order to remove the THF. The combined aqueous layers were extracted twice with 40-ml portions of the light petroleum. The two extracts were combined and washed six times with 2 N HCl. The light petroleum solutions were dried over magnesium sulfate, then poured into a 1-l round-bottomed flask, which was equipped for vacuum distillation with a 40-cm Vigreux column, condenser and receiver, cooled at $-75^{o}C$ (see Fig. 5). A tube filled with KOH pellets was placed between the receiver and the water pump. The apparatus was evacuated (10-20 mmHg) and the flask was gradually heated until the light petroleum began to pass over. The distillate was heated under reflux under nitrogen for 20 min (note 2), and was subsequently distilled in a partial vacuum. $CH_3C{\equiv}C\text{-}CH_2C{\equiv}CH$, b.p. *ca.* $55^{o}C$/100 mmHg, n_D^{22} 1.4474, was obtained in 74% yield.

Note 1. In the final stage of the distillation the remaining liquid is subjected to a relatively high temperature. This causes dimerization of the greater part of the yne-allene $RC{\equiv}C\text{-}CH{=}C{=}CH_2$, which is formed as a by product.

Note 2. During this operation the small amount of $CH_3C{\equiv}C\text{-}CH{=}C{=}CH_2$, present in the crude product, dimerized.

(51) *1-Ethoxy-1,4-hexadiyne*[69]

$$HC{\equiv}C\text{-}OC_2H_5 + C_2H_5MgBr \xrightarrow[\text{THF}]{} BrMgC{\equiv}C\text{-}OC_2H_5$$

$$CH_3C{\equiv}C\text{-}CH_2O\text{-}Ts + BrMgC{\equiv}C\text{-}OC_2H_5 \xrightarrow[\text{THF}]{\text{CuBr}} CH_3C{\equiv}C\text{-}CH_2\text{-}C{\equiv}C\text{-}OC_2H_5$$

Apparatus: 1-l flask, see Fig. 1.

To a solution of ethylmagnesium bromide in 350 ml of THF, prepared from 0.35 mol of ethyl bromide (see Chapter II, Exp. 6), was added 0.30 mol of ethoxyacetylene at $-15^{o}C$ (Chapter VI, Exp. 27). The temperature of the mixture was allowed to rise to $15^{o}C$ over a period of 30 min (occasional cooling may be necessary). The solution was then cooled to $0^{o}C$, 4 g of powdered copper(I) bromide were introduced and a solution of 0.30 mol of the tosylate of 2-butyn-1-ol (see Chapter VIII-3, Exp. 3 for a general procedure for tosylates) in 50 ml of THF was added in 5 min. During this addition the temperature was kept between 0 and $5^{o}C$ (dry--ice-acetone bath). Stirring at $10\text{-}15^{o}C$ was continued for an additional 45 min. The mixture was then poured into a solution of 10 g of KCN or NaCN and 40 g of NH_4Cl in 400 ml of water. After vigorous shaking the layers were separated and the aqueous phase was extracted four times with diethyl ether. The combined organic solutions were dried over $MgSO_4$ and concentrated in a water-pump vacuum. Distillation of the residue gave 1-ethoxy-1,4-hexadiyne, b.p. $50^{o}C/15$ mmHg, n_D^{20} 1.4550, in 87% yield.

(52) *1-Chloromethylthio-1-butyne*[27]

$$C_2H_5C{\equiv}CH \xrightarrow[\text{THF-hexane}]{\text{BuLi}} C_2H_5C{\equiv}C\text{-}Li \xrightarrow{S_8} C_2H_5C{\equiv}C\text{-}SLi$$

$$C_2H_5C{\equiv}C\text{-}SLi \xrightarrow[\text{THF-hexane}]{BrCH_2Cl} C_2H_5C{\equiv}C\text{-}S\text{-}CH_2Cl$$

Apparatus: 1-l flask, see Fig. 1 (no dropping funnel was used).

A solution of 0.20 mol of butyllithium in about 140 ml of hexane was cooled below -30^{0}C, then 0.25 mol of 1-butyne, dissolved in 30 ml of dry THF, was added in 10 min at -30^{0}C. Ten minutes later 140 ml of dry THF were added and subsequently 0.20 at of dry, powdered sulfur was introduced at -25^{0}C. The cooling bath was then removed and the temperature was allowed to rise to 0^{0}C. After stirring for 30 min at 0^{0}C the temperature was allowed to rise to 15^{0}C. Bromochloromethane (0.60 mol, large excess) was added in one batch; the reaction was weakly exothermic. The mixture was kept for 15 h at room temperature, then 300 ml of ice-water were added with vigorous stirring. After addition of 300 ml of redistilled pentane the THF was removed by washing the solution ten times with 200-ml portions of ice-water. The combined washings were extracted three times with small amounts of pentane. After drying the combined pentane layers over magnesium sulfate the solvents were removed by evaporation in a water-pump vacuum. Distillation of the residue in an oil-pump vacuum (< 0.5 mmHg) in the apparatus shown in Fig. 5 (receiver cooled below -30^{0}C), followed by redistillation in a water-pump vacuum, gave the acetylenic sulfide, b.p. 65^{0}C/12 mmHg, n_D^{20} 1.5205, in 62% yield.

In a similar way were prepared the following compounds:

$CH_3C{\equiv}C\text{-}S\text{-}CH_2Cl$, b.p. 58^{0}C/12 mmHg, n_D^{20} 1.5322, yield 55%;

t-$C_4H_9C{\equiv}C\text{-}S\text{-}CH_2Cl$, b.p. 70^{0}C/12 mmHg, n_D^{20} 1.4930, yield 64%;

$Me_3SiC{\equiv}C\text{-}S\text{-}CH_2Cl$, b.p. 78^{0}C/12 mmHg, $n_D^{18.5}$ 1.4988, yield 60%;

$PhC{\equiv}C\text{-}S\text{-}CH_2Cl$, b.p. 94^{0}C/0.005 mmHg, yield 64%;

$C_2H_5C{\equiv}C\text{-}SeCH_2Cl$, b.p. 35^{0}C/0.01 mmHg, n_D^{20} 1.5482, yield 72%.

In the last case the reaction with bromochloromethane was much faster, and the reaction time (at 20-25^{0}C) was 4 h.

(*53*) *2-Methyl-1,1-dimethylthio-1,2-butadiene*[157]

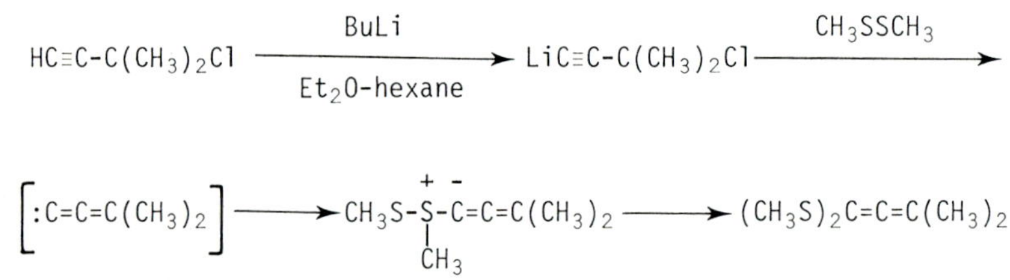

Apparatus: 1-l flask, see Fig. 1.

A mixture of 0.10 mol of freshly distilled 3-methyl-3-chloro-1-butyne (see Chapter VIII-3, Exp. 5) and 170 ml of dry diethyl ether was cooled to -100°C and 0.10 mol of butyllithium in about 70 ml of hexane was added at this temperature in 10 min. Five minutes later 0.10 mol of dimethyl disulfide was introduced within 1 min with cooling between -100 and -90°C. The cooling bath was subsequently removed and the temperature was allowed to rise. Above -25°C the clear light-brown solution became turbid and later a white precipitate was formed. When the temperature had reached 10°C, the reaction mixture was hydrolyzed by addition of 200 ml of water. The organic layer and one ethereal extract were dried over potassium carbonate and subsequently concentrated in a water-pump vacuum (bath temperature < 50°C). There remained 10.7-11.5 g (67-73%) of a light-brown oil, n_D^{23} 1.5591. The NMR spectrum corresponds to reasonably pure $(CH_3)_2C{=}C{=}C(SCH_3)_2$. Efforts to distil the compound led to partial dimerization.

(54) *Propynylcyclopentanol and propynylcyclohexanol*[8]

$$CH_3C{\equiv}CH + BuLi \xrightarrow[\text{THF-hexane}]{} CH_3C{\equiv}CLi$$

$$(CH_2)_nC{=}O + CH_3C{\equiv}CLi \xrightarrow[\text{LiBr}]{} (CH_2)_nC(OH)C{\equiv}CCH_3 \qquad n = 4,5$$

Apparatus: 1-l flask, see Fig. 1.

A solution of 0.21 mol of butyllithium in about 140 ml of hexane (note 1) was cooled below -40°C and 90 ml of dry THF were run in. Subsequently a cold (< -20°C) solution of 0.25 mol of propyne in 20 ml of dry THF was added with cooling below -20°C and a white precipitate was formed. A solution of 0.10 mol of anhydrous (note 2) lithium bromide in 30 ml of THF was added, followed by 0.20 mol of freshly distilled cyclopentanone or cyclohexanone, all at -30°C. The precipitate had disappeared almost completely after 20 min. The cooling bath was then removed and when the temperature had reached 0°C, the mixture was hydrolyzed by addition of 100 ml of a solution of 20 g of NH_4Cl in water. After shaking and separation of the layers four extractions with diethyl ether were carried out. The extracts were dried over magnesium sulfate and the solvents removed by evaporation in a water-pump vacuum. Careful distillation of the remaining liquids afforded the following carbinols:

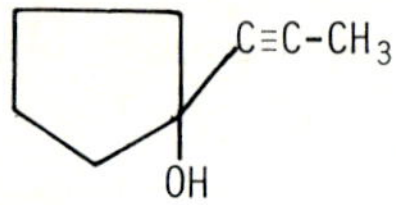 , b.p. 78°C/10 mmHg, n_D^{22} 1.4842, yield 82%;

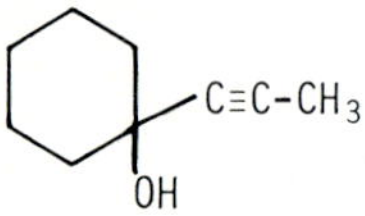 , b.p. 92°C/10 mmHg, yield 91% (this compound solidified in the receiver).

Note 1. The yields were considerably lower, when the metallation was performed with butyllithium or ethyllithium in diethyl ether (prepared from RBr and Li), to which THF was added as co-solvent. In liquid ammonia, using lithium amide as a base, the results were also unsatisfactory, especially with cyclopentanone. Other acetylenic carbinols derived from various acetylenes and ketones [except $CH_3COCH_2CH(OCH_3)_2$] could be obtained in excellent yields by the same procedure. The carbinol from cyclopentanone and LiC≡CH was obtained in a low yield, even when more LiBr was used.

Note 2. In the absence of LiBr the coupling was much slower and yields of the carbinols were poor and moderate, respectively. With ketones such as acetone and soluble acetylides the yields were also excellent when no salt was added.

(*55*) *2-Butyne-1-ol*[1]

$$HC{\equiv}CCH_2OH + 2\ LiNH_2 \longrightarrow LiC{\equiv}CCH_2OLi$$

$$LiC{\equiv}CCH_2OLi + CH_3I \longrightarrow CH_3C{\equiv}CCH_2OLi \xrightarrow{H_2O} CH_3C{\equiv}CCH_2OH$$

Apparatus: For the preparation of $LiNH_2$ and its reaction with propargyl alcohol, see Chapter I, Fig. 2; size of the flask, 5 l.

To a vigorously stirred suspension of 4 mol of lithium amide (see II, Exp. 11) in 2.5 l of liquid ammonia were added in 25 min 2 mol of propargyl alcohol (commercially available, purified before use by distillation at 100-120 mm). The suspension became very thin. Subsequently, the dropping funnel was combined with a gas inlet tube reaching about 1 cm beneath the surface of the ammonia. The vent on the splashing tube was removed. Methyl iodide (2 mol) was added to the vigorously stirred suspension over a period of 35-40 min, while nitrogen (1 l/min) was passed through the flask. 5 Min later (the flow of nitrogen had already been stopped) stirring was stopped and the ammonia was allowed to evaporate. The solid residue was dissolved in 1 l of ice-water. The resulting solution was subjected to continuous extraction with diethyl ether during 40 h (a perforator of sintered glass was used). The ethereal solution was dried over 100 g of magnesium sulfate. This was filtered off on a sintered-glass funnel and washed well with diethyl ether. After removing the diethyl ether *in vacuo* the remaining liquid was distilled through a 40-cm Vigreux column. A single receiver was used, this was cooled in ice water in order to trap the aqueous forerun. When the distillation temperature had nearly reached the boiling point of the product, the receiver was replaced with another one, and the distillation was continued. The aqueous forerun was dried over a small amount of magnesium sulfate after dilution with some diethyl ether. Working up afforded an additional small quantity of product. 2-Butyn-1-ol, b.p. 48°C/15 mmHg, n_D^{20} 1.4535, was obtained in 76-82% yield.

(*56*) *2-Heptyn-1-ol*[1]

$$HC{\equiv}C{-}CH_2OH + 2\ LiNH_2 \longrightarrow LiC{\equiv}C{-}CH_2OLi$$

$$LiC{\equiv}C{-}CH_2OLi + C_4H_9Br \longrightarrow C_4H_9C{\equiv}C{-}CH_2OLi \xrightarrow{H_2O} C_4H_9C{\equiv}C{-}CH_2OH$$

Apparatus: Chapter I, Fig. 2; size of the flask, 3 l.

To a vigorously stirred suspension of 2 mol of lithium amide in 2 l of liquid ammonia (see II, Exp. 11) was added in 15 min 1 mol of propargyl alcohol (commercial product, distilled in a partial vacuum before use). Subsequently, 1 mol of butyl bromide was added dropwise in 75 min. After an additional 1.5 h, stirring was stopped and the ammonia was allowed to evaporate. To the solid residue were added 500 ml of ice-water. After the solid mass had dissolved, six extractions with diethyl ether were performed. The (unwashed) combined extracts were dried over magnesium sulfate and then concentrated in a water-pump vacuum. Distillation of the residue through a 40-cm Vigreux column afforded 2-heptyn-1-ol, b.p. 83°C/12 mmHg, n_D^{20} 1.4550, in 76-83% yield.

(57) *2-Penten-4-yn-1-ol*[161]

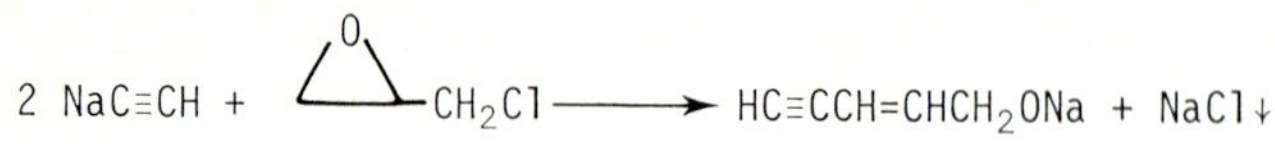

$$HC{\equiv}CCH{=}CHCH_2ONa + NH_4Cl \longrightarrow HC{\equiv}CCH{=}CHCH_2OH + NaCl\downarrow$$

trans

Apparatus: Chapter I, Fig. 2; size of the flask, 3 l; nitrogen (200 ml/min) was introduced during the experiment.
Warning: Use rubber gloves, the product causes blisters on the skin.

Epichlorohydrin (1 mol) was added dropwise over a period of 1.5 h to a solution of 2.2 mol of sodium acetylide in 1.5 l of liquid ammonia. During, as well as for a period of 1.5 h after, the addition the temperature of the mixture was kept at about -45°C. The cooling bath was removed after this period and the mixture was agitated vigorously for another 3 h. The thermometer and vent were removed, and 75 g of powdered ammonium chloride were added in 2-g portions with vigorous stirring. The ammonia was allowed to evaporate.

The remaining salt mass was dissolved in 500 ml of water. Eight to ten extractions with diethyl ether were carried out. The extracts (washing was not carried out) were dried over 100 g of magnesium sulfate. The siccative was sucked off on a sintered-glass funnel and thoroughly rinsed with diethyl ether. After removal of the diethyl ether in a water-pump vacuum, the residue was distilled at a very low pressure (0.1-0.5 mmHg and collected in a single receiver cooled to -40°C (see Chapter I, Fig. 5). The nonvolatile residue of this distillation was discarded. The contents of the receiver were redistilled at water-pump pressure, through a 40-cm Vigreux column. $HC{\equiv}CCH{=}CHCH_2OH$, purity varying between 90% and 95% in the various experiments, b.p. 68°C/12 mmHg, n_D^{20} 1.493-1.496, was obtained in 37-46% yield. Lower yields were obtained when the epichlorohydrin was added without cooling or at too fast a rate.

(58) *1-Penten-4-yn-3-ol and 4-hexen-1-yn-3-ol*[162]

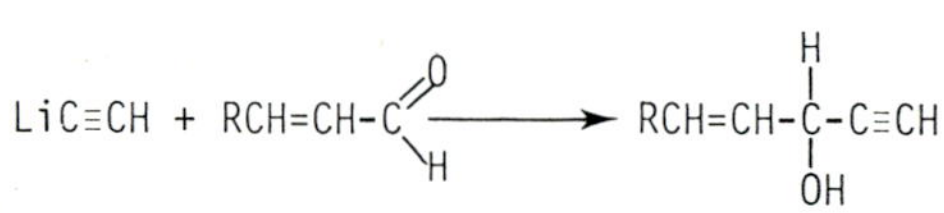

(R = H, CH_3)

Apparatus: 3-l round-bottomed three-necked flask with a combination of dropping funnel and gas inlet tube, reaching 1 cm beneath the surface of the NH_3, stirrer and a combination of thermometer and vent (see Chapter I, Fig. 1).

A concentrated solution of 3 mol of lithium acetylide in 1.4 l of liquid ammonia was prepared (see Ref. 1) from lithium amide and acetylene, in the presence of a small amount of triphenylmethane. The solution was cooled to -75°C (internal temperature) by means of a liquid nitrogen bath. During this cooling the mixture was vigorously agitated in order to prevent solidification of the ammonia on the walls. At the same time nitrogen was introduced at a rate of 400 ml/min. When the mixture had attained the prescribed temperature, addition of acrolein or crotonaldehyde (2.6 mol freshly distilled) was started, while a vigorous stream of nitrogen was passed through the flask (1.5-2 l/min). The aldehydes were added over 15-20 min with vigorous stirring. Continuous cooling was not necessary as the reactions were not very exothermic. After the addition the cooling bath was removed and the introduction of N_2 stopped. The mixture was allowed to stand for 30 min, then it was cautiously poured into a 5-l wide-necked round-bottomed flask. Ammonium chloride (3 mol) was introduced as quickly as possible in 1-g portions with manual swirling. The ammonia was evaporated by placing the flask in a water bath of 50-60°C. During this evaporation the flask was continuously swirled by hand in order to suppress bumping. To the slurry (still containing ammonia) which remained, crushed ice, just enough to dissolve the solids, was cautiously added (with swirling). The reaction flask was washed with a small amount (50 ml) of ice-water and this washing was added to the main portion. The mixture was extracted with diethyl ether: 10 times in the case of R = CH_3, at least 15 times in the case R = H (if desired, continuous extraction can be carried out). The unwashed extracts (note 1) were dried over 100 g of magnesium sulfate. This was filtered off on a sintered-glass funnel and thoroughly rinsed with ether. The greater part of the diethyl ether was distilled off at normal pressure through a 40-cm Vigreux column, avoiding bath temperatures higher than 90°C. The remaining liquid (note 2) was distilled very quickly through the same column, the (single) receiver being cooled in ice + ice-water (see Chapter I, Fig. 5). An aqueous forerun mainly consisting of carbinol was trapped in this manner. During the distillation of the acrolein-carbinol the bath temperature should not exceed 90-100°C: the viscous residue in this case was subjected to a distillation at very low pressure (< 0.5 mmHg), whereby the (single) receiver was cooled below -40°C, in this way a second crop of carbinol was obtained. Practically no residue remained after the distillation of the crotonaldehyde-carbinol .

HC≡C-CH(OH)-CH=CH_2, b.p. 40°C/15 mmHg, n_D^{20} 1.4522, was obtained in 74-78% yield. The product contained about 5% water (no other contaminants).

HC≡C-CH(OH)-CH=CHCH$_3$, b.p. 60°C/12 mmHg, n_D^{20} 1.4650, was obtained in 89% yield. This product contained only 2% water (no other contaminants).

Note 1. Washing, even with concentrated NH$_4$Cl, gives a considerable loss of product (especially in the case R = H).

Note 2. Prolonged heating might result in polymerization or decomposition of the products. A quick distillation is allowed, since the only "impurity" is water.

(*59*) *1-Phenyl-2-propyn-1-ol*

$$\text{HC}\equiv\text{CH} \xrightarrow[\text{hexane-THF}]{\text{BuLi}} \text{HC}\equiv\text{CLi} \xrightarrow[\text{LiBr}]{\text{PhCH=O}} \text{HC}\equiv\text{C-CH(OLi)-Ph} \xrightarrow{\text{H}_2\text{O}} \text{HC}\equiv\text{C-CH(OH)-Ph}$$

Apparatus: 1-l flask, see Fig. 1.

After the air in the flask had been completely replaced with nitrogen, it was cooled in a liquid nitrogen bath and a solution of 25 g of acetylene in 160 ml of dry THF was introduced. The solution had been prepared by dissolving acetylene (freed from acetone by means of a cold trap) in THF cooled at -80 to -90°C. A solution of 0.21 mol of butyllithium in about 150 ml of hexane was added in 5 min to the vigorously stirred solution. During this addition the temperature of the mixture was kept between -80 and -100°C by occasionally dipping the flask into the liquid nitrogen. To the white suspension were successively added at -80°C a solution of 10 g of anhydrous lithium bromide (note 1) in 30 ml of THF and 0.20 mol of freshly distilled benzaldehyde. The reaction mixture was kept for 3 h at -60°C, after which the temperature was allowed to rise to +10°C over a period of 2 h. The suspension became gradually thinner and disappeared completely at about 0°C. The yellow solution was poured into 150 ml of water. After addition of 20 g of ammonium chloride and vigorous shaking, the layers were separated. The aqueous layer was extracted twice with diethyl ether. The combined solutions were dried over magnesium sulfate and concentrated in a water-pump vacuum. The residue was distilled at low pressure giving the desired carbinol, (b.p. 40°C/0.1 mmHg), n_D^{20} 1.5505 in 66-70% yield. A small viscous residue remained in the distillation flask.

Note 1. Compare Chapter VIII-6, Exp. 1 for the effect of lithium bromide on the formation of carbinols, derived from propyne.

(60) 2,4-Hexadiyne-1,6-diol[163]

$$2\ HC{\equiv}CCH_2OH + \tfrac{1}{2}\ O_2 \xrightarrow[NH_4Cl]{CuCl} HOCH_2(C{\equiv}C)_2CH_2OH$$

Apparatus: 3-l round-bottomed flask, connected through vacuum tubes to a mercury manometer and a cylinder with oxygen (Fig. 6).

The flask was charged with a mixture of 37 g of finely powdered cuprous chloride, 120 g of ammonium chloride, 400 ml of water and 1 mol of freshly distilled (under about 100-mmHg pressure) propargyl alcohol. After the air in the flask had been completely replaced by oxygen (a vigorous stream of oxygen was blown into the flask during 1-2 min), a rubber stopper fitted with a glass tube was tightly placed on the flask. The system was charged with an overpressure of 0.5 atm (note 1) and the contents of the flask were vigorously agitated by means of a shaking apparatus.

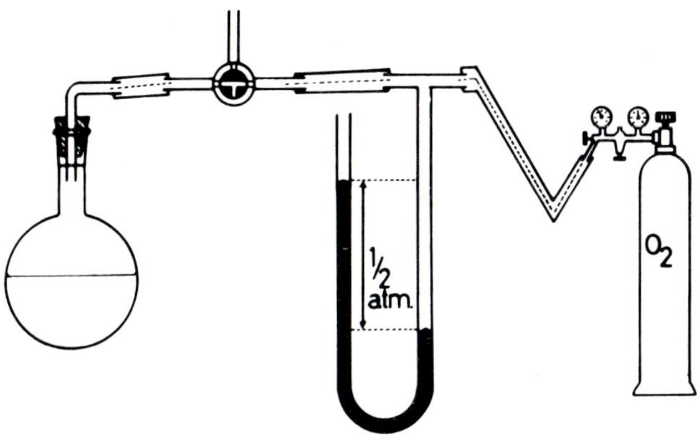

Fig. 6. Oxidative coupling of acetylenic carbinols

The mixture became warm (about 35°C) while the pressure dropped. After 30-60 min (depending on the efficiency of the shaking apparatus), the levels of the mercury in the manometer were equal. The flask was cooled (with swirling) in ice-water during some minutes, the system was again charged with 0.5 atm overpressure and shaking was continued. The diol began to separate from the solution after about

1.5 h of shaking. After levelling the pressure, the tap was turned on in order to neutralize small differences between the pressure of the system and the atmosphere. The stopper was removed and 0.5 mol of propargyl alcohol was added as quickly as possible, whereupon the stopper was again placed on the flask. After cooling the contents of the flask to room temperature, 0.5 atm of oxygen was added and shaking was continued. The operations described can be repeated several times (depending on the available time); after consumption of every charge of 0.5 atm of oxygen, the contents of the flask should be cooled to room temperature and after every *two* charges of oxygen, 0.5 mol of propargyl alcohol should be added. In the present experiment 3 mol of propargyl alcohol were converted. If no further amount of this alcohol is added, shaking should be continued for about 20 min after the green colour of the mixture has changed (note 2) and the consumption of oxygen has become considerably slower (as indicated by the dropping of the mercury level).

Working up was carried out by adding 35 ml of 36% HCl to the cooled suspension (note 3) and subsequently extracting the mixture 15 times with small (∿ 50 ml) portions of THF (note 4). A small amount of insoluble material was left behind. The brown THF-extracts were dried over potassium carbonate. This was sucked off on a sintered-glass funnel and rinsed with THF. Part of the THF was removed in a water-pump vacuum until the volume of the solution had become about 300 ml. 200 ml of diethyl ether were added to the warm solution and the mixture was kept for some hours in the refrigerator. The light-brown crystals (~ 97% purity) were sucked off and were rinsed with cold (-30°C) diethyl ether. The mother liquor was concentrated and again diluted with diethyl ether. Cooling at -25°C afforded a second crop of crystals. Repetition of this procedure brought the yield of fairly pure (about 95%) diol to 92%. Repeated crystallization of a small sample from diethyl ether gave a m.p. 108-109°C.

Note 1. The volume above the reaction mixture in the flask is about 2.5 l. In the case of an overpressure of 0.5 atm of oxygen, one charge corresponds to 1.25 l of 1 atm. According to the equation, 1 mol of $CH{\equiv}CCH_2OH$ consumes 5.5 l of oxygen: one charge of 0.5 atm thus is sufficient for about 0.25 mol of $HC{\equiv}CCH_2OH$. If no excess of this alcohol is present during the reaction, oxidation of CuCl may occur.

Note 2. The change in the green colour can be ascribed to the formation of Cu(OH)Cl, when the propargyl alcohol has been completely converted.

Note 3. The slightly soluble Cu(OH)Cl is converted into $CuCl_2$: a green solution is formed.

Note 4. Continuous extraction with diethyl ether can also be carried out.

(*61*) *2-Methyl-3,5-hexadiyn-2-ol*

$$\mathrm{HC{\equiv}C{-}C{\equiv}CH \xrightarrow[\text{THF}]{\text{BuLi}} HC{\equiv}C{-}C{\equiv}CLi \xrightarrow[\text{THF}]{(CH_3)_2C{=}O} HC{\equiv}C{-}C{\equiv}C{-}C(CH_3)_2OH}$$

(after hydrolysis)

Apparatus: 1-l flask, see Fig. 1.

The flask was cooled at -100°C, while the air was completely replaced with nitrogen. A solution of 0.12 mol of diacetylene in 100 ml of dry THF (V, Exp. 27) was placed in the flask, and 70 ml of hexane solution of 0.10 mol of butyllithium were added dropwise with vigorous stirring and cooling between -90 and -70°C. A white suspension was formed. Directly after the addition of the butyllithium 0.12 mol of dry acetone was added in one portion. The cooling bath was removed. After about 10 min the solution had become clear. Stirring was continued for 10 min, then a solution of 10 g of NH_4Cl in 100 ml of water was added with vigorous stirring. After separation of the layers three extractions with diethyl ether were carried out. The combined solutions were dried over magnesium sulfate and subsequently concentrated in a water-pump vacuum. The light-brown residue was distilled at a pressure $<$ 1 mmHg, keeping the temperature of the heating bath below 60°C. The carbinol, b.p. about 40°C/1 mmHg, n_D^{20} 1.4886 was obtained in 90% yield.

The compound immediately turned purple after distillation.

(*62*) *1-Pentyne and 1-hexyne*[179]

$$\mathrm{NaC{\equiv}CH + RBr \xrightarrow[\text{liq. } NH_3]{} RC{\equiv}CH}$$

Apparatus: 3-l flask, see Fig. 1.

A suspension of 2.3 mol of sodamide in about 1.5 l of liquid ammonia was prepared as described in Chapter II, Exp. 11. To this suspension was added 1 g of triphenylmethane. Acetylene (4-5 l/min) was introduced until the red colour of the triphenylmethane anion had disappeared. The flask was then placed in a cooling bath with dry-ice and acetone (-50°C) and the introduction of acetylene (1-l/min) was continued for an additional 10 min. Dry DMSO (200 ml) was then poured into the flask. The gas inlet tube was replaced with a combination of dropping funnel and gas inlet and the other outer neck of the flask was fitted with a combination of gas outlet and thermometer. A slow stream of nitrogen was passed through the flask. Propyl or butyl bromide (2.0 mol) was added dropwise with vigorous stirring. During this addition which was carried out in 45 min the temperature of the reaction mixture was maintained at about -40°C. After an additional 1 h 200 ml of a high--boiling light petroleum fraction (b.p. > 180°C at normal pressure) were added and the cooling bath was removed. The middle neck was fitted with a stopper, perforated by a glass tube (internal diameter at least 5 mm) just reaching until the salt precipitate. The supernatant liquid was transferred under a mixture of 1 kg of finely crushed ice and 50 ml of the extraction solvent (for this operation see Chapter I, Fig. 3 and accompanying text). To the salt slurry remaining in the flask were successively added with swirling 100 ml of extraction solvent and 200 ml of ice--water. After the remaining ice had melted the portions were combined and the layers were separated. The aqueous layer was extracted once with 50 ml of extraction solvent. After washing the combined extracts with 2 N HCl they were dried over magnesium sulfate. The alkynes were distilled off at normal pressure through a 40-cm Vigreux column: the distillation was stopped as soon as the extraction solvent began to reflux in the middle of the column. Redistillation afforded the pure alkynes in 75% or higher yields: $C_3H_7C{\equiv}CH$, b.p. 41°C, n_D^{20} 1.386 and $C_4H_9C{\equiv}CH$, b.p. 71°C, n_D^{20} 1.3990.

(63) 3-Cyclohexylidenepropenal (ref. 164)[*]

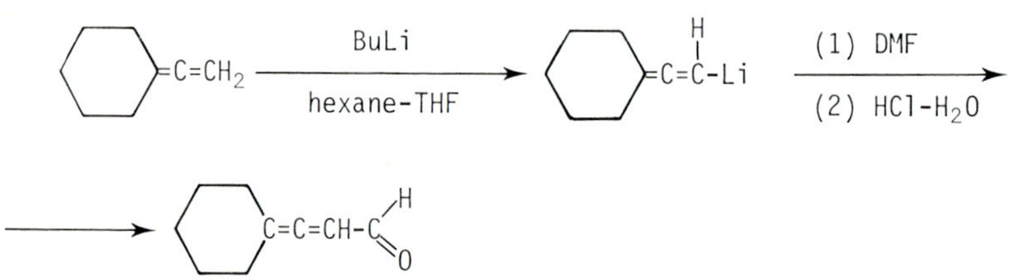

Apparatus: 1-l flask, see Fig. 1.

To a solution of 0.10 mol of butyllithium in about 70 ml of hexane were added with cooling below -20°C 100 ml of dry THF. Vinylidenecyclohexane (0.11 mol, see Chapter VI, Exp. 45) was added dropwise in 10 min with cooling between -50 and -60°C. After stirring for 1.5 h at -50°C the solution was cooled at -100°C and a mixture of 0.15 mol of dry dimethylformamide and 30 ml of THF was introduced in 10 min. During, and for 1.5 h after this addition, the temperature of the solution was kept between -80 and -85°C (note 1). The solution was then cooled to -100°C and was subsequently poured with vigorous stirring into 2.5 l of 0.1 N hydrochloric acid, cooled at 0°C. The mixture was allowed to stand for 2 h and then extracted seven times with redistilled pentane. The combined extracts were washed with ice--water, dried over magnesium sulfate and subsequently concentrated in a water-pump vacuum. Distillation of the residue through a short Vigreux column gave, after a first fraction of vinylidenecyclohexane, the allenic aldehyde, b.p. 94°C/12 mmHg, n_D^{19} 1.5204, in 52% yield. There was a viscous residue of about 1 g, probably resulting from polymerization of the allenic aldehyde (note 2). Redistillation gave about the same amount of residue.

In a closely similar way was prepared from $C_4H_9CH{=}C{=}CH_2$ (0.11 mol, see Chapter VI, Exp. 1), BuLi and $CH_3C({=}O)N(CH_3)_2$: $C_4H_9CH{=}C{=}CH{-}C({=}O){-}CH_3$, b.p. 80°C/10 mmHg, n_D^{20} 1.4680 in 50% yield. The main impurity was $H_2C{=}C{=}C(C_4H_9){-}C({=}O){-}CH_3$ (8 rel %).

Note 1. At higher temperatures the adduct decomposes and yields of the allenic aldehyde are lower.

Note 2. A similar experiment with $C_4H_9{-}CH{=}C{=}CH_2$ gave only traces of the allenic aldehyde and much resinous residue.

*We are indebted to Prof. S. Julia, Dr. J.-C. Clinet and Dr. G. Linstrumelle for submitting experimental details from the dissertation of Clinet (Paris, November 1980), prior to publication of full papers on this subject.

(64) *o-Trimethylsilyl phenylacetylene*[8]

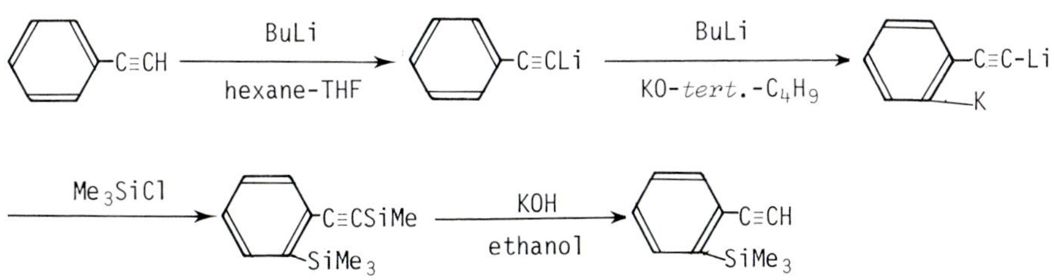

Apparatus: 1-l flask, see Fig. 1.

To a solution of 0.10 mol of phenylacetylene (commercially available, see also Ref. 1) in 100 ml of dry THF was added a solution of 0.21 mol of butyllithium in about 145 ml of hexane. During this addition the temperature was kept below -20°C. The obtained solution was cooled to -65°C and a solution of 0.12 mol of KO-*tert*.-C_4H_9 (commercially available, see Chapter IV, Exp. 4, note 2) in 100 ml of THF was added, while keeping the temperature below -55°C. After an additional 15 min the cooling bath was removed, the temperature was allowed to rise to -10°C and was kept at that level for 1 h (note 1). The reddish suspension was subsequently cooled to -50°C and 0.32 mol of trimethylchlorosilane was added in 10 min. The cooling bath was then removed and the temperature was allowed to rise to 10°C. The greater part of the hexane and THF was removed in a water-pump vacuum, using a rotary evaporator. To the remaining white salt mass was added a solution of 30 g of KOH in 175 ml of 96% ethanol. The mixture was heated at 60°C for 30 min, then 500 ml of ice-water were added. The organic products were extracted six times with redistilled pentane, then washed with water and dried over magnesium sulfate. After removing the solvent in a water-pump vacuum the residue was carefully distilled. The desired acetylene, b.p. 85°C/10 mmHg, n_D^{20} 1.5279, was obtained in 76-79% yield (note 1).

Note 1. When 0.04 mol of KO-*tert*.-C_4H_9 was used and stirring at -10°C was continued for 2.5 h, the yield of the acetylene was only 45%.

Chapter IV

BASE-PROMOTED INTERCONVERSIONS BETWEEN ACETYLENES AND CUMULENES

1. ISOMERIZATIONS WITH CATALYTIC AMOUNTS OF BASE

The composition of the products from the isomerization of an unsaturated compound under the influence of a catalytic amount of a base is governed by the relative thermodynamic stabilities of the starting compound and the product. Of particular synthetic interest are isomerizations in which there is an accumulation of an isomer in the isomerization sequence. Isolation of the desired intermediate in a reasonable state of purity is often a matter of careful selection of the base and the solvent. The following reactions are representative examples:

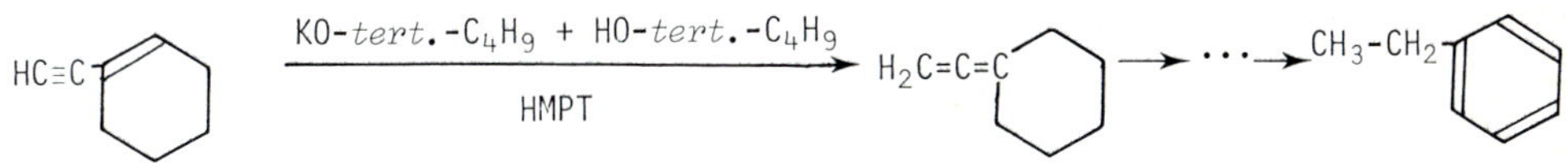

$$CH_3O\text{-}CH_2\text{-}C{\equiv}C\text{-}CH_2\text{-}OCH_3 \xrightarrow[\text{liq. } NH_3]{KO\text{-}tert.\text{-}C_4H_9} CH_3O\text{-}CH{=}C{=}CH\text{-}CH_2\text{-}OCH_3 \longrightarrow CH_3O(CH{=}CH)_2OCH_3$$

$$HC{\equiv}C\text{-}CH_2\text{-}C{\equiv}CH \xrightarrow[CH_3OH]{LiOPh} HC{\equiv}C\text{-}CH{=}C{=}CH_2 \longrightarrow HC{\equiv}C\text{-}C{\equiv}C\text{-}CH_3$$

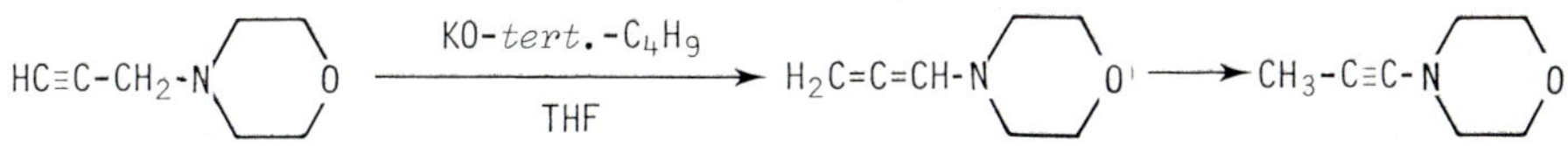

2. ISOMERIZATIONS VIA AN ANIONIC OR ORGANOMETALLIC INTERMEDIATE

Interaction between acetylenic or cumulenic derivatives with very strong bases often gives rise to the formation of an organometallic or anionic intermediate with a relatively high kinetic stability. Subsequent protolysis, e.g. by quenching with water, in general affords a mixture of isomers, which often have a lower thermodynamic stability than the starting compound. There are relatively few conversions which lead to the formation of (predominantly) one isomer. Some representative examples are the following:

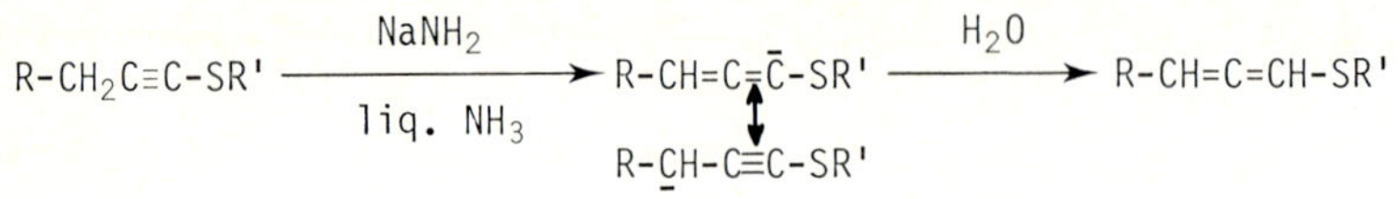

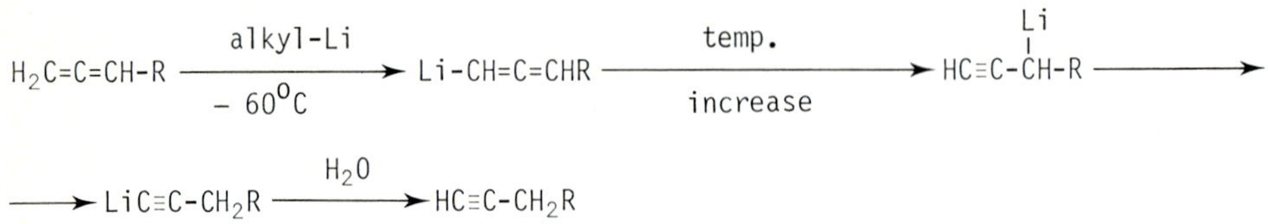

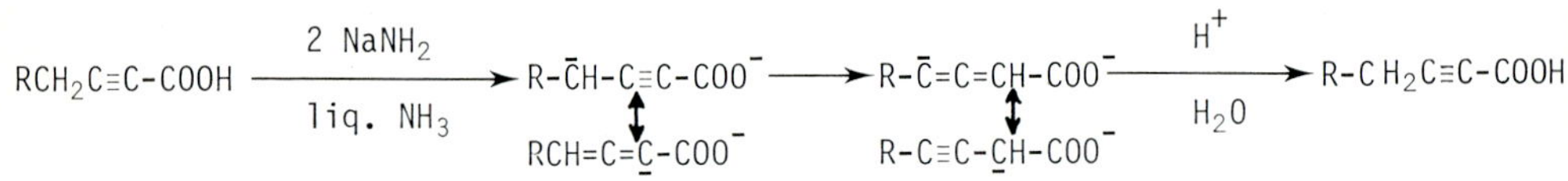

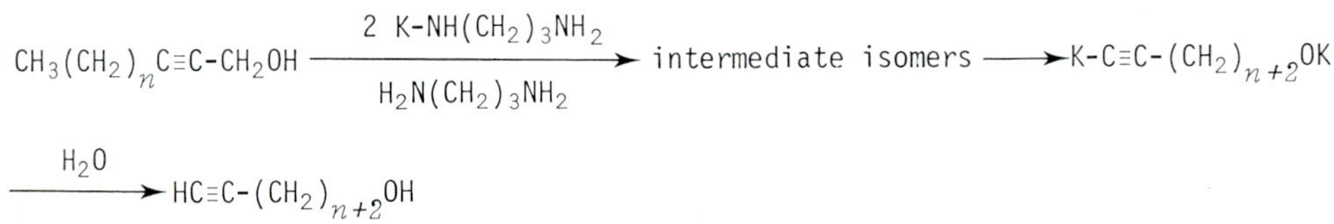

The last isomerization is remarkable in that the triple bond can shift through a long carbon chain to the terminus, where it is fixed as the (kinetically) stable acetylide. The reagent is a solution of potassium diamino-propylamide in 1,3-diaminopropane. In some cases alkali metal amides in liquid ammonia can also bring about "contra-thermodynamic" isomerizations; the reactions are successful only if the triple bond is in the 2-position.

The enyne system in the amines $H_2C{=}CH{-}C{\equiv}C{-}NR_2$ can be reversed by potassium amide in liquid ammonia. Addition of the enyne amines to an equivalent amount of this reagent gives the potassium acetylides, $K{-}C{\equiv}C{-}CH{=}CH{-}NR_2$, from which the "ynene" amines can be obtained in excellent yields by addition of solid ammonium chloride.

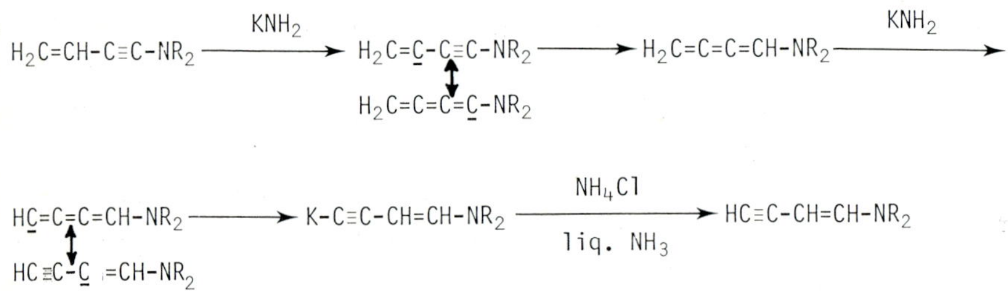

This isomerization, which must proceed through a 1,2,3-trienylamine, is not "contra-thermodynamic", since with a catalytic amount of potassium *tert.*-butoxide the same result is obtained. Enyne ethers, $H_2C{=}CH{-}C{\equiv}C{-}OR$, undergo a similar conversion into $HC{\equiv}C{-}CH{=}CH{-}OR$ upon interaction with alkali metal amides in liquid NH_3, followed by hydrolysis[8]. Enyne sulphides, $H_2C{=}CH{-}C{\equiv}C{-}SR$, and the hydrocarbons $H_2C{=}CH{-}C{\equiv}C{-}R$ (R = *tert.*-C_4H_9 or phenyl) give only tars or polymeric products under similar conditions[8].

3. EXPERIMENTS

(1) *1,2,4-Pentatriene*[77]

$$H_2C{=}CHCH_2C{\equiv}CH \xrightarrow[C_2H_5OH]{NaOH} H_2C{=}CH{-}CH{=}C{=}CH_2$$

Apparatus: 250-ml two-necked flask, provided with a reflux condenser and a thermometer, dipping into the liquid.

Powdered sodium hydroxide (8.0 g) was dissolved in 100 ml of 96% ethanol (note 1) and 19.8 g (0.3 mol) of freshly distilled allylacetylene[1] were added at 35°C. After swirling, the refractive index (note 2) of the solution was measured ($n_D^{20} \sim 1.387$). The solution was kept at 35°C for 45 min, by which time the refractive index then had reached its maximal value (about 1.396) (note 3). The product was distilled off, together with some alcohol and water, and collected in a receiver, cooled below 0°C. The distillation, which took about 15 min (note 3), was stopped when the temperature had reached 78°C. The distillate was transferred into a small separating funnel and shaken three times with 20-ml portions of cold (-10 to -5°C) concentrated ammonium chloride solution. The upper layer was dried over a small amount of magnesium sulfate. The refractive index (n_D^{20}) was 1.4704, corresponding to pure vinylallene. The yield was 86-90%. The compound can be stored for a limited period (a few weeks) at -25°C.

Note 1. With NaOH in *methanol* the isomerization was markedly slower.

Note 2. A small amount (0.1 - 0.2 ml) of the solution is sucked into a Pasteur pipette and the refractive index is measured immediately. The liquid should be transferred between the prisms in such a way that there is no opportunity for evaporation, as this would result in measuring too low a value.

Note 3. The ultimate result of the isomerization is a mixture of $CH_3C{\equiv}CCH{=}CH_2$ and $HC{\equiv}CCH{=}CHCH_3$ (about 9:1). As the refractive index of these compounds is intermediate between these of vinylallene and allylacetylene, longer reaction times will result in a decrease in n_D. Although the further conversion of vinylallene under the described reaction conditions is much slower, it is safer to keep the reaction time and the time used for the distillation as short as possible.

(2) *Allenylacetylene (compare Ref. 78)*

$$HC{\equiv}C\text{-}CH_2\text{-}C{\equiv}CH \xrightarrow[CH_3OH]{LiOPh} HC{\equiv}C\text{-}CH{=}C{=}CH_2 \quad (\longrightarrow HC{\equiv}C\text{-}C{\equiv}C\text{-}CH_3)$$

Apparatus: 100-ml round-bottomed flask and thermometer.

Lithium (0.10 mol) was dissolved in 100 ml of methanol and 0.11 mol of pure phenol was added (note 1). Of the solution of lithium phenolate obtained, 15 ml were mixed at room temperature with 5.3 g of freshly distilled 1,4-pentadiyne (see Chapter III, Exp. 49). The refractive index was determined directly after mixing and was n_D^{22} 1.381 (note 2). The mixture was then rapidly warmed to 48°C and kept at this temperature for 15 min. The n_D^{21} was now 1.391 (note 3). Water (100 ml) was then added and three extractions with 20-ml portions of high-boiling light petroleum (b.p. > 190°C) were carried out. The combined solutions were washed with 1 N hydrochloric acid and subsequently dried over magnesium sulfate. The extract was warmed in a distillation apparatus, using a water-pump vacuum of 10-15 mmHg. The vapour of the yne allene was collected in a receiver cooled at -75°C (for this procedure see Fig. 5 and accompanying text). As soon as the light petroleum began to pass over, nitrogen was admitted to the apparatus. This evacuation procedure was repeated once with the contents of the receiver, now with only slight external heating, so that the light petroleum did not boil. In the receiver were collected 3.5 g of product, consisting of a 9:1 mixture of the yne allene and the conjugated diyne. This was the best result obtained from several experiments, using various solvents and bases. Sodium phenolate gave considerably more 1,3-pentadiyne, and with lithium *p*-nitrophenolate as a base the isomerization to the yne allene was too slow.

Note 1. A slight excess of phenol was used. With a stoichiometric amount the solution may be too strongly basic, so that the further isomerization will become too fast.

Note 2. A very exact determination is not possible because of the dark brown colour of the solution.

Note 3. Heating for longer periods gave more 1,3-pentadiyne.

(3) *3-Vinylidene-1-cyclohexene*[81]

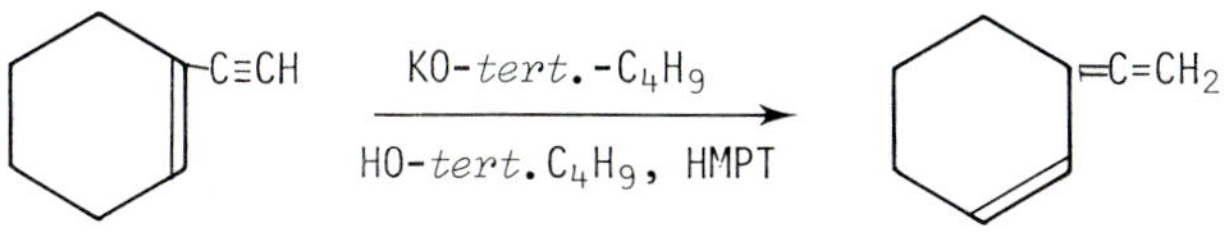

Apparatus: 250-ml flask, see Chapter I, Fig. 1.

In the flask were placed 40 g of dry, pure HMPT (note 1) and 3 g of finely powdered KO-*tert*.-C_4H_9 (see Exp. 4, note 2). The mixture was warmed at 50-60°C and after all of the solid material had passed into solution 7.5 g of *tert*.-butylalcohol and 9.5 g of ethynylcyclohexene (for the preparation of this compound see ref. 165) were added successively. The temperature of the mixture was adjusted to 55°C directly after these additions, and was kept for 10 min between 54 and 56°C (note 2) by occasional cooling or heating. The brown mixture was then poured into 200 ml of ice-water and five extractions with very small portions of redistilled pentane were carried out. The pentane extracts (total volume about 130 ml) were washed three times with water and subsequently dried over magnesium sulfate. The pentane solution was transferred into a 250-ml round-bottomed flask, which was equipped for distillation in a water-pump vacuum, using the apparatus of Chapter I, Fig. 5. The receiver was immersed in a bath at 0°C and the pentane was removed in a water-pump vacuum. The temperature of the heating bath was increased gradually. The allenic compound distilled between 35 and 40°C/17 mmHg and was obtained in yields between 80 and 85%. The n_D^{20} varied between 1.528 and 1.534; the product with the lowest value contained about 20% of ethylbenzene (the product of further isomerization) and the product with the highest value contained about 6% of ethylbenzene.

Note 1. For the purification of commercial HMPT see ref. 1, p. 140.

Note 2. Heating for longer periods at higher temperatures leads to the formation of more ethylbenzene.

(4) *Methoxyallene*[29]

$$HC{\equiv}C\text{-}CH_2OCH_3 \xrightarrow{KO\text{-}\textit{tert.}\text{-}C_4H_9} H_2C{=}C{=}CH\text{-}OCH_3$$

Apparatus: 500-ml flask, provided with a mechanical stirrer, a gas inlet and a reflux condenser.

The flask was charged with 2.0 mol of dry, freshly distilled (note 1) methyl propargyl ether (VIII-6, Exp. 7), potassium *tert.*-butoxide (note 2) (5 g) was added with stirring and the mixture was heated under reflux. The reaction was monitored by drawing small samples at intervals of about 15 min and determining the refractive index (note 3). Refluxing was stopped when the n_D^{20} had reached the maximum value (about 1.427). The flask was cooled to room temperature. Two stoppers were placed on the outer necks, some boiling stones were added and the stirrer was replaced with a 40-cm Vigreux column, connected to a descending condenser and receiver. A tube filled with KOH pellets was placed between the receiver and the water pump. The receiver was cooled in a dry-ice-acetone bath (-75°C) and the system was evacuated (10-20 mmHg). The distillation flask was placed in a water-bath at 15°C. When most of the allenic ether had evaporated, the bath temperature was increased to 45°C. After a further 15 min of warming nitrogen was admitted. The receiver contained pure methoxypropadiene, n_D^{20} 1.4264, yield 88-93%. The boiling point at 760 mmHg is 52°C.

Note 1. Traces of water, or peroxides, formed during long periods of storage may inactivate the KO-*tert.*-C_4H_9.

Note 2. The uncomplexed base, commercially available from Nobel-Dynamit, G.F.R., was used.

Note 3. If after 15 min the refractive index has not increased considerably, 2 g of *tert.*-butylalcohol should be added. This may activate the base.

(5) *tert.-Butoxyallene*[29]

$$HC{\equiv}CCH_2O\text{-}\textit{tert.}\text{-}C_4H_9 \xrightarrow{KO\text{-}\textit{tert.}\text{-}C_4H_9} H_2C{=}C{=}CH\text{-}O\text{-}\textit{tert.}\text{-}C_4H_9$$

Apparatus: 500-ml flask, see Fig. 1 (the dropping funnel was omitted).

In the flask were placed 1 mol of freshly distilled and dry (note 1) *tert.*-butyl propargyl ether[76] and 7 g of powdered KO-*tert.*-C_4H_9 (see Exp. 4, note 2). The mixture was warmed to 54°C (internal) and was kept at this temperature (note 2) by occasional cooling or warming until the refractive index (n_D^{20}) had increased to 1.4440 (about 70 min, note 3). The flask was then connected to a 40-cm Vigreux column, condenser and receiver, cooled at -75°C. Some boiling stones were added, a tube with KOH was placed between the water pump and the receiver, and most of the liquid was distilled at 15-20 mmHg using a bath at a temperature that was gradually increased to 55°C. When the distillation had stopped, nitrogen was admitted to the apparatus and the receiver was replaced with an empty one. The last traces of allenic ether were subsequently distilled from the brown mass at a pressure of 0.5 mmHg or lower, while the flask was warmed in a bath at 50°C and the receiver was cooled at -75°C. The yield of *tert.*-butoxyallene, n_D^{20} 1.4418, was 90-94%.

Note 1. If not kept under nitrogen in well closed bottles, *tert.*-butyl propargyl ether is gradually converted into a hydroperoxide, HC≡C-CH(OOH)O-*tert.*-C_4H_9, even during storage at -20°C. Small amounts of this peroxide as well as moisture will lead to inactivation of the base, so that the isomerization is very slow or does not take place. The presence of hydroperoxide appears from an increase in the n_D and from a KI-test: brown colour after shaking a small sample with an aqueous solution of KI. This colour disappears if shaking is continued for some minutes. The impure propargyl ether can be freed from hydroperoxide by adding some paraffin oil and subsequently distilling the product at $<$ 1 mmHg using a receiver, cooled at -70°C, and keeping the bath-temperature below 30°C.

Note 2. Especially at temperatures higher than 60°C further conversion of the allenic ether to $CH_3C{\equiv}C$-O-*tert.*-C_4H_9 takes place. This ether splits off isobutene under these conditions and the resulting ketene, $CH_3CH{=}C{=}O$, reacts with the allenic ether and the 1-alkynyl ether. The yields of *tert.*-butoxyallene are then much lower.

Note 3. Although the n_D^{20} of the pure allenic ether is only 1.4418, the final value for the reaction mixture is higher, as the KO-*tert.*-C_4H_9 partly dissolves in the allenic ether.

(6) *Phenoxyallene*[79]

$$HC{\equiv}C\text{-}CH_2OPh \xrightarrow[HO\text{-}tert.\text{-}C_4H_9,\ HMPT]{KO\text{-}tert.\text{-}C_4H_9} H_2C{=}C{=}CHOPh$$

Apparatus: 250-ml round-bottomed, one-necked flask and thermometer.

A mixture of 2 g of KO-*tert*.-C_4H_9 (see Exp. 4, note 2), 5 g of *tert*.-butylalcohol and 40 ml of pure HMPT was heated, with occasional swirling, at 50°C until all of the solid material had dissolved. A 10-g amount of HC≡C-CH_2OPh (b.p. 80°C/20 mmHg, n_D^{20} 1.5323, prepared from phenol, KOH in ethanol and propargyl bromide) was added and the mixture was heated at 60°C for 15 min. It was then poured into 200 ml of ice-water and the reaction products were extracted with diethyl ether. The ethereal extracts were washed with saturated NH_4Cl solution, dried over magnesium sulfate and then concentrated in a water-pump vacuum. There remained 9.5 g of 3:1 mixture (probably an equilibrium mixture) of phenoxyallene and the starting compound (n_D^{20} 1.5505). A 5-g amount of KOH was dissolved in 40 ml of ethanol and this solution was added to the mixture. Subsequently a solution of K_2HgI_4 in about 50 ml of water (prepared by shaking a mixture of 12 g of KI, 9 g of $HgCl_2$ and 50 ml of water until the red precipitate had dissolved) was added. The mixture was shaken for 5 min, after which 250 ml of water were added. The allenic ether was extracted with a 2:1 mixture of pentane and diethyl ether. The brown precipitate was filtered off. The filtrate was washed with water, dried over $MgSO_4$ and concentrated in a water-pump vacuum. The 7.2 g of residue (n_D^{21} 1.5588) was pure phenoxyallene.

(7) *1,4-Dimethoxy-1,2-butadiene*[80]

$$CH_3O\text{-}CH_2\text{-}C{\equiv}C\text{-}CH_2\text{-}OCH_3 \xrightarrow[liq.\ NH_3]{KO\text{-}tert.\text{-}C_4H_9} CH_3O\text{-}CH{=}C{=}CH\text{-}CH_2\text{-}OCH_3$$

Apparatus: 3-l wide-necked, round-bottomed flask.

In 400 ml of anhydrous liquid ammonia (note 1) (drawn from a cylinder) in the 3-l flask were dissolved 25 g of KO-*tert*.-C_4H_9 (see Exp. 4, note 2). 1,4-Dimethoxy-2-butyne (Chapter VIII-6, Exp. 8) (0.60 mol) was poured into the solution. The reaction mixture was allowed to stand (with occasional swirling) for 25 min, after which 50 g of powdered ammonium chloride were introduced in 5 min with manual swirling. The ammonia was driven off by placing the flask in a water bath at 40°C.

Water (300 ml) was then added and five extractions with 40-ml portions of pentane (redistilled) were performed. The combined solutions were freed from *tert.*-C_4H_9OH by shaking four times with 100-ml portions of water. The washings were combined and subsequently extracted twice with small amounts of pentane. The extracts were shaken two times with water and subsequently combined with the first extracts. After drying the solutions over potassium carbonate, the greater part of the pentane was distilled off on a water-bath at 70°C at normal pressure through a 40-cm Widmer column. The remaining liquid was fractionated carefully, affording the allenic ether, b.p. 38°C/15 mmHg, n_D^{20} 1.4540, in 86% yield.

Note 1. Water content < 0.1%.

(*8*) *1,4-Diethoxy-1,2-butadiene*[80]

$$C_2H_5OCH_2C{\equiv}CCH_2OC_2H_5 \xrightarrow[\text{liq. } NH_3]{KO\text{-}tert.\text{-}C_4H_9} C_2H_5O\text{-}CH{=}C{=}CH\text{-}CH_2OC_2H_5$$

Apparatus: 3-l wide-necked, round-bottomed flask.

Anhydrous liquid ammonia (note 1) (600 ml), drawn from a cylinder, was introduced into the flask. Potassium *tert.*-butoxide (see Exp. 4, note 2) (55 g) was then introduced cautiously with swirling. Subsequently 80 g of 1,4-diethoxybutyne [VIII-6, Exp. 8] was poured into the solution. A wad of cotton-wool was placed in the neck and the flask was allowed to stand for 70 min with occasional swirling. Powdered ammonium chloride (35 g) was then added in small portions with manual swirling. The flask was placed in a water-bath at 50°C and to the remaining mass 300 ml of ice-water were added. After the solid had passed into solution, five extractions with small portions of redistilled pentane were carried out and the combined extracts were washed five times with water and dried over potassium carbonate. The aqueous layers from the washings were combined and extracted once with 50 ml of pentane. This pentane layer was dried together with the original extract. The pentane was evaporated and the residue gave, after fractionation through a 30-cm Widmer column, the allenic ether, b.p. 59°C/15 mmHg, n_D^{22} 1.4492, in 87% yield.

Note 1. Water content < 0.1%.

(*9*) *1-Methoxy-4-ethoxy-1,2-pentadiene*[76]

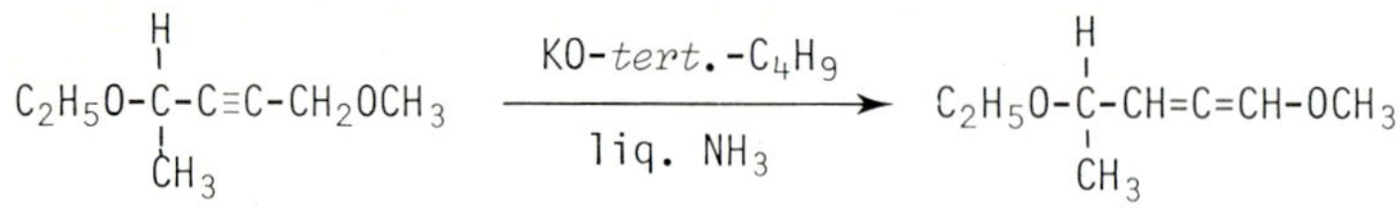

Apparatus: 1-l flask, see Fig. 2 (without dropping funnel).

To a solution of 22 g of KO-*tert.*-C_4H_9 (see Exp. 4, note 2) in 400 ml of anhydrous liquid ammonia were added 22 g of the bis-ether (note 1). After stirring for 4 h, 20 g of powdered ammonium chloride were introduced in small portions. The ammonia was removed by placing the flask in a water-bath at 40°C, then 200 ml of water were added and five extractions with small portions of redistilled pentane were carried out. The combined extracts were washed with water, dried over magnesium sulfate and then concentrated in a water-pump vacuum. The residue was carefully distilled through a 40-cm Vigreux column, giving the allenic bis-ether, b.p. 52°C/15 mmHg, n_D^{21} 1.4456 in 80% yield.

Note 1. Prepared by adding butyllithium in hexane to a solution of $HC{\equiv}CCH_2OCH_3$ in THF at -10°C and subsequently alkylating the solution of $LiC{\equiv}CCH_2OCH_3$ obtained with $CH_3CH(Cl)OC_2H_5$ at -10 to 0°C. The α-chloroether was prepared by introducing the required amount of dry gaseous HCl into $H_2C{=}CH{-}OC_2H_5$ at-25°C. $C_2H_5OCH(CH_3)C{\equiv}CCH_2OCH_3$ has a b.p. of 59°C/15 mmHg and n_D^{15} 1.4320.

(*10*) *1,4,4-Triethoxy-1,2-butadiene*[76]

$$C_2H_5O{-}CH_2C{\equiv}C{-}CH(OC_2H_5)_2 \xrightarrow[\text{DMSO}]{\text{KO-}tert.\text{-}C_4H_9} C_2H_5O{-}CH{=}C{=}CH{-}CH(OC_2H_5)_2$$

Apparatus: 250-ml flask and thermometer.

To a mixture of 15 g of the triethoxyacetylene (see Chapter III, Exp. 40) and 40 ml of dry DMSO was added with manual swirling 1.5 g of KO-*tert.*-C_4H_9 (see Exp. 4, note 2). The addition was carried out in five equal portions at intervals of 3 min. The temperature of the mixture gradually rose to about 30°C. After the addition the solution was warmed for an additional 10 min at 35°C. The isomerization product was isolated by adding 150 ml of an aqueous solution of 20 g of NH_4Cl and subsequently extracting five times with diethyl ether. The ethereal solutions were washed with water and dried over potassium carbonate and then concentrated in a water-pump vacuum. Distillation of the residue through a 20-cm Vigreux column gave the allene, b.p. 92°C/15 mmHg, n_D^{20} 1.4467, in 82% yield.

(*11*) *4-Methyl-1-methoxy-1,2,4-pentatriene*[81]

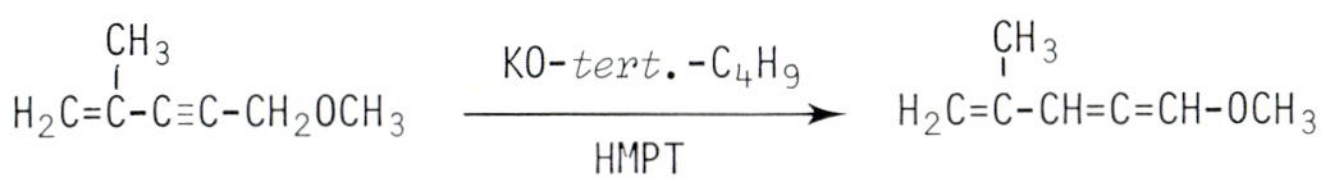

Apparatus: 250-ml flask, see Fig. 1.

A mixture of 1 g of KO-*tert*.-C_4H_9, 0.6 g of HO-*tert*.-C_4H_9 and 40 ml of dry, pure HMPT was warmed at 40-50°C until all of the solid material had disappeared. The solution obtained was cooled to 15°C and 15 g of the enyne ether (note 1) were added in 2 min. The temperature of the mixture was kept at *ca*. 25°C and after the addition this temperature was maintained for an additional 10 min. The flask was then equipped for distillation by means of an oil pump, using a pressure of 1 mmHg or lower. The volatile product was collected in a single receiver, cooled at -75°C. The contents of the receiver were washed three times with ice-water in a small separating funnel (50-100 ml) and subsequently dried over magnesium sulfate (just enough to cause disappearance of the turbidity). The liquid was then distilled in a water-pump vacuum using a 20-cm Vigreux column and a single receiver, cooled in ice (note 2). The allenic ether (b.p. 20-25°C/15 mmHg, n_D^{15} 1.4906) was obtained in 83% yield. The compound dimerizes (or polymerizes) rapidly at room temperature, and should be stored at -80°C.

Note 1. The enyne ether (b.p. 40°C/25 mmHg, n_D^{15} 1.4614) was obtained in 75% yield by converting the commercially available enyne $HC{\equiv}C\text{-}C(CH_3){=}CH_2$ (see also Ref. 1) into the Grignard compound by refluxing with C_2H_5MgBr in diethyl ether and subsequently adding $ClCH_2OCH_3$ (commercially available) at -20 to -10°C. Aqueous work-up, drying of the ethereal solution, removal of the diethyl ether by distillation at normal pressure and distillation of the residue gave the enyne ether.

Note 2. The allenic ether is extremely sensitive towards oxygen. All operations during the work-up must be carried out under nitrogen. After termination of the distillation nitrogen should be admitted to the distillation apparatus. Too strong heating during the distillation involves the risk of an explosion.

(*12*) *N,N-Dimethylaminoallene*[166]

$$HC{\equiv}CCH_2N(CH_3)_2 \xrightarrow[\text{HMPT}]{\text{KO-}tert.\text{-}C_4H_9.\text{HO-}tert.\text{-}C_4H_9} H_2C{=}C{=}CH\text{-}N(CH_3)_2$$

Apparatus: 100-ml round-bottomed flask and thermometer; swirling by hand.

A clear solution of 0.01 mol of KO-*tert.*-C_4H_9 and 0.01 mol of HO-*tert.*-C_4H_9 in 10 ml of pure, dry HMPT (note 1) was prepared by warming the mixture at 50^oC. The solution was then cooled to 17^oC and added to 5.0 g of the propargylamine (see Chapter VIII-5, Exp. 1), in the 100-ml flask. The air in the flask was quickly replaced with nitrogen and the mixture was kept at 23^oC for 10 min (occasional cooling in a bath of 15^oC may be necessary). During this period the flask was occasionally swirled by hand. The flask was then connected to a distillation apparatus (see Fig. 5), consisting of a 30-cm Vigreux column, condenser and receiver, cooled at -78^oC. Between the receiver and the water pump was placed a tube filled with KOH pellets. The system was evacuated and the distillation flask gradually warmed at about 80^oC. After termination of this operation nitrogen was admitted. The contents of the receiver consisted of reasonably pure (*ca.* 95%) dimethylaminoallene, n_D^{20} 1.4658, yield 94%. The NMR spectrum indicated the presence of about 1% of starting compound and 4% of $CH_3C{\equiv}CN(CH_3)_2$ (note 2). The product can be stored under pure nitrogen at -25^oC for at least 24 h.

Note 1. Prepared by addition as much of a solution of 1.5 g of lithium in 75 ml of liquid NH_3 to 1 l of the commercial product as to cause a uniformly blue solution, for a least 1 min. The liquid is subsequently distilled two times at 10-20 mmHg.

Note 2. All operations of the isolation procedure must be carried out without any delay. The distillation apparatus must be perfectly dry, since allenic amines are extremely water-sensitive.

The equilibrium mixture consists of about 20% of $CH_3C{\equiv}C\text{-}N(CH_3)_2$ and 80% of $H_2C{=}C{=}CH\text{-}N(CH_3)_2$. The isomerization method is therefore unsuitable for the preparation of the *yneamine*. In the case of the isomerization of $HC{\equiv}CCH_2N(C_2H_5)_2$ the equilibrium ratio allenic amine: yneamine is about 1:4. As the allene can be removed by simply heating the mixture (polymerization), $CH_3C{\equiv}CN(C_2H_5)_2$ can be successfully prepared from $HC{\equiv}CCH_2N(C_2H_5)_2$. The procedure, which is published in detail in Ref. 1, involves treating of $HC{\equiv}CCH_2N(C_2H_5)_2$ (1 mol) with KO-*tert.*-C_4H_9 (2-4 g) in DMSO (80-100 ml) at about 40^oC for 20 min and subsequently distilling off the volatile products at 10-15 mmHg. The distillate is heated for about 1 h at 100°C and subsequently subjected to vacuum distillation. The yield of $CH_3C{\equiv}C\text{-}N(C_2H_5)_2$ is about 75%.

(*13*) *N-Morpholinoallene*[166]

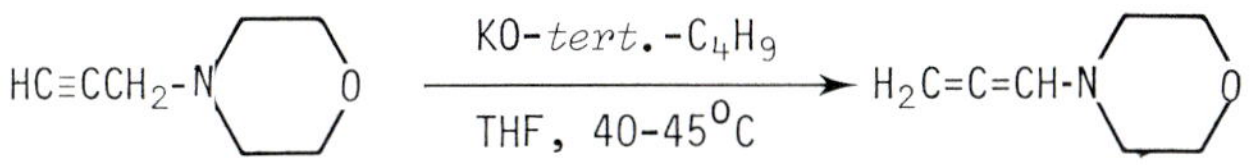

Apparatus: 200-ml round-bottomed, three-necked flask provided with a gas inlet, a thermometer and a gas outlet; magnetic stirring.

In the flask were placed 10.0 g of the propargylic amine (see Chapter VIII-5, Exp. 1). The air in the flask was replaced with nitrogen and a solution of 0.01 mol of KO-*tert.*-C_4H_9 in 10 g of THF (free from hydroperoxide) was added. The mixture was warmed at about 40°C. A weakly exothermic reaction was observed and the temperature rose to about 45°C. After 1-2 min the gel originally present, had disappeared almost completely and a brown solution had formed. The refractive index of the solution (note 1) was measured after intervals of about 2 min. After the maximum value (n_D^{20} about 1.464) had been reached, heating at 40°C was continued for another 1 to 2 min. *Tert.*-butylalcohol (0.01 mol) was then added (note 2) and the mixture was cooled to 20°C. The THF was removed in a water-pump vacuum and the residue was distilled (b.p. 30-40°C) in a high vacuum (pressure $<$ 0.5 mmHg), the (single) receiver being cooled at 0°C. Towards the end of the distillation the temperature of the heating bath was increased to 60-70°C in order to minimise the hold-up. The yield of the allene, n_D^{20} 1.5125, was 82-87%. The NMR spectrum indicated the presence of about 4% of $CH_3C{\equiv}C$-Morph. The product rapidly turned yellow upon exposure to the air and polymerized at room temperature within a few hours.

Note 1. A small sample was taken by means of a Pasteur pipette and the liquid was placed on the prism. Care should be taken that no evaporation of THF takes place: this will give too high a n_D.

Note 2. The further conversion into the yneamine $CH_3C{\equiv}C$-Morph is stopped almost completely by the addition of HO-*tert.*-C_4H_9 which forms the 1:1 complex with KO-*tert.*-C_4H_9. If the isomerization with KO-*tert.*-C_4H_9 is carried out in DMSO, an equilibrium mixture of about 80% of the allenic amine and 20% of the yneamine is formed after 1-2 min at 30°C.

A similar procedure with propargylpiperidine (reaction temperature 45-55°C) led to a mixture of 92% of the allene and 8% of the yneamine. During the high-vacuum distillation the receiver was cooled at -30°C. The allenic amine, b.p. 30°C/0.5 mmHg, n_D^{20} 1.5068, was obtained in a yield of 82%. The equilibrium mixture obtained in the isomerization under the influence of KO-*tert*.-C_4H_9 in DMSO consisted of about 70% of allene and 30% of yneamine. Base-catalysed isomerization is therefore not a suitable method to prepare the *yneamine*.

(*14*) *1-Phenylthio-1-propyne*[79]

$$C_6H_5SNa + HC{\equiv}CCH_2Cl \xrightarrow[C_2H_5OH]{} C_6H_5S\text{-}CH_2C{\equiv}CH \xrightarrow[C_2H_5OH]{NaOC_2H_5} CH_3C{\equiv}C\text{-}SC_6H_5$$

Apparatus: 1-l round-bottomed, three-necked flask, provided with a dropping funnel, a mechanical stirrer and a reflux condenser.

A solution of 0.30 mol of sodium ethoxide was prepared by dissolving 0.30 mol of sodium in 300 ml of 100% ethanol. Thiophenol and propargyl chloride (0.30 mol) were successively added in 15 min without external cooling. Five minutes later a solution of sodium ethoxide in 100 ml of 100% ethanol, prepared from 3 g of sodium, was run in. The suspension was heated under reflux for an additional 1 h. The mixture was cooled to room temperature and subsequently poured into 500 ml of ice--water. The product was isolated by extracting five times with diethyl ether, washing the extract with water, drying over magnesium sulfate and concentrating the extract in a water-pump vacuum. Distillation through a short column gave 1-phenylthio-1-propyne, b.p. 100°C/1 mmHg, n_D^{20} 1.6092, in 88% yield.

(*15*) *Ethylselenoallene*[79]

$$HC{\equiv}CCH_2Se\text{-}C_2H_5 \xrightarrow[HOC_2H_5]{KOC_2H_5} H_2C{=}C{=}CH\text{-}Se\text{-}C_2H_5 \quad (\longrightarrow CH_3C{\equiv}C\text{-}Se\text{-}C_2H_5)$$

Apparatus: 250-ml flask, see Fig. 1.

A solution of potassium ethoxide in ethanol was prepared by adding 4.0 g of KO-*tert.*-C_4H_9 (see Exp. 4, note 2) to 24 g of 96% ethanol. The solution was heated at 60°C and 25.2 g of ethyl propargyl selenide (see Chapter VIII-6, Exp. 5) were added. The mixture was warmed at 62°C (internal) for 35 min. The refractive index (initially 1.430) was then 1.440. The solution was poured into 200 ml of ice-water and the products were extracted with pentane. After washing the pentane solution with water it was dried over magnesium sulfate. Removal of the solvent in a water-pump vacuum, followed by distillation, gave a mixture of 15% of $HC{\equiv}C\text{-}CH_2SeC_2H_5$, 82% of $H_2C{=}C{=}CH\text{-}SeC_2H_5$ and a trace (3%) of $CH_3C{\equiv}C\text{-}SeC_2H_5$, in 94% yield (b.p. 42°C/15 mmHg, n_D^{18} 1.5520).

Note 1. Lengthening of the reaction time will lead to a decrease of the amount of starting compound, but more of the end product $CH_3C{\equiv}C\text{-}Se\text{-}C_2H_5$ will be present.

(16) *Allenyl ethyl ketone*[82]

$$HC{\equiv}CCH_2\text{-}\overset{\displaystyle O}{\overset{\|}{C}}\text{-}C_2H_5 \xrightarrow[H_2O]{NaHCO_3} H_2C{=}C{=}CH\text{-}\overset{\displaystyle O}{\overset{\|}{C}}\text{-}C_2H_5$$

Apparatus: 100-ml round-bottomed flask with a stopper, and shaking machine (vibrator), or a 100-ml round-bottomed flask with a mechanical stirrer.

A mixture of 15 g of the crude propargyl ketone (purity about 80%, see Chapter VIII-6, Exp. 6), 5 g of sodium hydrogen carbonate and 10 ml of water was shaken vigorously at room temperature or agitated vigorously with a mechanical stirrer. The isomerization was followed by determining the refractive index of the upper layer (shaking or stirring was interrupted for a few minutes). When this had reached its maximum value (about 1.464) (note 1) shaking or stirring was stopped and the mixture was extracted five times with small portions of diethyl ether. The extracts were dried (without washing) over magnesium sulfate and after removing the solvents in a water-pump vacuum the ketone was distilled (b.p. 35-40°C/ /15 mmHg, n_D^{20} 1.4662), yield 85%. The product contained about 15% of an unknown impurity.

Note 1. The reaction time at room temperature was 2-3 h. At 40°C the isomerization was finished within 30 min. With a dilute (1 g per 10 ml) aqueous solution of K_2CO_3, only about 5 min were required.

(*17*) *2,4,5-Hexatrienenitrile*

$$HC{\equiv}CCH{=}CH{-}CH_2Br \xrightarrow[H_2O,\ C_2H_5OH]{CuC{\equiv}N,\ KC{\equiv}N} \begin{matrix} HC{\equiv}CCH{=}CHCH_2C{\equiv}N \\ + \\ H_2C{=}C{=}CH{-}CH{=}CH{-}C{\equiv}N \end{matrix} \xrightarrow{K_2CO_3} H_2C{=}C{=}CH{-}CH{=}CH{-}C{\equiv}N$$

Apparatus: 1-l flask, see Fig. 1.

In the flask were placed 5.0 g of copper(I) cyanide, 25 ml of water, 50 ml of ethanol and 0.20 mol of 5-bromo-3-penten-1-yne (note 1). The mixture was heated to 60°C and a solution of 0.22 mol of potassium cyanide in 40 ml of water was added in 20 min, whilst keeping the temperature between 60 and 65°C (occasional cooling may be necessary). After the addition the mixture was heated for an additional 10 min at 80°C, then cooled to room temperature and 200 ml of water were added. The reaction products were extracted seven times with diethyl ether. The combined extracts were washed once with concentrated NH_4Cl solution and dried over magnesium sulfate. Removal of the diethyl ether by evaporation in a water-pump vacuum gave 17.5 g of a residue, consisting of equal amounts of 3-hexen-5-yne- and 2,4,5-hexatrienenitrile. This mixture was dissolved in 50 ml of diethyl ether and 3 g of potassium carbonate were added. The temperature rose in a few minutes from 20 to about 30°C. The mixture was shaken for an additional 15 min (by hand or mechanically) and was subsequently poured with swirling into 50 ml of 2 N hydrochloric acid saturated with ammonium chloride. After separation of the layers, two extractions with diethyl ether were carried out. The combined ethereal solutions were washed once with 50 ml of concentrated NH_4Cl solution and dried over magnesium sulfate. Removal of the solvent and subsequent distillation through a 30-cm Vigreux column afforded the pure nitrile (similar amounts of *cis*- and *trans*-isomers), b.p. 55°C/15 mmHg, n_D^{20} 1.5372, in 78% yield.

Note 1. Prepared by addition at -10°C of the required amount of PBr_3 to an ethereal solution of $HC{\equiv}CCH{=}CHCH_2OH$ (III, Exp. 57), to which 5 ml of pyridine had been added. After standing for 2 h at 20°C the mixture was poured into water. The bromide, b.p. 40°C/15 mmHg, n_D^{20} 1.5413, was obtained in 77% yield.

(18) *Neopentylacetylene*[15]

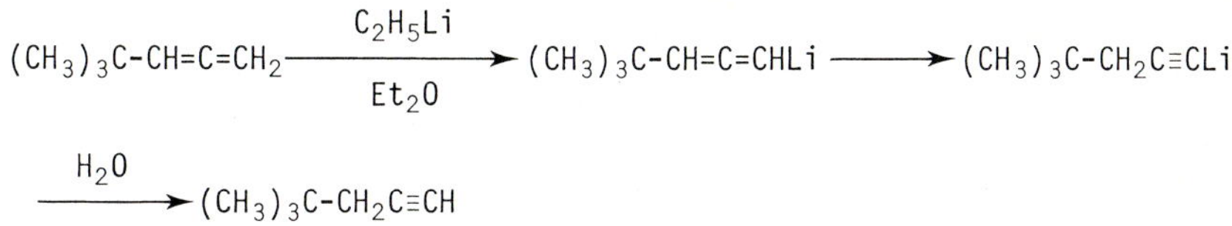

Apparatus: 500-ml flask, see Fig. 1.

A solution of $(CH_3)_3C\text{-}CH{=}C{=}CLi$, obtained by addition at $-60^{\circ}C$ of 0.20 mol of *tert.*-butylallene (see Chapter VI, Exp. 2) to a solution of 0.25 mol of ethyllithium in about 200 ml of diethyl ether (see Chapter II, Exp. 1) was warmed to $25^{\circ}C$ and held at this temperature for 15 min. Subsequently the solution was cooled to below $0^{\circ}C$ and 50 ml of saturated NH_4Cl solution were added dropwise with vigorous stirring, keeping the temperature below $20^{\circ}C$. The upper layer was separated off and the aqueous layer was extracted twice with 25-ml portions of diethyl ether. The combined solutions were dried over a small amount of magnesium sulfate. Slow distillation through a 40-cm Widmer column gave neopentylacetylene (b.p. $76^{\circ}C$/760 mmHg, n_D^{20} 1.3992) in 74% yield (towards the end of the distillation a partial vacuum was applied in order to minimize the hold-up).

(19) *4-Phenyl-1-butyne*[15]

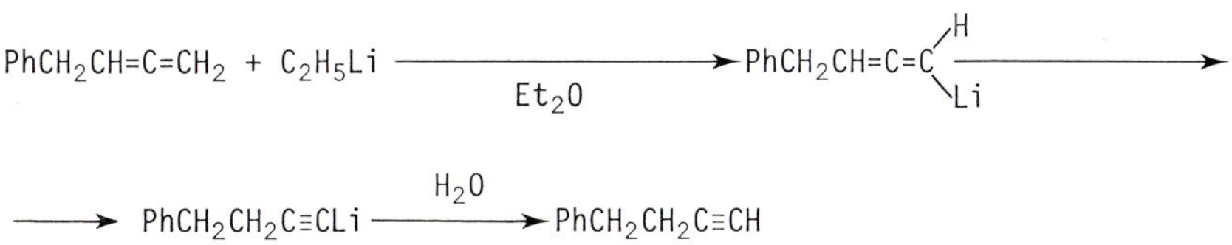

Apparatus: 250-ml flask, see Fig. 1.

To a solution of 0.05 mol of 4-phenyl-1,2-butadiene (see Chapter V, Exp. 19) was added in 10 min at -25 to $-35^{\circ}C$ a solution of 0.10 mol of ethyllithium in 80 ml of diethyl ether (see Chapter II, Exp. 1). After the addition the cooling bath was removed and the reaction mixture was warmed to $30^{\circ}C$ in about 15 min and held at this temperature for an additional 15 min. The brown solution was then cautiously poured into 200 ml of ice-water. After separation of the layers four extractions with diethyl ether were carried out. The combined ethereal solutions were dried over magnesium sulfate and concentrated in a water-pump vacuum. Distillation of the residue in a small apparatus gave 4-phenyl-1-butyne, b.p. 72°C/15 mmHg, n_D^{20} 1.5244, in 78% yield.

(*20*) *1-NN-diethylamino-1-buten-3-yne*[153]

$$H_2C{=}CH{-}C{\equiv}C{-}N(C_2H_5)_2 \xrightarrow[\text{liq. } NH_3]{KNH_2} KC{\equiv}C{-}CH{=}CH{-}N(C_2H_5)_2 \xrightarrow{NH_4Cl}$$

$$\longrightarrow HC{\equiv}C{-}CH{=}CH{-}N(C_2H_5)_2$$

Apparatus: 1-l flask, see Fig. 2.

To a filtered solution (see Chapter II, Exp. 12) of 0.20 mol of potassium amide in 250 ml of liquid ammonia was added in 10 min a mixture of 0.15 mol of the enyne amine (see Chapter V, Exp. 15) and 50 ml of dry diethyl ether. A thin suspension was formed gradually. After stirring for 1 h 12 g of powdered ammonium chloride were introduced in small portions. The ammonia was then evaporated by immersing the flask in a water-bath at 40°C, care being taken that no moisture could enter the flask (note 1). When the stream of escaping ammonia had become very faint, the water-bath was removed. After drying the outer walls of the flask with a cloth, the product was extracted from the salt by shaking 5-8 times with small portions of dry diethyl ether and carefully decanting the solutions from the salt. The combined solutions were concentrated in a water-pump vacuum and the residue was distilled through a 30-cm Vigreux column, giving the ynene amine, b.p. 78°C/15 mmHg, n_D^{20} 1.5277, in 92% yield (*E/Z* ratio *ca.* 94:6).

In a similar way was prepared from $HC{\equiv}C{-}CH{=}CH{-}N(CH_3)_2$, b.p. 74°C/23 mmHg, n_D^{20} 1.5419, was prepared in 80% yield from $H_2C{=}CH{-}C{\equiv}C{-}N(CH_3)_2$ (see Chapter V, Exp. 14) (*E/Z* ratio *ca.* 94:6).

Note 1. The product is water-sensitive.

(*21*) *6-Heptyn-1-ol*[83]

$$K + NH_3 \xrightarrow{Fe^0} KNH_2$$

$$KNH_2 + H_2N(CH_2)_3NH_2 \longrightarrow KNH(CH_2)_3NH_2$$

$$2\ KNH(CH_2)_3NH_2 + C_4H_9C{\equiv}C\text{-}CH_2OH \longrightarrow KC{\equiv}C(CH_2)_5OK$$

$$KC{\equiv}C(CH_2)_5OK \xrightarrow{H_2O} HC{\equiv}C(CH_2)_5OH$$

Apparatus: 1-l one-necked, round-bottomed flask.

A filtered solution of 0.55 mol of potassium amide was prepared (as described in Chapter II, Exp. 12). The solution was subsequently concentrated to a volume of about 40 ml by placing the flask in a water-bath at 30-40°C. 1,3-Diaminopropane (dried over powdered KOH and subsequently distilled)(230 ml) was added cautiously and the flask was warmed in a water-bath at 40°C. When the stream of escaping ammonia had become very faint, the flask was connected, via a tube, filled with KOH pellets, with the water pump. The bath temperature was lowered to about 30°C and the flask was evacuated until the suspension had disappeared almost completely, which took about 1 h. Nitrogen was then admitted to the flask and 0.20 mol of 2-heptyn-1-ol (see III, Exp. 56) was added with swirling. During this addition the temperature rose to 60-70°C. The dark brown solution was heated for 1 h at 70°C and, after cooling to 20°C, 500 ml of cold (0°C) saturated ammonium chloride solution was added with swirling (note 1). Ten extractions with small portions of diethyl ether were carried out and the combined extracts were dried over magnesium sulfate and concentrated in a water-pump vacuum. Careful distillation afforded the desired alcohol, b.p. 89°C/15 mmHg, n_D^{20} 1.4531, in 81% yield.

Note 1. A considerable part of the diamine can be recovered by distilling it off from the reaction mixture, using a bath temperature not exceeding 50°C. The residue upon treatment with aqueous ammonium chloride gives the acetylenic alcohol in good yield.

(22) *3-Pentyn-1-oic acid and 3-heptyn-1-oic acid*[11]

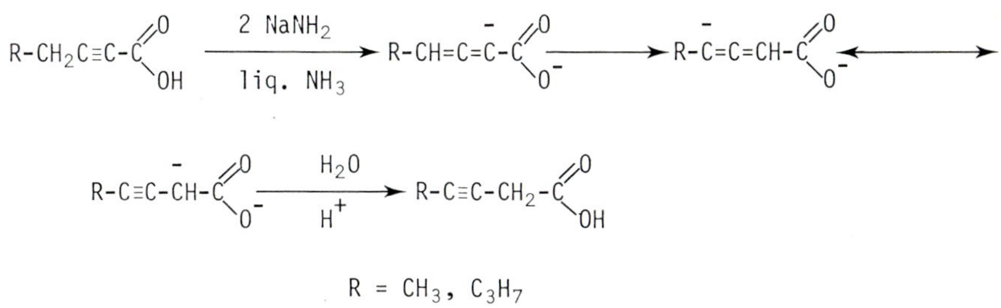

R = CH_3, C_3H_7

Apparatus: 3-l wide-necked flask with mechanical stirrer, see Fig. 4.

A suspension of 0.4 mol of sodium amide in 400 ml of liquid ammonia was prepared in the usual way (see Chapter II, Exp. 11). To the suspension was added in 5 min 0.10 mol of the carboxylic acid (prepared in almost 100% yield by introducing an excess of carbon dioxide into a solution of alkynyllithium in THF--hexane, compare Ref. 1 and Chapter III, Exp. 35) dissolved in 50 ml of THF. A vigorous reaction ensued : the greyisch black solution that was originally formed very soon changed into a white suspension. After 30 min, 40 g of powdered ammonium chloride were added in 5 min with vigorous stirring. The ammonia was driven off by placing the flask in a water-bath at 50^{0}C. During this operation the flask was swirled manually to suppress bumping. To the solid mass that remained were added 300 ml of diethyl ether. Heating at 50^{0}C was continued until the flow of escaping ammonia had become hardly perceptible (a rubber stopper with a narrow gas outlet should be placed on the flask). The flask was then cooled to room temperature and 200 g of finely crushed ice were added. Subsequently the mixture was acidified by cautiously adding 5 N hydrochloric acid to the swirled mixture to give a pH of about 1. The solution was saturated with ammonium chloride and then extracted ten times with small portions of diethyl ether. The combined extracts (no washing was carried out) were dried over magnesium sulfate and concentrated in a water-pump vacuum. The residue thus obtained contained some water, which was removed by warming for 1 h at 30^{0}C and 0.5 mmHg. The remaining light-brown solids were > 96% pure according to the NMR spectra. The yields were about 90%. Very pure samples can be obtained by crystallization from pentane (see Ref. 11).

(*23*) *Methylthiopropadiene*[23]

$$CH_3SSCH_3 \xrightarrow[\text{liq. } NH_3]{2\text{ Na}} CH_3SNa$$

$$HC{\equiv}C{-}CH_2Cl + CH_3SNa \xrightarrow[\text{liq. } NH_3]{} HC{\equiv}C{-}CH_2SCH_3 \xrightarrow[\text{liq. } NH_3]{NaOC_2H_5} CH_3C{\equiv}C{-}SCH_3$$

$$CH_3C{\equiv}C{-}SCH_3 \xrightarrow[\text{liq. } NH_3]{NaNH_2} H_2C{=}C{=}\overset{-}{C}{-}SCH_3 \xrightarrow{H_2O} H_2C{=}C{=}CH{-}SCH_3$$

Apparatus: 5-l round-bottomed, wide-necked flask with a mechanical stirrer for the preparation of $CH_3C{\equiv}C\text{-}SCH_3$; 2-l round-bottomed flask and a rubber stopper, perforated by a glass tube, internal diameter 4-5 mm; the glass tube, when placed on the flask, reached nearly to the bottom of the flask; the upper end of the glass tube was connected to a plastic tube.

a. Preparation of 1-methylthio-1-propyne, $CH_3C{\equiv}C\text{-}SCH_3$

Anhydrous liquid ammonia (2 l) was transferred from a cylinder into the reaction flask. Sodium (47 g) was introduced in pieces of 1-2 g. Ten minutes after the addition of the sodium, dimethyl disulfide (commercially availabe) was added dropwise from a dropping funnel with a long stem. The reaction was very vigorous and addition of ammonia to maintain the volume of the reaction mixture at 2 l, was necessary. The addition of dimethyl disulphide, which took 30 min, was stopped as soon as the blue colour of the sodium had disappeared: slightly more than 1 mol appeared to be required. Propargyl chloride (2.0 mol) was subsequently added in 40 min to the vigorously agitated mixture. Ammonia was added to bring the volume of the mixture at 1.5 l. Twenty minutes after the addition of propargyl chloride a solution of 0.3 mol of sodium ethoxide in 150 ml of liquid ammonia was cautiously poured into the reaction mixture. This solution had been prepared in a 500-ml round-bottomed flask by cautious addition of 0.4 mol of absolute ethanol to 0.3 mol of sodium in liquid ammonia. Stirring was continued for 1 h after the addition of sodium ethoxide. A rubber stopper with a narrow hole was placed on the flask and the ammonia was allowed to evaporate. To the remaining salty mass was added 1 l of water and, after dissolution of the salt, the upper layer was separated as sharply as possible. The aqueous layer was extracted twice with 50-ml portions of pentane. The organic layer and the pentane extracts were separately dried over a small amount of magnesium sulfate. After practically all pentane had been distilled off from the extract at normal pressure (using a 40-cm Vigreux column), the remaining liquid and the first organic fraction were combined and distilled in a partial vacuum. 1-Methylthio-1-propyne, b.p. between 60 and $70^{0}C/100$ mmHg, n_D^{20} 1.5020, was obtained in 82% yield.

b. Preparation of methylthiopropadiene, $H_2C{=}C{=}CH\text{-}SCH_3$

A suspension of 0.35 mol of sodium amide in 400 ml of liquid ammonia was prepared in the usual way (*c.f.* Chapter II, Exp. 11) (note 1). To this suspension was added with swirling 0.30 mol of 1-methylthio-1-propyne, in portions of about 10 g, waiting about 10 s after the addition of each portion. Swirling was continued for 1.5 min after the addition of the last portion. Immediately thereafter the end of the plastic tube was placed on the bottom end of a 3-l conical flask, containing 700 g of crushed ice and the stopper with the glass tube was placed on the flask. The brown ammoniacal solution was transferred by the pressure of the ammonia through the tubes under the ice (see Fig. 3). During this procedure, which took 1-2 min, the flask with the ice was swirled continuously. Pentane (b.p. 35-40°C) (150 ml) was added and the flask was filled with nitrogen and placed in a water-bath at 40°C. After the remaining ice had melted, the mixture was poured into a separating funnel and shaken vigorously under nitrogen (note 2). The upper layer was separated off and the aqueous layer was extracted 4-5 times with very small portions of pentane under nitrogen. After drying the extracts over magnesium sulfate, the pentane was distilled off under nitrogen through a 40-cm Widmer column; the temperature of the heating bath was increased gradually but was kept below 75°C (note 3). When the distillation had stopped, the temperature of the bath was lowered to 30°C. The condenser was connected to a single receiver and the distillation apparatus was evacuated. When the pressure had dropped to 30 mmHg, the receiver was placed in a cooling bath at -20°C. The volatile allenic sulfide was trapped in the receiver (note 4). The temperature of the heating bath was gradually increased to 60-70°C and, after termination of the distillation, nitrogen was admitted to the apparatus (note 2). Methylthioallene, n_D^{20} 1.5369, was obtained in 78-84% yield.

Note 1. If more than 100 mg of the iron salt is used as catalyst for the conversion of lithium, the separation of the layers during the work-up may be difficult, owing to the formation of iron hydroxide gel.

Note 2. Allenic sulfides undergo autoxidation when exposed to air.

Note 3. Prolonged heating at higher temperatures gives rise to cyclo-dimerization.

Note 4. If distillation at 760 mmHg were possible, one would expect a b.p. between 110 and 120°C.

In a similar way $H_2C{=}C{=}CH{-}SC_2H_5$, b.p. 30°C/12 mmHg, n_D^{20} 1.5210, was prepared in 86% yield from $CH_3C{\equiv}C{-}SC_2H_5$[1]. After the greater part of the solvent had been distilled off at normal pressure (bath temperature below 70°C), the remaining liquid was distilled at water-pump pressure and collected in a single receiver cooled at 0°C.

(24) *tert.-Butylthioallene*[23]

$$tert.\text{-}C_4H_9SH + Na \xrightarrow[\text{liq. } NH_3]{} tert.\text{-}C_4H_9SNa$$

$$HC{\equiv}C\text{-}CH_2Cl + tert.\text{-}C_4H_9SNa \xrightarrow[\text{liq. } NH_3]{} HC{\equiv}C\text{-}CH_2\text{-}S\text{-}tert.\text{-}C_4H_9 \xrightarrow[\text{liq. } NH_3]{NaOC_2H_5}$$

$$\longrightarrow CH_3C{\equiv}C\text{-}S\text{-}tert.\text{-}C_4H_9$$

$$CH_3C{\equiv}C\text{-}S\text{-}tert.\text{-}C_4H_9 \xrightarrow[\text{liq. } NH_3]{NaNH_2} H_2C{=}C{=}\overset{-}{C}\text{-}S\text{-}tert.\text{-}C_4H_9 \xrightarrow{H_2O}$$

$$\longrightarrow H_2C{=}C{=}CH\text{-}S\text{-}tert.\text{-}C_4H_9$$

Apparatus: 5-l round-bottomed, wide-necked flask with a mechanical stirrer for the preparation of $CH_3C{\equiv}C\text{-}S\text{-}tert.C_4H_9$; 1-l round-bottomed flask and a rubber stopper, perforated by a glass tube of 4-5 mm I.D., connected to a plastic tube, for the conversion into the allenic sulfide. The glass tube extended to 2 mm from the bottom of the flask.

a. Preparation of $CH_3C{\equiv}CS\,tert.\text{-}C_4H_9$

Anhydrous liquid ammonia (3.5 l) was drawn from a cylinder and introduced into the reaction flask. Sodium (23.5 g) was cut into 1-g pieces and dissolved in the ammonia. Ten minutes later *tert.*-butylmercaptan was added from a dropping funnel with a long stem. The addition was carried out in 30 min, and was stopped as soon as the blue colour of the dissolved sodium had disappeared: slightly more than 1 mol of the thiol had been added. Propargyl chloride (1.00 mol) was subsequently introduced in 30 min. During this period a solution of 0.5 mol of sodium ethoxide in 500 ml of liquid ammonia was prepared in a 1-l round-bottomed flask by cautious addition of 0.55 mol of ethanol (100%) to a solution of 0.5 mol of sodium in about 650 ml of liquid ammonia. The solution obtained was added to the reaction mixture 20 min after all of the propargyl chloride had been introduced. Stirring was continued for an additional 3 h. A stopper with a gas outlet was then placed on the flask and the ammonia was allowed to evaporate overnight. To the residue were added 500 ml of water. After dissolution of the solid three or four extractions with small amounts of diethyl ether were carried out. The combined ethereal solutions were dried over magnesium sulfate, after which the greater part of the diethyl ether was distilled off at normal pressure through a 40-cm Vigreux column. The remaining liquid was distilled, affording the acetylenic sulfide, b.p. 50°C/ /17 mmHg, n_D^{25} 1.4810, in 88% yield.

b. Preparation of $H_2C{=}C{=}CH{-}S{-}$*tert.*$-C_4H_9$

To a suspension of 0.30 mol of sodium amide in 350 ml of liquid ammonia (see II Exp. 11) was added with swirling 0.20 mol of *tert.*-butylthiopropyne in five equal portions with intervals of about 10 s. Before adding the acetylenic sulfide the end of the plastic tube was placed on the bottom of a 3-l flask containing 600 g of crushed ice. Three minutes after addition of the last portion of the acetylenic sulfide the stopper and glass tube were placed on the flask, so that the brown ammoniacal solution was forced under the ice. During this procedure, which took about 2 min, the flask with the ice was swirled continuously. The flask was placed in a water-bath at 40-50°C. When the remaining ice had disappeared, six extractions with pentane or diethyl ether were carried out and the combined extracts were dried over magnesium sulfate. After the greater part of the solvent had been distilled off at normal pressure through a 40-cm Vigreux column (bath temperature not exceeding 65°C), the remaining liquid was distilled in a water-pump vacuum. The allenic sulfide, b.p. 48°C/17 mmHg, n_D^{25} 1.4992, was obtained in 87% yield. The product had a purity of about 95%, the impurities being the isomers.

(25) *Phenylthioallene*[23]

$$CH_3C{\equiv}C{-}SPh \xrightarrow[\text{liq. } NH_3]{NaNH_2} H_2C{=}C{=}\overset{-}{C}SPh \xrightarrow{H_2O} H_2C{=}C{=}CH{-}SPh$$

Apparatus: For the hydrolysis procedure see Chapter I, Fig. 3; in the procedure below a 1-l *one*-necked flask was used and mechanical stirring was not carried out.

A suspension of 0.40 mol of sodium amide in 300 ml of liquid ammonia was prepared as described in Chapter II, Exp. 11. To the suspension was added with swirling a mixture of 0.25 mol of $CH_3C{\equiv}C{-}S{-}Ph$ (see Chapter IV, Exp. 14) and 40 ml of THF in about 2 min (note 1). Swirling was continued after the addition. Three minutes later (note 1) the stopper with glass tube was placed on the flask. The brown solution was forced through the glass tube and the plastic tube, connected to it under 400 g of finely crushed ice, which was contained in a 3-l conical flask (see Chapter I, Fig. 3, and accompanying description of this operation). The flask was placed for a few minutes in a water-bath until the remaining ice had disappeared. Five extractions with a 1:1 mixture of diethyl ether and pentane were then carried out as rapidly as possible (preferably under nitrogen in view of the oxygen-sensitivity of the allenic sulphide). The extracts were dried and then concentrated in a water-pump vacuum. The residue was distilled through a 20-cm Vigreux column, affording phenylthioallene (b.p. *ca.* 70°C/0.01 mmHg, n_D^{20} 1.6205) in 80% yield. The product contained small amounts (< 5%) of $HC{\equiv}C{-}CH_2{-}SPh$.

Note 1. If the addition is carried out too slowly and/or the reaction mixture is allowed to stand for too long, part of the allenic anion may be converted into $^{-}C{\equiv}C{-}CH_2{-}SPh$, which gives $HC{\equiv}C{-}CH_2{-}SPh$ upon hydrolysis. If shorter reaction times are applied, the conversion of the $CH_3C{\equiv}C{-}SPh$ appears to be incomplete.

(*26*) *1-Methylthio-1,2-pentadiene*[23]

$$C_3H_7C{\equiv}C{-}SCH_3 \xrightarrow[\text{liq. } NH_3]{NaNH_2} C_2H_5CH{=}C{=}\overset{-}{C}{-}SCH_3 \xrightarrow{H_2O} C_2H_5CH{=}C{=}CH{-}SCH_3$$

Apparatus: 1-l one-necked, round-bottomed flask; for the hydrolysis procedure see Chapter I, Fig. 3.

A solution of 0.50 mol of sodium amide in 400 ml of liquid ammonia was prepared as described in Chapter II, Exp. 11. A mixture of 0.30 mol of 1-methylthio-1-pentyne (prepared from pentynyllithium and dimethyl disulfide[63]) and 50 ml of dry diethyl ether was added in five equal portions at intervals of 30 s. After addition of the last portion the reaction mixture was swirled for another 3 min. The stopper, perforated by the glass tube, was placed on the flask, after the end of the plastic tube had been placed under 600 g of crushed ice in a 2-l conical flask (see Fig. 3). On gentle swirling, the reaction mixture was forced through the tube under the ice. During this procedure, which took 1-2 min, the flask with ice was also swirled continuously. The air in the flask was replaced with nitrogen (note 3). After melting of the ice the mixture was transferred into a separating funnel and five extractions with pentane (b.p. 36-40°C) were carried out as quickly as possible. The combined pentane solutions were dried over magnesium sulfate, subsequently concentrated in a water-pump vacuum and the residue was distilled through a 30-cm Vigreux column. The allenic sulfide, b.p. 53°C/17 mmHg, n_D^{19} 1.5212, was obtained in 86% yield.

Note 1. An impression of the water content can be obtained by adding at intervals very small pieces of sodium until the blue coulour persists. A good quality of ammonia should not contain more than 0.1% of water.
Note 2. Detailed instructions for working with liquid ammonia are given in Ref. 1.
Note 3. Allenic sulfides are oxygen-sensitive.

(*27*) *1-NN-Dimethylamino-4-methoxy-1,2-butadiene*[155]

$$(CH_3)_2N\text{-}CH_2C{\equiv}CCH_2OCH_3 \xrightarrow[\text{THF, 35-40}^{\circ}\text{C}]{\text{KO-}tert.\text{-}C_4H_9\ +\ \text{HO-}tert.\text{-}C_4H_9} (CH_3)_2N\text{-}CH{=}C{=}CH\text{-}CH_2OCH_3$$

Apparatus: 200-ml flask, see Fig. 1.

To a solution of 9.6 g of KO-*tert.*-C_4H_9 (see Exp. 4, note 2) in 55 ml of dry THF were added successively 1.2 g of *tert.*-butylalcohol and 0.30 mol of 1-dimethylamino-4-methoxy-2-butyne (see Chapter V, Exp. 14). The mixture was warmed at 30-35°C for 40 min (note 1) (occasional cooling may be necessary at the beginning to maintain this temperature range). The greater part of the THF was then removed by means of a rotary evaporator, keeping the bath temperature below 25°C. To the viscous residue were added 120 ml of dry, redistilled pentane and after shaking (under nitrogen, note 2) the precipitate was filtered off by suction filtration on a dry (note 2) sintered-glass funnel. The precipitate was rinsed with pentane. After concentration of the filtrate in a water-pump vacuum the residue was carefully distilled in order to separate the small amount of elimination product $(CH_3)_2N\text{-}C{\equiv}C\text{-}CH{=}CH_2$ (*c.f.* Chapter V, Exp. 14) from the desired product. This passed over at 66°C/25 mmHg (n_D^{20} 1.4814) and consisted of 89-92% of the allenic amine. The contaminants were the starting compound (about 8%) and a trace of the allenic ether $(CH_3)_2N\text{-}CH_2\text{-}CH{=}C{=}CH\text{-}OCH_3$. The yields varied from 62 to 75%.

Note 1. Heating for a longer period or at higher temperatures results in the formation of more $(CH_3)_2N\text{-}C{\equiv}C\text{-}CH{=}CH_2$ and lower yields of the desired product.
Note 2. The product is oxygen- and water-sensitive.

(*28*) *1-Methoxy-4-NN-dimethylamino-1,2-butadiene*[156]

$$CH_3O\text{-}CH_2C{\equiv}C\text{-}CH_2\text{-}N(CH_3)_2 \xrightarrow[\text{THF-hexane}]{\text{BuLi}} CH_3O\text{-}\overset{\overset{\displaystyle Li}{|}}{C}H\text{-}C{\equiv}C\text{-}CH_2N(CH_3)_2 \xrightarrow{H_2O}$$

$$\longrightarrow CH_3O\text{-}CH{=}C{=}CH\text{-}CH_2N(CH_3)_2 + \text{starting compound}$$

Apparatus: 1-l flask, see Fig. 1.

A solution of 0.22 mol of butyllithium in 150 ml of hexane was cooled below -40°C and 140 ml of dry THF were added. Subsequently 0.20 mol of 1-dimethylamino-4-methoxy-2-butyne (see Chapter V, Exp. 14) were added in 10 min with cooling between -35 and -45°C. After an additional 15 min 100 ml of an aqueous solution of 25 g of ammonium chloride were added with vigorous stirring. After separation of the layers four extractions with diethyl ether were carried out. The solutions were dried over potassium carbonate and then concentrated in a water-pump vacuum. Distillation of the residue gave a mixture of 8-10% of starting compound and 90-92% of the allenic ether, b.p. 50°C/12 mmHg, n_D^{20} 1.4648, in 82% yield (note 1).

Note 1. The same procedure with 2-alkynylethers, Alkyl-C≡C-CH_2OR, gave lower relative amounts of the corresponding allenic ethers (varying between about 60:40 for R = CH_3 to 75:25 for R = *tert*.-C_4H_9 in favour of the allenic isomer).

Chapter V

SYNTHESIS OF ACETYLENES AND CUMULENES BY ELIMINATION REACTIONS

1. 1,2-DEHYDROHALOGENATION

One of the commonest preparative methods for acetylenic compounds is the abstraction of hydrogen halide from an organic halide with a sufficiently strong base. A wide variety of bases is available. The procedure generally involves addition of bromine to an alkene, followed by either stepwise dehydrobromination with two different bases or, in some cases, double dehydrobromination with the same base. The same holds for 1,1-dihalo compounds, which can be prepared by addition of secondary or tertiary alkyl halides to vinyl chloride under the influence of $AlCl_3$ or by Wittig coupling of aldehydes with polyhalogenmethanes. A number of practical examples is given in Ref. 1 and in a recent well-documented review[3b]. Sodium amide in liquid ammonia has been shown to give good results in many cases[1, 40]. Some chemists prefer to work with base-organic solvent pairs such as KO-*tert*.-C_4H_9-DMSO[41], BuLi in diethyl ether or THF[39, 42] and lithium dialkylamide--diethyl ether or THF[43]. The combination of KO-*tert*.-C_4H_9 and THF can be recommended if partial dehydrohalogenation is desired. Starting from cyclooctene, for example, 1-bromocyclooctene can be obtained via the dibromide. Reaction of 1-bromocyclooctene with sodium amide in liquid ammonia affords some cyclooctyne, but the yield and purity are unsatisfactory. Lithium diisopropylamide in a mixture of hexane and THF can be successfully applied to dehydrobrominate 1-bromocyclooctene, giving cyclooctyne in high yields[44]. In a recent method[45] dehydrohalogenation of several halogen compounds is carried out with KO-*tert*.-C_4H_9 in light petroleum fractions of suitable volatility, using 18-crown-6 as a catalyst.

2. 1,4- AND 1,6-ELIMINATIONS

As indicated in the general scheme below, butatrienes are the first products from base-induced 1,4-elimination of hydrogen and a suitable leaving group. The butatriene in general very readily undergoes isomerization into enynes, if sufficiently "acidic" protons are available (see Chapter 11 in Ref. 3a). In aprotic media cumulenic ethers are fixed as their lithio derivatives if an excess of alkyllithium is applied[21].

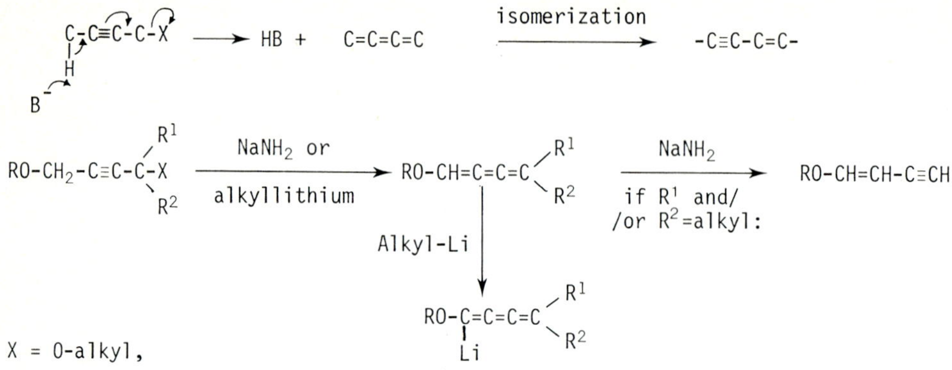

X = O-alkyl, O-CH(CH$_3$)OC$_2$H$_5$ or OSiMe$_3$

Cumulenic ethers with the 4-positions blocked by alkyl groups can be obtained from bis-ethers, RO-CH$_2$C≡C-C(R^1)(R^2)OR, and sodium amide in liquid NH$_3$, applying the inverse-order technique[46]. 1,4-Elimination with alkyllithium, however, is more easily performed[21]. Further examples of 1,4-eliminations are represented by the following schemes[8, 33]:

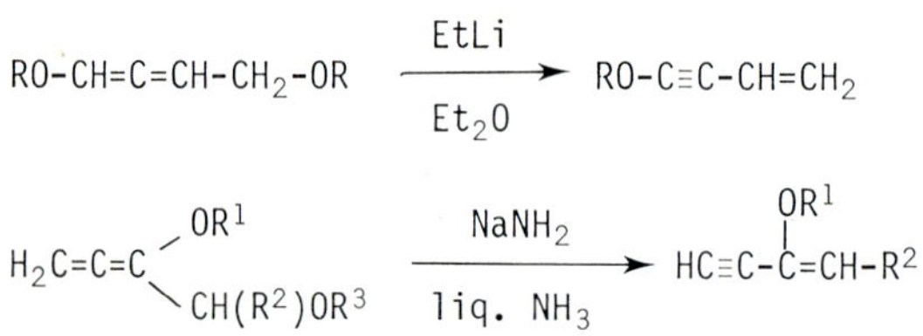

It is evident that in both cases the elimination is induced by the (kinetically favoured) abstraction of an sp^2-proton.

Other 1,4-eliminations of synthetic interest are the formation of enyne amines[47] and enyne sulfides[48] from the easily accessible Mannich coupling products of propargylethers and 1,4-dialkyl- or -diarylthio-2-butynes, respectively:

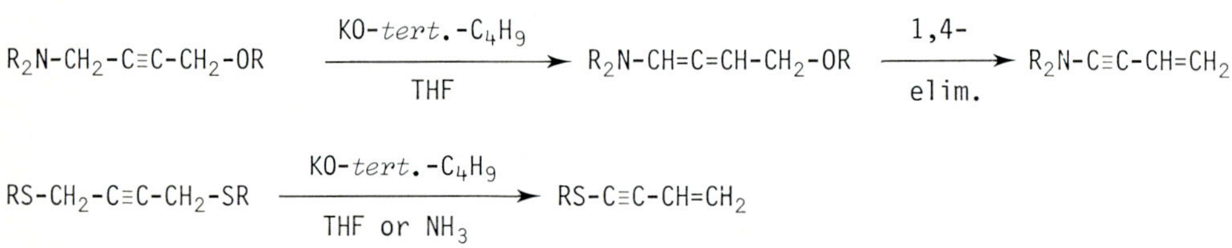

1,6-Eliminations are scarcely known. An interesting example is the formation of hexatriyne from 1,6-dichloro-2,4-hexadiyne[8, 49]. Whereas during interaction of the latter compound and KOH or alkali metal amide in liquid ammonia much carbonaceous material is produced, KO-*tert.*-C_4H_9 in THF gives the extremely unstable polyyne in the surprisingly high yield[8] of about 90%.

$$Cl\text{-}CH_2C{\equiv}C\text{-}C{\equiv}C\text{-}CH_2Cl \xrightarrow[\text{THF, -90}^{o}\text{C}]{\text{KO-}\textit{tert.}\text{-C}_4\text{H}_9} H\text{-}C{\equiv}C\text{-}C{\equiv}C\text{-}C{\equiv}C\text{-}H$$

The intermediate in this double 1,6-elimination may be the pentaene $Cl\text{-}CH{=}C{=}C{=}C{=}C{=}CH_2$, but routes via other highly unsaturated compounds cannot be ruled out. Since the elimination is a very clean reaction, prior isolation of the explosive triyne is not always necessary: 2-ethynylselenophen, for example, can be obtained in about 50% yield by adding a THF solution of hexatriyne to a solution of sodium selenide in methanol[8] (see Chapter VII-3, Exp. 3).

Treatment of conjugated enyne systems, $H\text{-}\overset{\diagdown}{\underset{\diagup}{C}}\text{-}C{=}C\text{-}C{\equiv}C\text{-}\overset{|}{C}H\text{-}OR$, with a suitable base in principle may give rise to 1,6-elimination of HOR with primary formation of the system C=C-C=C=C=C, which subsequently may undergo isomerization to a dienyne[8]. An example is:

$$CH_3CH{=}CH\text{-}C{\equiv}C\text{-}CH(OC_2H_5)_2 \xrightarrow[\text{DMSO}]{\text{KO-}\textit{tert.}\text{-C}_4\text{H}_9\text{-HO-}\textit{tert.}\text{-C}_4\text{H}_9} [H_2C{=}CH\text{-}CH{=}C{=}C{=}CH\text{-}OC_2H_5]$$

$$\longrightarrow H_2C{=}CH\text{-}C{\equiv}C\text{-}CH{=}CH\text{-}OC_2H_5$$

3. 1,1-DEHALOGENATION OF GEMINAL DIHALOCYCLOPROPYL DERIVATIVES

Treatment of geminal dihalocyclopropyl compounds with a strong base such as butyllithium has been for several years the most versatile method for cumulenes. The dihalo compounds are easily obtained by addition of dihalocarbenes to double-bond systems[50-52]. If the dihalocyclopropanes are reacted at low temperatures with alkyllithium, a cyclopropane carbenoid is formed, which in general decomposes above -40 to -50°C to afford the cumulene. Although at present a number of alternative methods are available[10], the above-mentioned synthesis is the only suitable one for cyclic cumulenes [e.g. 1,2-cyclononadiene and 1,2,3-cyclodecatriene] and substituted non-cyclic cumulenes [e.g. $(CH_3)_2C{=}C{=}C{=}C(CH_3)_2$].

4. 1,2-DEHALOGENATION AND 1,4-DEHALOGENATION

The recent discovery that propadiene can be converted into a lithio compound, which is stable enough at low temperature to allow functionalization[28], has made it an important starting compound for the synthesis of derivatives. A fortunate circumstance is that allene can be obtained in > 80% yield in an extremely simple way, namely by treatment of the cheap dichloropropene with powdered zinc in a mixture of water and ethanol[53]:

$$H_2C{=}C(Cl)CH_2Cl \xrightarrow[C_2H_5OH\text{-}H_2O,\ 70\text{-}80^{\circ}C]{Zn} H_2C{=}C{=}CH_2$$

Homologues in principle can also be prepared from $RCH(Br)\text{-}C(Br){=}CH_2$ but for the synthesis of these starting compounds three steps have to be carried out[10]. The dehalogenation procedure has also been used in the preparation of di- and trifluoroallenes[10].

For some volatile aliphatic 1,2,3-trienes simple preparative methods have been developed[54-55]. These involve 1,4-elimination of halogen or halogen and sulfinate with powdered zinc in high-boiling solvents, especially DMSO or HMPT:

$$R\text{-}\underset{\displaystyle X}{\underset{|}{C}H}\text{-}C{\equiv}C\text{-}CH_2\text{-}Y \xrightarrow{Zn} R\text{-}CH{=}C{=}C{=}CH_2$$

The extremely unstable butatriene, $H_2C{=}C{=}C{=}CH_2$, can be obtained in more than 90% yield from the easily accessible 1,4-dichloro-2-butyne[54] and zincin DMSO. The reaction is performed in a water-pump vacuum at 80-90°C, collecting the cumulene in a trap cooled at -190°C. For the homologues the yields are lower, perhaps because their volatilities are lower, so that the products do not escape immediately from the solution in which partial polymerization occurs at the elevated temperatures.

The alternative method of elimination of chlorine and a sulfinate group is particularly useful for the preparation of *tert.*-butylbutatriene, since the corresponding dichloride cannot be obtained in good yields. The required sulfinate can be synthesized in a high overall yield[55], starting with propargyl chloride:

$$HC{\equiv}C\text{-}CH_2Cl \xrightarrow[(2)\ RCH{=}O]{(1)\ BuLi} RCH(OH)\text{-}C{\equiv}C\text{-}CH_2Cl \xrightarrow[Et_3N,\ -50^{\circ}C]{CH_3S({=}O)Cl} R\underset{\displaystyle O\text{-}\underset{\displaystyle \overset{\|}{O}}{S}\text{-}CH_3}{\underset{|}{C}H}\text{-}C{\equiv}C\text{-}CH_2Cl$$

5. EXPERIMENTS

(1) *tert.-Butylacetylene*[41]

$$\textit{tert.}\text{-}C_4H_9CH{=}CH_2 \xrightarrow[Et_2O]{Br_2} \textit{tert.}\text{-}C_4H_9CHBrCH_2Br \xrightarrow[DMSO]{KO\text{-}\textit{tert.}\text{-}C_4H_9} \textit{tert.}\text{-}C_4H_9\text{-}C{\equiv}CH$$

Apparatus: 1-l flask (see Fig. 1) for the addition of bromine; 1-l two-necked flask, provided with a dropping funnel and a 40-cm Vigreux column, for the dehydrohalogenation; the Vigreux column was connected to a condenser and receiver, cooled at -75°C.

To a mixture of 0.40 mol of neohexene (commercially available) and 200 ml of dry diethyl ether 0.35 mol of bromine was added with cooling between -40 and -60°C. The diethyl ether and excess of neohexene were then completely removed by evaporation in a water-pump vacuum.In the second flask was placed a solution of 90 g of commercial KO-*tert.*-C_4H_9 (see Chapter IV, Exp. 4, note 2) in 250 ml of DMSO. The dibromo compound was added in five portions during 15 min from the dropping funnel; after the addition of each portion the flask was swirled gently in order to effect homogenization. Much heat was evolved and part of the *tert.*-butylacetylene passed over. After the addition the flask was heated for 30 min in a bath at 80-100°C. The mixture was then cooled to 30-40°C and the dissolved acetylene was "sucked over" in a water-pump vacuum (10-20 mmHg) and condensed in a receiver cooled at -78°C. The contents of the receiver were shaken three times with 20-ml portions of ice-water in a small separating funnel or dropping funnel in order to remove small amounts of *tert.*-butylalcohol. After drying over a very small amount of magnesium sulfate the product was distilled to afford *tert.*-butylacetylene, b.p. 38°C/760 mmHg, n_D^{20} 1.3750, in 86-92% yield.

(2) *Cyclooctyne*[44]

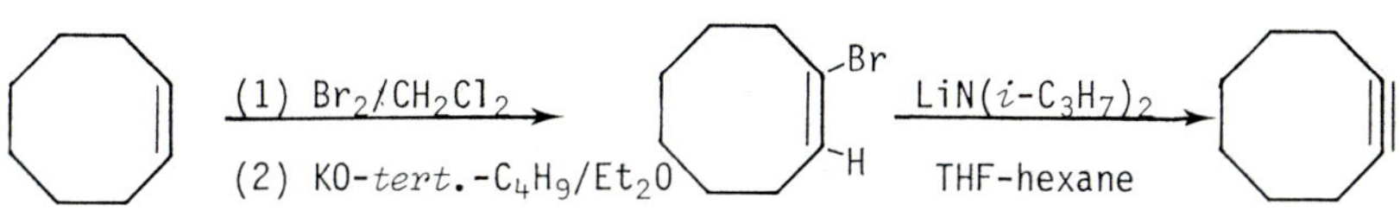

Apparatus: 1-l round-bottomed, three-necked flask with a dropping funnel, a mechanical stirrer and a thermometer, combined with a vent, for the addition of bromine and the dehydrobromination to 1-bromocyclooctene; 1-l flask (see Fig. 1) for the preparation of cyclooctyne.

a. 1-Bromocyclooctene

To a solution of 1 mol of freshly distilled cyclooctene in 300 ml of dry dichloromethane was added, with cooling between -30 and -50°C, sufficient bromine (about 1 mol) to give a distinct persistent brown colour. The solution obtained was concentrated in a water-pump vacuum, until all of the dichloromethane had been removed. The residue was dissolved in 4C0ml of dry diethyl ether and a solution of 1.50 mol of KO-*tert*.-C_4H_9 (see ChapterIV, Exp. 4, note 2) in 150 ml of THF was added in 20-30 min, while keeping the temperature at about 0°C. The cooling bath was then removed and vigorous stirring was continued for an additional 2.5 h at 5°C. The mixture was treated with 500 ml of ice-water and the aqueous layer was extracted three times with diethyl ether. The combined organic solutions were dried over magnesium sulfate and subsequently concentrated in a water-pump vacuum. Distillation of the residue through a 40-cm Vigreux column gave 1-bromocyclooctene, b.p. 90°C/18 mmHg, n_D^{20} 1.5187, in 70-74% yield. Re-treatment of the residue with KO-*tert*.-C_4H_9 afforded an additional small amount (*ca*. 4%) of 1-bromocyclooctene.

b. Cyclooctyne (note 1)

To a solution of 0.50 mol of butyllithium in about 350 ml of hexane were added successively 200 ml of dry THF and 0.50 mol of dry diisopropylamine (dried over powdered KOH) with cooling at -35°C. Ten minutes later 1.00 mol of 1-bromocyclooctene was added in one portion at -35°C. After keeping the mixture for 15 min at -25°C the temperature was allowed to rise gradually to +15°C over 1 h: occasional cooling was applied if the temperature rose too rapidly. Stirring at 15°C was continued for an additional 1.5 h. The brown solution was then poured (note 1) into 500 ml of ice-water and 4 N hydrochloric acid was added with stirring and cooling at 0-10°C until the aqueous layer had become neutral. The aqueous layer was extracted 3-5 times with small portions of redistilled pentane and the combined solutions were dried over magnesium sulfate and concentrated in a water-pump vacuum. Careful fractionation of the residue through a 40-cm Widmer column afforded cyclooctyne, b.p. 48°C/16 mmHg, n_D^{20} 1.4876, in 76-84% yield (based on converted 1-bromocyclooctene). About 0.45 mol of 1-bromocyclooctene was recovered.

Note 1. Cyclooctyne has a very penetrating and unpleasant odour. The precautions usually taken when working with thiols should therefore be adopted.

(3) 1-NN-diethylamino-2-phenylthioacetylene[127]

$$F_3C\text{-}CH_2OH \xrightarrow[\text{diethyl ether}]{\text{tosyl chloride, KOH}} F_3C\text{-}CH_2O\text{-}Ts \xrightarrow[\text{DMF}]{\text{PhSNa}} F_3C\text{-}CH_2\text{-}SPh$$

$$\xrightarrow[\text{diethyl ether}]{LiN(C_2H_5)_2} (C_2H_5)_2N\text{-}C{\equiv}C\text{-}SPh$$

Apparatus: 1-l flask, see Fig. 1.

Starting from 0.20 mol of trifluoroethanol the tosylate was prepared in the usual way (see Chapter VIII-3, Exp. 3) in almost quantitative yield. An oil-immersion of sodium hydride, corresponding to 0.20 mol of NaH, was transferred into the reaction flask. The flask was filled with nitrogen, 50 ml of dry pentane were added and, after swirling, the flask was allowed to stand for 1 min and the pentane was then carefully poured off. This procedure was repeated once, using the same amount of pentane. The sodium hydride was subsequently suspended in 175 ml of dry DMF and 0.20 mol of redistilled thiophenol was added in 15 min. The mixture was warmed in a water-bath at 40°C. After 30 min the bath was removed and the crude tosylate of trifluoroethanol, dissolved in 50 ml of DMF, was added in 15 min. The temperature of the mixture rose to 60-70°C. After the exothermic reaction had subsided and the temperature began to drop, the mixture was warmed at about 60°C for an additional 20 min. Ice-water (500 ml) was then added and the product was extracted five times with diethyl ether. The combined ethereal solutions were washed with 2 N NaOH and subsequently dried over magnesium sulfate. After removing the solvent by evaporation in a water-pump vacuum the remaining liquid was distilled through a 30-cm Vigreux column to give trifluoroethyl phenyl sulfide, b.p. 70°C/15 mmHg, n_D^{20} 1.4880, in 72% yield.

The sulfide (0.10 mol) was mixed with 0.32 mol of dry diethylamine (dried over KOH powder and subsequently distilled) and 250 ml of dry diethyl ether. To the mixture was added in 20 min, with cooling between -30 and -40°C, a solution of 0.32 mol of butyllithium in about 215 ml of hexane. After the addition the cooling bath was removed and the temperature was allowed to rise. After the mixture had been stirred for 2 h the temperature was increased to 30°C (water-bath) and stirring was continued for an additonal 3 h. The brown reaction mixture was then poured into a cold (0°C) solution of 5 g of sodium hydrogen carbonate in 500 ml of water. The organic layer and three ethereal extracts were combined, washed with ice-water and subsequently dried over potassium carbonate. The residue remaining after evaporation of the solvent in a water-pumpvacuum was distilled through a short column, giving the acetylenic derivative, b.p. 110°C/0.5 mmHg, n_D^{20} 1.5865, in 73% yield.

(4) *N-Morpholinoacetylene*[70]

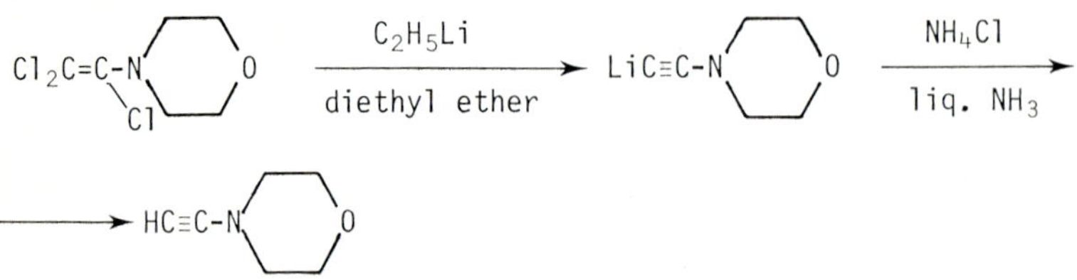

Apparatus: 500-ml flask (see Fig. 1) for the dehalogenation of the trichlorovinylamine (note 1).

To a mixture of 0.05 mol of the trichlorovinylamine and 100 ml of dry diethyl ether was added in 25 min with cooling at -25°C a solution of 0.15 mol of ethyllithium in about 130 ml of diethyl ether (see Chapter II, Exp. 1), a white precipitate being formed. The thermometer and gas outlet were then replaced with a reflux condenser and the mixture was warmed under reflux for 15 min. It was then cooled to -30°C and cautiously poured with manual swirling into a solution of 6 g of ammonium chloride in 100 ml of dry (water content $< 0.1\%$) liquid ammonia in a 1-l wide-necked, round-bottomed flask. A curved gas outlet was then placed on the flask, which was immersed in a water bath at 40°C. When the stream of escaping ammonia vapour had become very faint, the flask was connected with the water pump via a tube filled with KOH pellets. The bath was removed and the flask was evacuated for 30 min. Boiling dry diethyl ether (100 ml) was then added to the salty mass and the mixture was warmed for 5 min in a water-bath at 30°C. The salt was filtered off on a sintered-glass funnel (G-2) and, after pressing, rinsed five times with dry diethyl ether (30°C). The filtrate was concentrated in a water-pump vacuum and the residue was distilled in a small apparatus, affording the ynamine, b.p. 60°C/10 mmHg, n_D^{22} 1.4776, in 50-58% yield (note 2).

Note 1. The preparation of the starting compound for the yneamine was performed in a way similar to that in the literature and gave slightly lower yields.

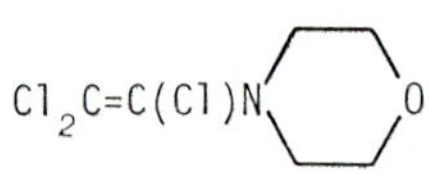

had b.p. *ca.* 100°C/1 mmHg and n_D^{23} 1.5224

Note 2. In connection with the extreme water-sensivity of the yneamine, moisture must be excluded during the work-up.

(5) *Diethynyl sulfide*[71]

$$2\ H_2C{=}CHCl + SCl_2 \xrightarrow[CH_2Cl_2]{} (Cl_2CHCH_2)_2S$$

$$(Cl_2CHCH_2)_2S \xrightarrow[\text{methanol}]{KOH} (ClCH{=}CH)_2S$$

$$(ClCH{=}CH)_2S + 4\ NaNH_2 \xrightarrow[\text{liq. } NH_3]{} (NaC{\equiv}C)_2S \xrightarrow{H_2O} (HC{\equiv}C)_2S$$

a. 2,2,2',2'-Tetrachlorodiethyl sulfide

Apparatus: 3-l round-bottomed, three-necked flask, provided with a dropping funnel, a mechanical stirrer and a reflux condenser, cooled with dry-ice and acetone (cold finger). The upper end of this condenser was connected with a drying tube filled with anhydrous calcium chloride.

To a solution of 7.5 mol of vinyl chloride in 1.5 l of dry dichloromethane were added in 4 h with cooling in a large ice-bath 3 mol of freshly distilled (b.p. 58-62^{0}C/760 mmHg) sulfur dichloride[74]. After the addition stirring was stopped and the flask was allowed to stand for 15 h in the ice-bath. The condenser was replaced with a drying tube containing $CaCl_2$ (refluxing of the vinyl chloride had stopped completely after the addition of sulfur dichloride).

b. 2,2-Dichlorodivinyl sulfide

Apparatus: 3-l round-bottomed, three-necked flask, provided with a dropping funnel, a mechanical stirrer and a thermometer, combined with a vent.

The remaining faintly yellow to almost colourless solution was concentrated in a water-pump vacuum. The residue was placed, together with 500 ml of methanol, in the reaction flask, which was cooled in an ice-bath. A cold (0-5^{0}C) solution of 8 mol of KOH in 1.5 l of methanol was added at a rate such that the temperature could be kept between 30 and 40^{0}C (about 1 h). The cooling bath was then removed and the mixture was warmed for an additional 1.5 h at 45^{0}C. The suspension was poured into 5 l of ice-water and the organic layer was separated off. The upper layer was extracted three or four times with small amounts of redistilled pentane. The organic layers were combined, dried over anhydrous magnesium sulfate and then concentrated in a water-pump vacuum. Distillation through a 40-cm Vigreux column gave (crude) 2,2'-dichlorodivinyl sulfide, b.p. 53-75°C/12 mmHg (mixture of cis- and trans-isomers) in 67-75% yield (calculated on sulfur dichloride).

c. Diethynyl sulfide

Apparatus: 2-l flask, see Fig. 1.

A suspension of 2.2 mol of sodium amide in 1.2 l of liquid ammonia was prepared (see Chapter II, Exp. 11). The suspension was occasionally cooled in a bath of liquid nitrogen while a stream of nitrogen was passed through the flask. From the dropping funnel was added with vigorous stirring 0.5 mol of crude 2,2'-dichlorodivinyl sulfide at a rate such that the temperature of the ammoniacal reaction mixture could easily be kept between -50 and -55°C (about 20 min) (if during the addition a crust of solid ammonia is formed on the bottom of the flask, the cooling bath should be temporarily removed). After addition of the dichloro compound the cooling bath was removed and the temperature was allowed to rise until the ammonia started to boil. The dropping funnel, stirrer and thermometer were removed. On the outer necks were placed two rubber stoppers and on the middle neck a rubber stopper, perforated by a glass tube (I.D. 7 mm), extending to within 2 mm of the bottom of the flask. The black reaction mixture was hydrolyzed by pressing it under a mixture of 800 g of finely crushed ice and 200 ml of high-boiling light petroleum (b.p. > 180°C) in a 4-l conical flask. During this operation the reaction flask was gently swirled by hand (in order to suspend the sodium choride) and the flask with the ice was vigorously swirled (for a general description of this hydrolysis procedure, see Chapter I, Fig. 3 and the accompanying text; note 1). The flask containing the hydrolysed reaction mixture was kept in a water-bath at 40°C until the remaining ice had disappeared, then the layers were separated (note 2). The aqueous layer was extracted three times with small amounts of light petroleum and the combined organic solutions were washed with cold 1 N HCl and subsequently dried over magnesium sulfate. The brown extract was placed in a 1-l round-bottomed flask, connected with a 40-cm Vigreux column, condenser and receiver, cooled at -75°C (see Fig. 5). Between the receiver and the water pump was placed a tube containing KOH pellets. After the system had been evacuated (10-20 mmHg) the distillation flask was gradually heated until the light petroleum began to pass over. Nitrogen was then admitted and the contents of the receiver, diethynyl sulfide and a small amount of light petroleum, were subjected to a similar evacuation procedure. This was terminated by admitting nitrogen as soon as the small amount of light petroleum began to reflux in the lower part of the column. The receiver now contained fairly pure (*ca.* 95%) diethynyl sulfide, n_D^{20} 1.5090, yield 52-63%. The compound rapidly turned brown at room temperature, but could be kept, without any decomposition, for several days at -80°C.

Note 1. Especially in the case of diethynyl sulfide the performance of this operation requires considerable experimental experience: it should preferably be carried out by two persons.

Note 2. The separation may be difficult owing to the presence of carbonaceous material.

(6) *2-Ethoxy-2-penten-4-yne*[72]

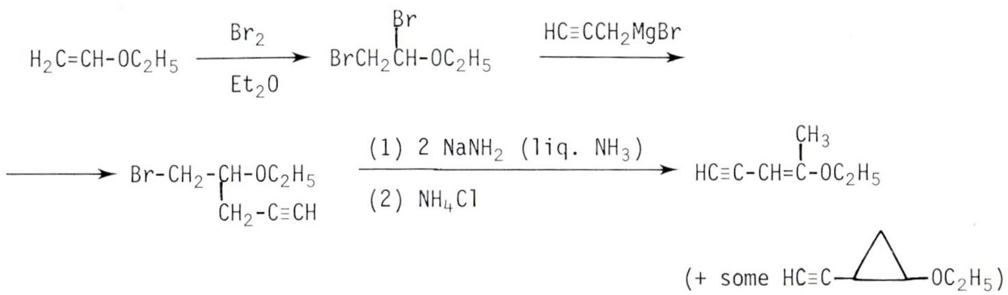

Apparatus: 2-l flask (Fig. 1) for the addition of bromine and the coupling with propargylmagnesium bromide; 6-l flask (Fig. 4) for the reaction in liquid ammonia.

To a solution of 1.2 mol of freshly distilled ethyl vinyl ether in 200 ml of dry diethyl ether was added 1.0 mol of bromine with efficient cooling (at -30 to -60°C, liquid nitrogen bath). The solution obtained was subsequently added in 30 min to a solution of propargylmagnesium bromide in about 800 ml of diethyl ether, prepared from 1.15 mol of propargyl bromide (see Chapter II, Exp. 8). During this addition the mixture was kept between -15 and -5°C. Thirty minutes later the reaction mixture was poured into 500 ml of an aqueous solution of 100 g ammonium chloride. After shaking, the layers were separated and the aqueous layer was extracted twice with small portions of diethyl ether. The combined ethereal solutions were dried over magnesium sulfate and subsequently concentrated in a water-pump vacuum by means of a rotary evaporator to a volume of about 200 ml. This residue was added dropwise in 15 min to a suspension of 2.5 mol of sodium amide (see Chapter II, Exp. 11) in 2 l of liquid ammonia. The reaction was very vigorous and a very dark reaction mixture was formed. After the addition the ammonia was driven off by placing the flask in a water-bath at 50°C. To the remaining semi-solid mass were added with vigorous stirring 400 ml of an aqueous solution of 100 g of ammonium chloride. The mixture was extracted (4-6 times) with a 1:1

mixture of diethyl ether and pentane. The combined solutions were washed with water and dried over magnesium sulfate. After the greater part of the solvents had been distilled off at normal pressure through a 40-cm Vigreux column, the residue was distilled through the same column. The enyne ether, b.p. 42°C/17 mmHg, n_D^{18} 1.4627, was obtained in 67% yield. There was a considerable high-boiling residue. The distilled product contained varying amounts (10-15%) of a contaminant, probably the cyclopropyl derivative, indicated in the reaction scheme.

(7) *1-Ethoxy-1,2,3-butatriene*[21]

$$C_2H_5O\text{-}CH_2C{\equiv}CCH_2\text{-}OC_2H_5 \xrightarrow[\text{diethyl ether}]{2\ C_2H_5Li} C_2H_5O\text{-}\underset{}{\overset{Li}{\overset{|}{C}}}{=}C{=}C{=}CH_2 \xrightarrow{H_2O}$$

$$\longrightarrow C_2H_5O\text{-}CH{=}C{=}C{=}CH_2$$

Apparatus: 1-l flask, see Fig. 1.

A mixture of 0.25 mol of 1,4-diethoxy-2-butyne (see Chapter VIII-6, Exp. 8) and 100 ml of dry diethyl ether was cooled at -45°C. A solution of 0.55 mol (note 1) of ethyllithium (note 2) (see Chapter II, Exp. 1) in about 450 ml of diethyl ether was added in 30 min. The temperature of the mixture was kept between -45 and -35°C. After the addition the mixture was stirred for an additional 20 min at -40°C. The light brown solution was then poured into 500 ml of ice-water, which had been saturated with nitrogen (note 3). After shaking the upper layer was separated off and the aqueous layer was extracted twice with small portions of diethyl ether. The combined solutions (washing was not carried out) were dried over magnesium sulfate (just enough to give a clear solution). The solution was decanted from the drying agent and poured into a 1-l round-bottomed flask. Some boiling stones were added and the flask was connected to a 40-cm Vigreux column, descending condenser and receiver, cooled in ice. The diethyl ether was evaporated in a water-pump vacuum, using a water-bath at 15-20°C. The residue was distilled from the same flask, care being taken that the bath temperature did not rise above 60-70°C. During the final stage of the distillation, water from the bath was poured along the column in order to minimize the hold-up. Most of the product passed over between 25 and 40°C/15 mmHg. After cooling to room temperature nitrogen was admitted to the distillation apparatus. The cumulenic ether, n_D^{21} 1.5127, was obtained in 76% yield.

Note 1. If only one equivalent of ethyllithium is used, the conversion of the bis--ether is not complete. The necessity for two equivalents can be explained by assuming that the rates of 1,4-elimination of ethanol and subsequent 1-lithiation are comparable.

Note 2. Commercial butyllithium in hexane as solvent or butyllithium in diethyl ether, prepared from butyl bromide and lithium, can also be used in principle, but we prefer to use ethyllithium because hexane is not easily separable from the rather volatile cumulenic ether and during the reaction of butyl bromide with lithium some octane is formed which cannot be separated from ethoxybutatriene.

Note 3. Cumulenic ethers are extremely oxygen-sensitive: all operations during the work-up must be carried out under nitrogen.

(*8*) *4-Methyl-1-methoxy-1,2,3-pentatriene*[21]

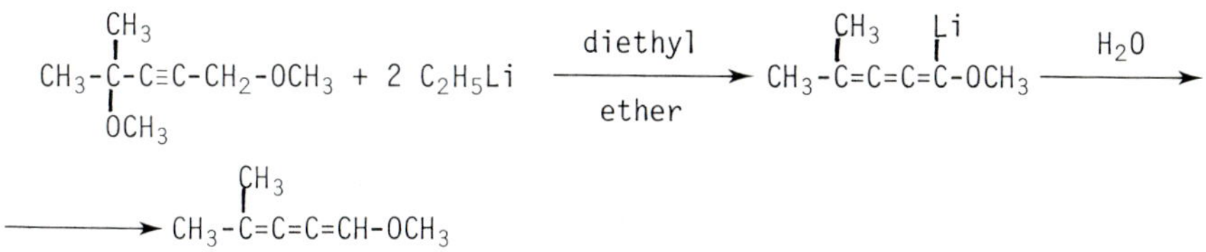

Apparatus: 1-l flask, see Fig. 1.

To a mixture of 0.15 mol of the acetylenic bis-ether (see Chapter VIII-6, Exp.1) and 100 ml of dry diethyl ether was added in 20 min with cooling between -40 and $-50^{o}C$ a solution of 0.30 mol of ethyllithium (note 1) in about 250 ml of diethyl ether (see Chapter II, Exp. 1). After stirring for an additional 15 min at $-45^{o}C$ the mixture was cautiously poured into 200 ml of ice-water and 50 g of crushed ice (note 2). After vigorous shaking the layers were separated and the aqueous layer was extracted twice with small portions of diethyl ether. The combined ethereal solutions were dried over magnesium sulfate. After concentration of the extracts in a water-pump vacuum the remaining liquid was distilled through a 30-cm Vigreux column to give the cumulenic ether, b.p. $40^{o}C$/11 mmHg, n_D^{20} 1.5190, (note 3) in 88% yield.

Note 1. Butyllithium in hexane can be used in principle, but the yield is lower because during the evaporation of the hexane some of the cumulenic ether is entrained.

Note 2. In connection with the oxygen-sensivity of cumulenic ethers the work-up must be carried out under nitrogen.

Note 3. The refractive index is much lower than that of the product obtained in Exp. 11 (1.5276). GLC and NMR results show the presence of small amounts (5-7%) of impurities, probably having a much lower refractive index.

(9) 1,4-Diethoxy-1,2,3-butatriene[21]

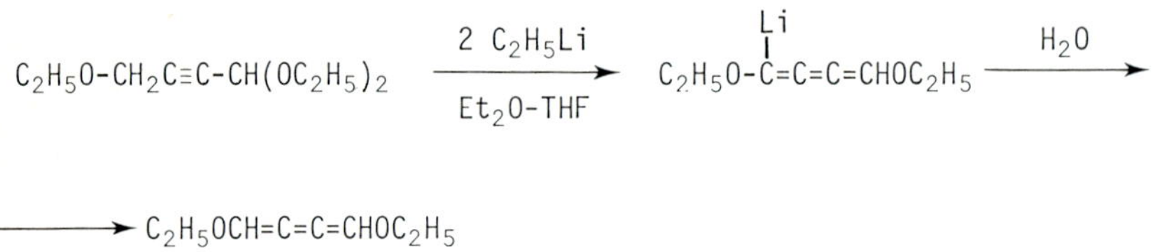

Apparatus: 1-l flask, see Fig. 1.

To a mixture of 0.20 mol of 1,1,4-triethoxy-2-butyne (see Chapter III, Exp. 40), 50 ml of dry THF and 50 ml of dry diethyl ether was added at -45 to -50°C a solution of 0.42 mol of ethyllithium in about 280 ml of diethyl ether (see Chapter II, Exp. 1). Stirring at -50°C was continued for 30 min, then the reaction mixture was poured into 300 ml of saturated ammonium chloride solution. After shaking, the layers were separated and the aqueous layer was extracted twice with small portions of diethyl ether. The combined ethereal solutions were dried over magnesium sulfate and concentrated in a water-pump vacuum and the residue was distilled at about 1 mmHg through a 30-cm Vigreux column. The cumulenic bis-ether, b.p. 50°C/0.2 mmHg, n_D^{20} 1.5094, was obtained in 84-88% yield.

Note 1. Because of the extreme sensivity of the cumulenic ether towards oxygen, all operations during the work-up must be carried out under nitrogen.

(10) 1-tert.-Butoxy-1,2,3-pentatriene[8]

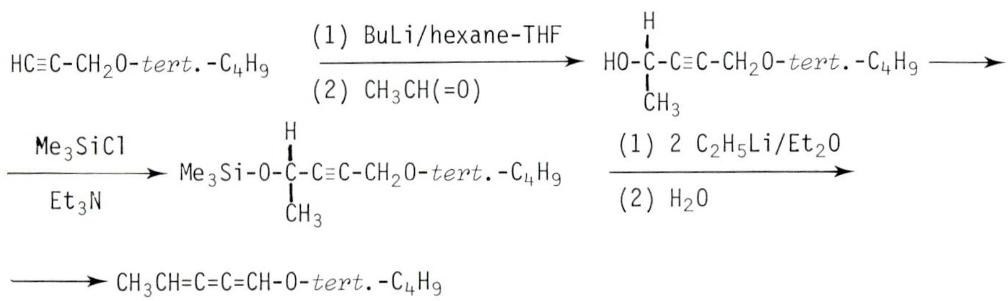

Apparatus: 1-l flask (see Fig. 1) for all conversions.

a. Preparation of $Me_3Si-O-CH(CH_3)-C{\equiv}C-CH_2O$-*tert*.-$C_4H_9$

To a solution of 0.40 mol of butyllithium in about 280 ml of hexane were added 300 ml of dry THF at -20 to -40°C. Subsequently 0.40 mol of freshly distilled *tert*.-butyl propargyl ether[76] was added, keeping the temperature below -30°C. Freshly distilled acetaldehyde (0.40 mol) was then added at the same temperature during about 15 min. The cooling bath was removed and, after an additional 15 min, 200 ml of an aqueous solution of 30 g of ammonium chloride were introduced. After separation of the layers the aqueous layer was extracted twice with diethyl ether and the combined solutions were dried over magnesium sulfate and concentrated in a water-pump vacuum. The residue was distilled through a 30-cm Vigreux column, giving the carbinol, b.p. 110°C/15 mmHg, n_D^{20} 1.4545, in 82% yield.

The distillate was dissolved in a mixture of 350 ml of dry diethyl ether and 45 g of dry triethylamine (dried over powdered KOH). Trimethylchlorosilane (45 g) was added in 20 min with cooling at about 10°C. After standing for 1 h at room temperature the precipitate was sucked off on a dry sintered-glass funnel and rinsed with pentane. The filtrate was concentrated in a water-pump vacuum. The small amount of salt which precipitated during this operation was removed by a second suction filtration. Subsequent distillation afforded the trimethylsilyl ether, b.p. 100°C/15 mmHg, n_D^{20} 1.4330, in 94% yield.

b. Preparation of the cumulenic ether

To a solution of 0.25 mol of the trimethylsilyl ether in 120 ml of dry diethyl ether was added in 20 min at -35°C 0.50 mol of ethyllithium in about 400 ml of diethyl ether (see Chapter II, Exp. 1). After an additional 30 min at -30°C the reaction mixture was poured into a solution of 40 g of ammonium chloride in 300 ml of water. After shaking, the upper layer was separated off and dried over magnesium sulfate and the aqueous layer was extracted twice with diethyl ether. The ethereal solution of the cumulenic ether was concentrated in a water-pump vacuum and the residue carefully distilled through a 30-cm Vigreux column at 1 mmHg. The product passed over at about 55°C, had n_D^{20} 1.5118, and was obtained in a yield of 87%. Distillation at water-pump pressure (b.p. 72°C/15 mmHg) gave some losses due to polymerization.

Note 1. Cumulenic ethers are very oxygen sensitive and all operations during the work-up must therefore be carried out under nitrogen.

(11) 4-Methyl-1-methoxy-1,2,3-pentatriene[46]

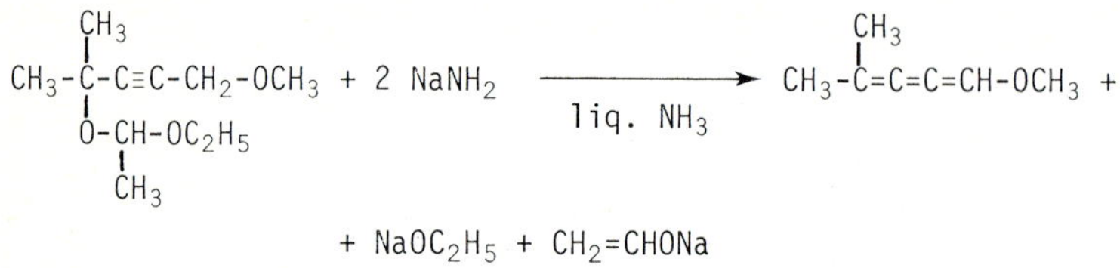

Apparatus: 3-l round-bottomed, three-necked flask, provided with a mechanical stirrer. During the addition of the sodium amide suspension the other necks were open.

In a 1-l round-bottomed flask was prepared a suspension of 1.15 mol of sodium amide in 500 ml of liquid ammonia (see Chapter II, Exp. 11). This suspension was added (note 1) in ten portions at intervals of about 2 min to a vigorously stirred mixture of 0.50 mol of the acetylenic acetal (note 2) in 300 ml of anhydrous liquid ammonia and 100 ml of dry diethyl ether. After an additional 1 h 30 g of solid ammonium chloride were added in small portions. The ammonia was then evaporated by placing the flask in a water-bath at 40^{0}C, stirring being continued. When the stream of escaping ammonia had become very faint, nitrogen was passed through the flask and 300 ml of ice-water were added. The cumulenic ether was extracted three times with small amounts of redistilled pentane (b.p. 36-40^{0}C). The combined extracts were dried over magnesium sulfate and subsequently concentrated in a water--pump vacuum. Careful fractionation of the remaining liquid through a 40-cm Widmer column gave the cumulenic ether, b.p. 40^{0}C/11 mmHg, n_D^{20} 1.5276, in 54-65% yield. The residue consisted mainly of starting compound (note 3). Based on converted material the yield was about 90%.

Note 1. Some experimental skill is required. The sodium amide has to be suspended as homogeneously as possible prior to pouring it into the reaction flask: this can be done by vigorous manual swirling, just before the addition. The suspension can be added most convenietnly through a funnel placed on one of the necks of the reaction flask.

Note 2. Prepared from the carbinol, $(CH_3)_2C(OH)C{\equiv}C\text{-}CH_2OCH_3$, and an excess of ethyl vinyl ether in the presence of a catalytic amount of *p*-toluenesulfonic acid at 0-10^{0}C[1]. The glassware used for the distillation of the acetal was rinsed with a dilute solution of Et_3N in acetone prior to distillation.

Note 3. Cunulenic ethers are extremely air-sensitive. All operations during the work-up must be performed under nitrogen. After termination of the distillation nitrogen must be admitted to the distillation apparatus.

(12) 1-Ethoxy-3-buten-1-yne[8]

$$C_2H_5O\text{-}CH{=}C{=}CH\text{-}CH_2OC_2H_5 \xrightarrow[Et_2O]{C_2H_5Li} C_2H_5O\text{-}C{\equiv}C\text{-}CH{=}CH_2$$

Apparatus: 1-l flask, see Fig. 1.

To a solution of 0.30 mol of the allenic ether (see Chapter IV, Exp. 8) in 100 ml of dry diethyl ether was added with cooling at -60 to -70°C a solution of 0.30 mol of ethyllithium (see Chapter II, Exp. 1, note 1) in 250 ml of diethyl ether. The addition took 30-40 min. The temperature of the reaction mixture was then allowed to rise gradually to -20°C over 30 min and the brown reaction mixture was poured into 200 ml of water. After shaking, the upper layer was separated off and the aqueous layer was extracted four times with small portions of diethyl ether. The combined ethereal solutions were washed with water and dried over magnesium sulfate. The greater part of the diethyl ether was distilled off at normal pressure through a 40-cm Vigreux column, keeping the bath temperature below 65°C (note 2), (see Fig. 5). The distillation flask was then cooled to 20-30°C and the remaining diethyl ether was removed in a water-pump vacuum, keeping the receiver immersed in a bath at -10°C. The volatile enyne ether passed over between 20 and 35°C/12 mmHg, had n_D^{20} 1.4594, and was obtained in 62-70% yield. A small amount (about 5%) of $C_2H_5\text{-}OCH{=}C{=}C{=}CH_2$ was present.

Note 1. Butyllithium in hexane cannot be used, since separation of the volatile enyne ether from the hexane would result in serious losses of product.

Note 2. At higher temperatures some of the product might decompose into ethene and vinylketene.

(13) 3-Methoxy-3-buten-1-yne and 3-methoxy-3-penten-1-yne[33]

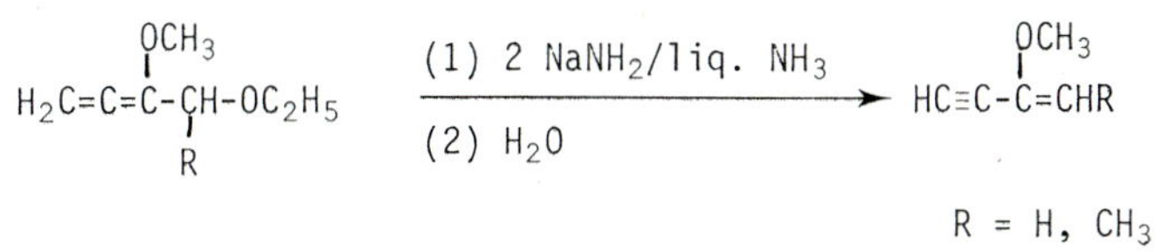

Apparatus: 1-l flask, see Fig. 2.

A suspension of 0.70 mol of sodium amide in 400 ml of liquid ammonia was prepared as described in Chapter II, Exp. 11. To this suspension was added in 15 min 0.30 mol of the allenic bis-ether (see Chapter III, Exp. 13 and 14). The reaction was very vigorous. The ammonia was then evaporated by placing the flask in a water--bath at 40°C. Towards the end of this operation 200 ml of redistilled pentane were added and warming was continued untill the stream of escaping ammonia had become very faint. More pentane was added to bring the volume of the suspension at about 150 ml, the dropping funnel was then combined with a gas inlet and a slow stream of nitrogen was passed through the flask, which was completely immersed in a bath of ice-water. Ice-water (200 ml) was cautiously added with vigorous stirring. The layers were separated and the aqueous layer was extracted three times with small portions of pentane. The combined pentane solutions were dried over magnesium sulfate and the greater part of the pentane was distilled off at normal pressure through a 40-cm Vigreux column. Distillation of the residue in a partial vacuum gave $HC{\equiv}C\text{-}C(OCH_3){=}CH_2$, b.p.29-31°C/65 mmHg, n_D^{20} 1.4419, yield 82% and $HC{\equiv}C\text{-}C(OCH_3){=}CHCH_3$, b.p. 35-39°C/42 mmHg, n_D^{20} 1.4545, yield 92% (63:37 mixture of *E*- and *Z*-isomers).

(*14*) *1-NN-dimethylamino-3-buten-1-yne*[47]

$$CH_3O\text{-}CH_2C{\equiv}CH + HO\text{-}CH_2N(CH_3)_2 \xrightarrow[\text{dioxane}]{Cu(OAc)_2} CH_3O\text{-}CH_2C{\equiv}C\text{-}CH_2N(CH_3)_2$$

$$\xrightarrow[\text{THF}]{KO\text{-}\mathit{tert}.\text{-}C_4H_9} H_2C{=}CH\text{-}C{\equiv}C\text{-}N(CH_3)_2$$

a. Preparation of $CH_3OCH_2C{\equiv}CCH_2N(CH_3)_2$ (Mannich coupling of methyl propargyl ether)

Apparatus: 1-l three-necked, round-bottomed flask with a dropping funnel, a mechanical stirrer and a thermometer, combined with a vent; at a later stage the thermometer and vent were replaced with a reflux condenser.

In the flask were placed 60 g of powdered paraformaldehyde, 100 ml of dioxane and 3 g of copper(II) acetate and 0.3 mol of liquid dimethylamine was added at -20°C. The temperature was allowed to rise gradually to 40-45°C with occasional cooling and when the reaction had subsided, the mixture was cooled to 20°C and a second portion of 0.3 mol of the amine was added. When this had reacted, the remainder of the 2.0 mol of dimethylamine was added in the same way. The mixture was then heated to 80°C, after which the thermometer was replaced with a reflux condenser. Methyl propargyl ether (2.0 mol) (see VIII-6, Exp. 8) was added in 30 min from the dropping funnel and stirring and heating at 100-110°C (bath temperature) were continued until the yellow precipitate had disappeared completely (1.5-2 h). The dropping funnel and reflux condenser were then replaced with two stoppers and the mechanical stirrer with a 40-cm Vigreux column. The latter was connected to a descending condenser and a receiver, cooled at -75°C. After cooling, the contents of the flask were distilled at 10-15 mmHg and collected in one receiver. The distillate was shaken with 60 g of potassium carbonate, and subsequently separated from the solid mass by suction filtration. The drying agent was rinsed with diethyl ether. Careful distillation through a 40-cm Widmer column first gave dioxane (this was discarded before continuing the distillation) and then the coupling product, b.p. 60°C/15 mmHg, n_D^{20} 1.445, in 92% yield.

b. Preparation of the enyne amine $H_2C{=}CH{-}C{\equiv}C{-}N(CH_3)_2$

Apparatus: 1-l round-bottomed flask with a gas inlet, a mechanical stirrer and a thermometer, combined with a gas outlet.

A solution of 0.80 mol of KO-*tert*.-C_4H_9 (see Chapter IV, Exp. 4, note 2) in 150 g of dry THF was heated at 50°C. The Mannich product (0.40 mol) was added in one portion. The temperature rose in a few minutes to 60°C or higher and a thick precipitate was formed. The thermometer and vent were replaced with a reflux condenser and the mixture was heated under reflux for 30 min. It was then cooled to 30°C and 150 ml of dry, redistilled pentane were added. The solid material was filtered off through a sintered-glass funnel and, after pressing, rinsed with pentane. The pentane and the greater part of the THF and *tert*.-C_4H_9OH were distilled off from the filtrate through a 40-cm Vigreux column (bath temperature not higher than 110°C) in a slow stream of nitrogen. When the distillation had stopped, the residue in the distillation flask was cooled to 30-40°C and the distillation was continued at 40 mmHg, using a 40-cm Widmer column. After THF and *tert*.-C_4H_9OH had distilled, the enyne amine passed over between 45 and 55°C, n_D^{20} 1.5034, yield 78%.

(*15*) *1-NN-Diethylamino-3-buten-1-yne*[47]

$$CH_3OCH_2C{\equiv}CH + HO{-}CH_2{-}N(C_2H_5)_2 \xrightarrow{Cu(OAc)_2} CH_3O{-}CH_2C{\equiv}C{-}CH_2N(C_2H_5)_2$$

$$\xrightarrow[THF]{KO\text{-}tert.\text{-}C_4H_9} H_2C{=}CH{-}C{\equiv}C{-}N(C_2H_5)_2$$

Apparatus: 1-l three-necked, round-bottomed flask with a dropping funnel, a mechanical stirrer and a reflux condenser.

a. Mannich coupling of methyl propargyl ether

A mixture of 45 g of dry paraformaldehyde and 1.60 mol of diethylamine was warmed until a vigorous reaction started (occasional cooling may be necessary), so that the amine refluxed temporarily. When this reaction had subsided, 3 g of copper(II) acetate were added and the mixture was heated in a bath at 100°C. Methyl propargyl ether (see VIII-6, Exp. 8) (1.50 mol) was added in five equal portions at intervals of 5 min through the reflux condenser. A yellow precipitate was formed. After the addition, heating at about 115°C (bath temperature) was continued for a further 2 h. The reaction mixture was then distilled at 10-15 mmHg from a 2-l flask fitted with a 20-cm Vigreux column. This column was connected to a descending condenser and receiver, cooled at -75°C (note 1). The distillate was warmed to 20°C and then vigorously shaken with 15 g of anhydrous potassium carbonate. After separation of the aqueous layer which had formed, the product was shaken with 20 g of potassium carbonate and then decanted from the drying agent, which was rinsed with small portions of diethyl ether. These washings were combined with the dried product and careful distillation through a 40-cm Vigreux column gave the desired Mannich product, b.p. 77°C/15 mmHg, n_D^{20} 1.4500, in 78-86% yield.

b. Elimination of methanol

To a solution of 0.50 mol of KO-*tert.*-C_4H_9 (uncomplexed base, commercially available) and 180 ml of dry THF was added at room temperature 0.25 mol of the Mannich product. A weakly exothermic reaction took place. The mixture was heated in a bath at 60°C and the THF started to reflux (occasional cooling in an ice-water bath may be necessary). When the refluxing had ceased, the mixture was heated for 30 min in a bath at 70°C, then cooled to room temperature and 300 ml of redistilled, dry pentane were added. The precipitate of potassium methoxide and

KO-*tert.*-C_4H_9-HO-*tert.*-C_4H_9 was rapidly sucked off through a sintered-glass funnel and the precipitate on the filter was rinsed with pentane. The greater part of the pentane and THF was distilled off from the filtrate through a 30-cm Vigreux column (bath temperature not higher than 75°C). The remaining liquid was carefully fractionated through a 40-cm Widmer column. The enyne amine, b.p. 49°C/15 mmHg, n_D^{19} 1.4938, was obtained in 82% yield.

Note 1. Aqueous work-up and extraction with diethyl ether can also be carried out but will take longer. In the proposed procedure the distilled reaction product is collected in a cooled receiver; if no cooling is applied, the required pressure of 10-15 mmHg cannot be realised because of the presence of volatile components and water in the reaction mixture.

(*16*) *1-Methylthio-3-butén-1-yne*[48]

$$2\ Na + CH_3SSCH_3 \xrightarrow{\text{liq. } NH_3} 2\ CH_3SNa$$

$$ClCH_2C{\equiv}CCH_2Cl + 2\ CH_3SNa \xrightarrow{\text{liq. } NH_3} CH_3S\text{-}CH_2C{\equiv}C\text{-}CH_2\text{-}SCH_3$$

$$CH_3S\text{-}CH_2C{\equiv}C\text{-}CH_2\text{-}SCH_3 \xrightarrow[\text{liq. } NH_3]{\text{KO-}tert.\text{-}C_4H_9} CH_3S\text{-}C{\equiv}C\text{-}CH{=}CH_2\ (+\ KSCH_3)$$

Apparatus: 6-10-l wide-necked, round-bottomed flask with a mechanical stirrer; see Chapter I, Fig. 4.

In the flask were placed 2.3 l of liquid ammonia (water content < 0.1%), 25 g of sodium in 1-g pieces were introduced and after 15 min (note 1) 0.50 mol of dimethyl disulfide (commercially available) was introduced in 1-2-g portions by means of a pipette. During the addition, which was carried out in about 20 min (extremely vigorous reaction), the mixture was stirred vigorously. After completion of the addition the blue colour should have disappeared completely (if not, more drops of dimethyl disulfide should be added). 1,4-Dichloro-2-butyne (0.50 mol) (VIII-2,Exp. 8) was then added in 15 min with vigorous stirring. After ten min 0.75 mol of powdered KO-*tert.*-C_4H_9 (see Chapter IV, Exp. 4, note 2) was introduced very rapidly (within 3 min) with vigorous stirring (note 2). Three minutes thereafter 150g of finely crushed ice were added during about 1 min (note 2).

The greater part of the ammonia was then evaporated by placing the flask for about 30 min in a water-bath at 50°C. To the remaining mixture were added 500 ml of ice-water, then five extractions with redistilled pentane were carried out. The combined extracts were washed three times with water and dried over magnesium sulfate. The greater part of the pentane was distilled off at normal pressure through a 40-cm Vigreux column. The remaining liquid was distilled through the same column, giving the enyne sulfide, b.p. 34°C/12 mmHg, n_D^{20} 1.5560, in 83% yield.

Note 1. If the sodium is not allowed to dissolve completely, the undissolved sodium may be covered with sodium methanethiolate during the addition of dimethyl disulfide and it takes at least 1 h for all of the sodium to be converted.

Note 2. The potassium methanethiolate, liberated in the elimination, very readily adds to the enyne sulfide and rapid working is therefore necessary. Addition of ice prior to the work-up causes inactivation of the potassium methanethiolate by solvation.

(17) *Hexatriyne*[8]

$$ClCH_2C{\equiv}C{-}C{\equiv}C{-}CH_2Cl \xrightarrow[\text{THF}]{\text{KO-}\textit{tert.}\text{-}C_4H_9} HC{\equiv}C{-}C{\equiv}C{-}C{\equiv}CH$$

Apparatus: 1-l flask, see Fig. 1.

A solution of 0.10 mol of 1,6-dichloro-2,4-hexadiyne (VIII-2, Exp. 9) in 150 ml o dry THF was cooled to -90°C and a solution of 0.35 mol of KO-*tert.*-C_4H_9 (see Chapter IV, Exp. 4, note 2) in 120 ml of THF was added in about 15 min with vigorous stirring. The reaction was very exothermic and efficient cooling in a liquid nitrogen bath was necessary to keep the temperature between -90 and -70°C (note 1). After the addition, the mixture was stirred for an additional 20 min at -65°C. High-boiling light petroleum (b.p. > 190°C) (150 ml) was then added to the dark mixture, followed by 200 ml of 2 N HCl. The layers were separated (note 2) as soon as possible, taking care that their temperature remained between -5 and -10°C. The aqueous layer was extracted twice with 30-ml portions of light petroleum. The combined light petroleum solutions were washed ten times with 150-ml portions of 2 N HCl (cooled at -10°C) in order to remove the THF and *tert.*-C_4H_9OH, and were subsequently dried by swirling for a few minutes with a small amount of magnesium sulfate. The brown extract was decanted from the magnesium sulfate and transferred into a 1-l round-bottomed flask. The flask was equipped for vacuum distillation with a 40-cm Vigreux column, condenser and single receiver cooled at -75°C (see Fig. 5), between the receiver and the water pump was placed a tube filled with KOH pellets. The apparatus was evacuated (10-15 mmHg) and the flask slowly warmed until the light petroleum began to reflux in the middle of the column. Pure nitrogen was then admitted. The receiver (note 3) contained white crystals of hexatriyne, yield 84-90%, which turned violet upon brief contact with the air. Immediately after weighing (note 4) a sufficient amount of the desired solvent was added and the NMR spectrum of the solution was recorded. Small amounts (< 5%) of light petroleum appeared to be present as the only impurity.

Note 1. Care should be taken that no solid crust of THF is formed on the bottom of the flask; if it is, the bath should be removed temporarily and addition of the base interrupted.

Note 2. In view of the instability of hexatriyne, all operations during work-up must be carried out under nitrogen without any delay.

Note 3. Do not forget to weigh the receiver before starting the distillation!

Note 4. The compound is highly explosive in a concentrated state. Weighing of the still cold receiver should be carried out without delay.

(*18*) *1-Ethoxy-1,5-hexadien-3-yne*[8]

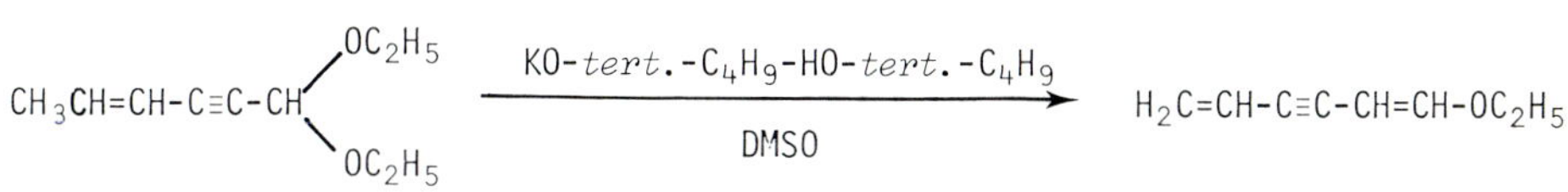

Apparatus: 500-ml flask, see Fig. 1.

A mixture of 0.12 mol of KO-*tert.*-C_4H_9, 0.12 mol of HO-*tert.*-C_4H_9 and 75 ml of dry DMSO was warmed in a one-necked, round-bottomed flask at about 50°C until all solid material had dissolved. After cooling to room temperature the solution obtained was transferred intothe dropping funnel. In the flask were placed 50 ml of dry DMSO and 0.10 mol of the enyne acetal (note 1). The solution of the base was added dropwise over a period of 30 min, while keeping the temperature of the reaction mixture between 28 and 32°C. After an additional 20 min, 200 ml of ice--water were added to the dark solution and the product was extracted five times with a 1:1 mixture of diethyl ether and redistilled pentane (note 2). The combined organic solutions were washed twice with a concentrated solution of ammonium chloride, then dried over magnesium sulfate. The residue remaining after removing the solvents by evaporation in a water-pump vacuum was distilled through a 25-cm Vigreux column, giving the dienyne ether, b.p. 50°C/1 mmHg, n_D^{20} 1.5273, in 87% yield (*cis* to *trans* ratio *ca.* 1).

Note 1. Prepared in 82% yield by refluxing an ethereal solution of $CH_3CH{=}CH{-}C{\equiv}C{-}MgBr$[1] for 8 h with an equivalent amount of $HC(OC_2H_5)_3$ and subsequently hydrolysing the reaction mixture with aqueous ammonium chloride solution. $CH_3CH{=}CH{-}C{\equiv}C{-}CH(OC_2H_5)_2$ has b.p. 87°C/15 mmHg (*cis* to *trans* ratio *ca.* 3:2). For the preparation of $CH_3CH{=}CH{-}C{\equiv}CH$ and its Grignard derivative, see Ref. 1.

Note 2. In view of the oxygen-sensitivity of the product, all operations during the work-up must be carried out under nitrogen.

(19) *4-Phenyl-1,2-butadiene*[51, 52]

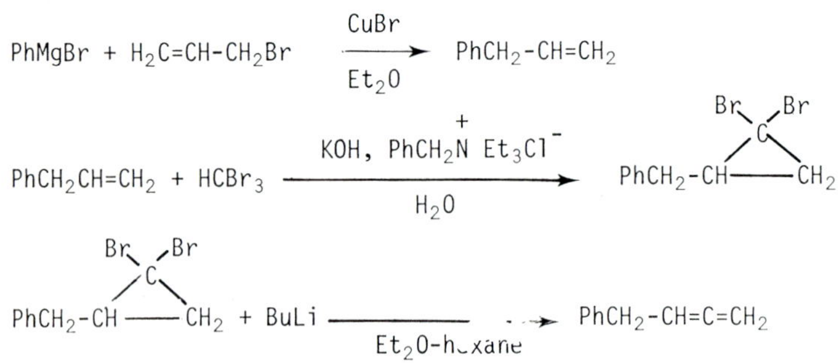

a. Preparation of allylbenzene

Apparatus: 1-l round-bottomed, three-necked flask, provided with a dropping funnel, combined with a gas inlet, a mechanical stirrer and a reflux condenser.

To a refluxing solution of phenylmagnesium bromide in 650 ml of diethyl ether, prepared from 1.15 mol of bromobenzene (see Chapter II, Exp. 5) was added 1.00 mol of allyl bromide at a rate such that refluxing was maintained (about 30 min). Thirty minutes after refluxing had stopped, a trace of copper(I) bromide was added to terminate the conversion. The reaction mixture was cautiously poured on to 500 g of finely crushed ice, then 200 ml of 4 N hydrochloric acid were added. After the remaining ice had melted the layers were separated and the aqueous layer was extracted three times with diethyl ether. The combined ethereal solutions were washed with saturated NaCl solution and dried over magnesium sulfate. The greater part of the diethyl ether was distilled off at normal pressure through a 40-cm Vigreux column and distillation of the residue gave allylbenzene, b.p. 44°C/15 mmHg, $n_D^{21.5}$ 1.5102, in 90% yield.

b. Addition of dibromocarbene to allylbenzene[50]

Apparatus: 2-l three-necked, round-bottomed flask with a thermometer, a mechanical stirrer and a vent.

To a solution of 250 g of NaOH in 275 ml of water were added at 25°C 0.5 g of triethylbenzyl ammonium chloride, 5 ml of ethanol, 0.50 mol of allylbenzene and 1.00 mol of bromoform and the mixture was vigorously stirred. The temperature rose to about 45°C in 10-15 min and was kept at that level by occasional cooling. Stirring was continued for 10 h after the exothermic reaction had subsided. After addition of 1 l of ice-water the product was extracted with diethyl ether (for the first extraction a sufficient amount has to be used to obtain an upper layer). The combined ethereal solutions were washed with water and dried over magnesium sulfate. After the diethyl ether had been removed by evaporation in a water-pump vacuum, the residue was subjected to a high-vacuum distillation (< 0.5 mmHg), keeping the temperature of the heating bath below 100°C. The residue remaining after the unconverted bromoform had been distilled off (b.p. < 60°C/0.1-0.5 mmHg) weighed 120 g and consisted of reasonably pure adduct.

c. Preparation of 4-phenyl-1,2-butadiene (Apparatus: 1-l flask, see Fig. 1)

The 120 g of residue were dissolved in 350 ml of dry diethyl ether and a solution of 0.35 mol of butyllithium in about 280 ml of hexane was added dropwise during 30 min, while maintaining the temperature at about -60°C. After the addition the temperature was allowed to rise to -25°C and stirring at that temperature was continued for an additional 30 min. The mixture was then poured with swirling into 1 l of ice-water and the upper layer and two extracts of the aqueous layer were combined and dried over magnesium sulfate. The solvents were removed in a water-pump vacuum by means of a rotary evaporator. Distillation of the residue gave the allene, b.p. 75°C/15 mmHg, n_D^{21} 1.5396, in 60% yield (based on allylbenzene).

(20) 1,2-Cyclononadiene[51, 52]

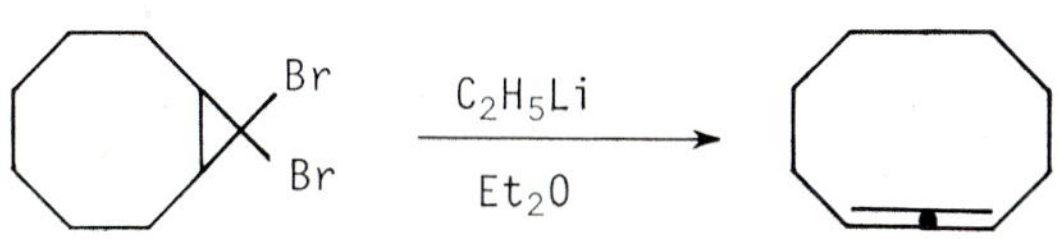

Apparatus: 1-l three-necked, round-bottomed flask, see Fig. 1.

A mixture of 0.20 mol of the adduct from cyclooctene and dibromocarbene (note 1) and 250 ml of dry diethyl ether was cooled to -65°C. A solution of 0.23 mol of ethyllithium (note 2) in 200 ml of diethyl ether (see Chapter II, Exp. 1) was added in 15 min with cooling between -60 and -50°C. The reaction was very exothermic (note 3). After the addition the cooling bath was removed and the temperature was allowed to rise to about -10°C and the reaction mixture was poured into 200 ml of ice-water. The aqueous layer was extracted twice with diethyl ether. After drying, the solvent was removed in a water-pump vacuum and the remaining liquid was distilled through a 40-cm Vigreux column. 1,2-Cyclononadiene, b.p. 62°C/22 mmHg, n_D^{20} 1.5059, was obtained in 86% yield.

Note 1. Prepared from cyclooctene, bromoform and KO-*tert.*-C_4H_9 (see Ref. 52) or by the phase-transfer catalyst method[50].

Note 2. Butyllithium in hexane can also be used.

Note 3. Liquid nitrogen may be used for cooling. In the case of dry-ice-acetone the addition to the dibromo compound will take more than 15 min.

(21) *1,2,3-Cyclodecatriene*[51, 52]

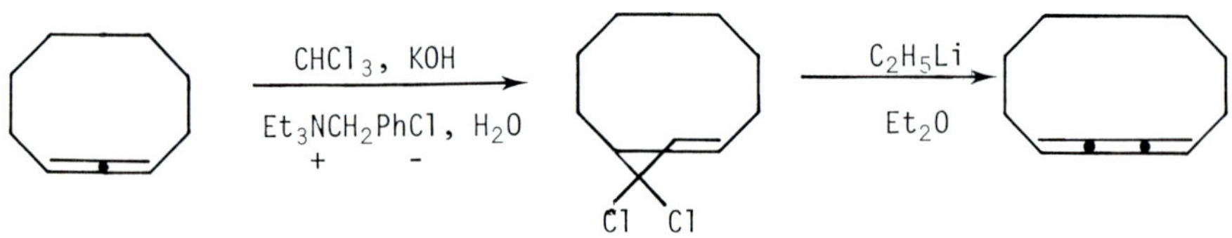

Apparatus: 250-ml three-necked flask (see Fig. 1) for the addition of dichlorocarbene and the dechlorination.

a. Addition of dichlorocarbene

To 75 ml of a 50% aqueous solution of KOH were added 0.25 mol of chloroform, 0.2 g of triethylbenzylammonium chloride and 0.10 mol of freshly distilled 1,2-cyclononadiene (see this chapter, Exp. 20) and the mixture was stirred vigorously for 10-12 h. Water (200 ml) was added and the products were extracted with diethyl ether. The extracts were dried over magnesium sulfate, concentrated in a water-pump vacuum and the residue was distilled through a short Vigreux column. The adduct, b.p. 80°C/0.15 mmHg, n_D^{20} 1.5300, was obtained in 75% yield.

b. Preparation of 1,2,3-cyclodecatriene

A solution of 0.06 mol of ethyllithium (see Chapter II, Exp. 1) in about 50 ml of diethyl ether was added at -50 to -55°C in 15 min to a solution of 0.05 mol of the dichlorocarbene adduct in 40 ml of diethyl ether. After this addition the temperature was allowed to rise to -25°C and the colour of the mixture changed from light yellow to green. After addition of 50 ml of a saturated ammonium chloride solution (note 1) and shaking, the upper layer was separated and dried over magnesium sulfate. The aqueous layer was extracted twice with diethyl ether or pentane and the combined solutions were concentrated in a water-pump vacuum. The last traces of solvent couldbe removed by applying a high vacuum. During the removal of the solvents the bath temperature was kept at about 10°C. The residue (yield > 90%) consisted of reasonably pure 1,2,3-cyclodecatriene (^{1}H NMR). An attempt to distil the compound resulted in polymerization.

Note 1. Cumulenes are extremely air-sensitive: all operations during the work-up must be carried out under nitrogen.

(22) *2,5-Dimethyl-2,3,4-hexatriene (tetramethylbutatriene)*[51, 52] *and 2,4-dimethyl--2,3-pentadiene (tetramethylallene)*

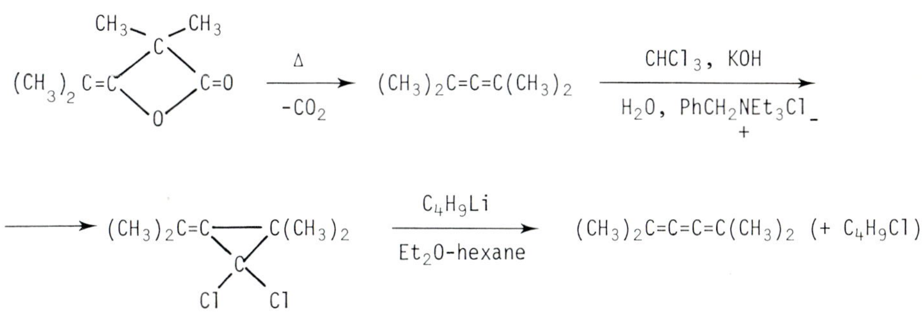

Apparatus: Ketene generator (500 ml distillation flask) (see Ref. 20, p. 529) for the preparation of tetramethylallene; 1-l round-bottomed, three-necked flask, provided with a mechanical stirrer, a thermometer and a vent for the addition of dichlorocarbene; 500-ml flask (see Fig. 1) for the dechlorination.

a. Preparation of tetramethylallene[153]

The ketene generator was connected to two cold traps (-80° C). Between the traps and the water pump was placed a tube filled with calcium chloride lumps. In the distillation flask was placed 1.0 mol of the commercially available 3-hydroxy-2,2,4--trimethyl-3-pentenoic acid β-lactone and after the apparatus had been evacuated (10-20 mmHg), the flask was heated in an oil-bath at 95-105°C and the voltage was adjusted (about 50 V) so that complete decomposition of the lactone vapour occurred (no reflux). After 60-80 min the electrical heating of the glowing spiral was terminated and the heating bath was removed. After the generator had cooled, nitrogen was admitted and the contents of the traps were distilled under normal pressure. Tetramethylallene, b.p. 86°C/760 mmHg, n_D^{20} 1.4405, was obtained in 80% yield. The residue in the reaction flask of the ketene apparatus weighed about 15 g and consisted mainly of starting compound.

b. Addition of dichlorocarbene

To 250 ml of a 50% aqueous solution of KOH were added 0.45 mol of tetramethylallene, 0.90 mol of chloroform and 0.10 g of triethylbenzylammonium chloride and the mixture was agitated vigorously without external cooling. After 4 h the mixture was cooled to 30°C and 200 ml of ice-water were added. After extraction with diethyl ether and drying the light yellow solution over magnesium sulfate the solvent was removed by evaporation in a water-pump vacuum. Distillation (note 1) gave the adduct, b.p. 59°C/15 mmHg, n_D^{23} 1.4812, in 82% yield.

c. Preparation of tetramethylbutatriene

A solution of 0.10 mol of the dichlorocarbene adduct in 130 ml of dry diethyl ether was cooled to -35°C. A solution of 0.13 mol of butyllithium in about 80 ml of hexane was added dropwise in 30 min, while maintaining the temperature of the reaction mixture close to -30°C. After this addition the cooling bath was removed, the temperature was allowed to rise to -10°C and a concentrated aqueous solution (50 ml) of ammonium chloride was added with vigorous stirring. The upper layer was separated off and dried (without washing) over magnesium sulfate. The diethyl ether, hexane and butyl chloride were removed by evaporation in a water-pump vacuum, keeping the bath temperature below 20°C. The residue was distilled through a short column and collected in a single receiver cooled at 0°C. The product passed over at 40°C/18 mmHg and solidified in the receiver. The yields varied from 60 to 80% (note 2).

Note 1. In view of the strong foaming, a 500-ml distillation flask should be used.

Note 2. Traces of oxygen cause polymerization of the cumulene and all operations must therefore be carried out under nitrogen. Under pure nitrogen at $-80^{\circ}C$ the compound can be kept unchanged for several days.

(23) *Propadiene*[53]

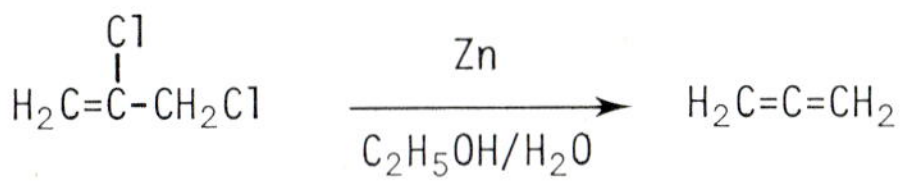

Apparatus: 1-l three-necked, round-bottomed flask with a dropping funnel, a gas--tight mechanical stirrer and a very efficient condenser. The top of the condenser was connected to a cold trap ($-80^{\circ}C$).

In the flask were placed 120 g of powdered zinc (Merck), 250 ml of 96% ethanol and 40 ml of water, and in the dropping funnel 1 mol of 2,3-dichloro-1-propene (commercially available). The mixture was stirred at a rate such that the zinc powder was suspended completely. After the mixture had been heated to about $80^{\circ}C$ (gentle refluxing of the ethanol), the heating bath was removed and dropwise addition of the dichloride was started. The addition, which was carried out at a rate such that the ethanol refluxed gently, took about 1.5 h. The reaction was terminated by heating the mixture under reflux for an additional 30 min. The trap was connected to an empty one, cooled at $-80^{\circ}C$, and was subsequently placed in a water-bath at $0^{\circ}C$. When the greater part of the allene had evaporated and the evolution of gas had subsided, the bath temperature was increased to about $35^{\circ}C$. The remaining mixture of ethanol and the dichloropropene was discarded. The yield of pure allene was about 85%.

Note 1. Allene has a b.p. of $-34.5^{\circ}C$. In view of the high volatility, precise weighing of amounts for small scale experiments (0.05-0.10 mol) is difficult. It is therefore better to prepare a stock solution by adding a sufficient volume (50-100 ml) of the pre-cooled ($-50^{\circ}C$) reaction solvent (mostly THF) to a fixed amount of allene (e.g. 0.5 mol). The required amount of allene can be obtained by taking a corresponding part of the solution.

(*24*) *Butatriene*[54]

$$ClCH_2C\equiv CCH_2Cl \xrightarrow[\text{DMSO}]{\text{Zn}} H_2C=C=C=CH_2$$

Apparatus: 1-l flask with a dropping funnel, a gas-tight mechanical stirrer and a very efficient reflux condenser; the top of the condenser was connected with a trap. A tube containing anhydrous $CaCl_2$ was placed between the trap and the water pump. The connection of the trap was made in such a way that the cumulene vapour could enter the large annular space (the long inner tube being connected to the water pump).

The flask was charged with 70 ml of dry DMSO, 35 g of zinc powder and 10 g of sodium iodide, and in the dropping funnel were placed 0.20 mol of 1,4-dichloro--2-butyne (see VIII-2, Exp. 8). After the system had been evacuated to 10-20 mmHg, stirring was started and the flask was heated (free flame) until the DMSO began to reflux. The trap was immersed in liquid nitrogen and the dichlorobutyne was added in 15 min from the dropping funnel. The reaction was vigorous and occasional cooling was sometimes necessary in order to moderate refluxing. After the addition heating, under reflux (occasional heating) was continued for 20 min. Nitrogen was admitted and the trap was placed in a bath of dry-ice and acetone. The yield of pure butatriene was generally higher than 90%. The compound should be used directly after its preparation. If diluted with an inert organic solvent it can be stored for 12-24 h at -80°C under nitrogen. If the undiluted compound is warmed to about 0°C, it may explode violently.

(*25*) *1,2,3-Pentatriene*[54]

$$ClCH_2C\equiv C\text{-}CH(CH_3)\text{-}Cl \xrightarrow[\text{DMSO}]{\text{Zn, NaI}} H_2C=C=C=CH\text{-}CH_3$$

Apparatus: 500-ml three-necked, round-bottomed flask, provided with a dropping funnel, a gas-tight mechanical stirrer and an efficient reflux condenser; a trap and a tube filled with anhydrous calcium chloride were placed between the condenser and the water pump in such a way that during the reaction the vapour of the product entered the large annular space of the trap (see Exp. 24); the drying tube was placed between the trap and the water pump.

In the flask were placed 70 ml of dry DMSO, 40 g of powdered zinc and 8 g of sodium iodide, and in the dropping funnel 0.20 mol of 1,4-dichloro-2-pentyne (see Chapter VIII-2, Exp. 2). The system was evacuated by means of the water pump and the trap was then placed in liquid nitrogen. The flask was heated until the DMSO began to reflux in the lower part of the condenser. The heating bath was removed and the dichloride was added dropwise in 10 min (note 1). After the addition, the mixture was heated for 45 min under gentle reflux. Pure nitrogen was then admitted and the trap was placed in a bath at -75°C. The NMR spectrum (note 2) showed the product (n_D^{20} 1.514) to be reasonably pure. Traces of DMSO (swept along with the cumulene) were sometimes present. The increase in weight of the trap corresponded to a yield of *ca.* 70%. The cumulene polymerizes completely within a few hours at room temperature, but can be stored at -80°C under nitrogen for 12-24 h.

Note 1. The heat developed by the reaction is just enough to cause gentle refluxing, provided that the mixture is not stirred too vigorously. If refluxing stops during the addition, external heating must be applied.

Note 2. Traces of oxygen will induce polymerization of the cumulene. The NMR tube must therefore be filled with nitrogen before putting the sample in it. Low-temperature NMR gives the most representative results.

(*26*) *5,5-Dimethyl-1,2,3-hexatriene*[55]

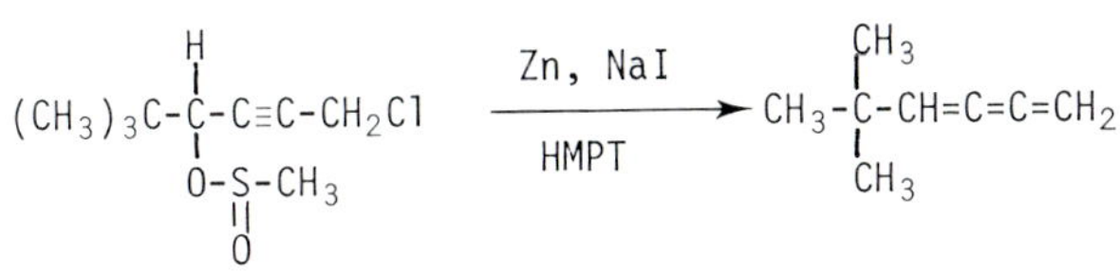

Apparatus: 250-ml distillation flask with a neck about 10 cm long (note 1), connected to a condenser and receiver (250 ml flask), cooled at -75°C. Between the water pump and the receiver was placed a tube filled with lumps of anhydrous calcium chloride.

In the flask were succesively placed 0.10 mol of the sulfinate (note 2), 25 ml of dry, pure HMPT (note 3), 4 g of powdered sodium iodide, 40 g of zinc dust and some boiling stones. After swirling for a few seconds the flask was connected with the other parts of the distillation apparatus, the system was evacuated immediately by means of the water pump (note 4) and the flask was then heated cautiously (free flame). A vigorous reaction started suddenly and the cumulene and part of the HMPT passed over. When the distillation had stopped completely and the remainder of the HMPT in the distillation flask no longer boiled, pure nitrogen was admitted to the apparatus. The receiver was immediately connected to a 20-cm Vigreux column, condenser and smaller receiver (50- or 100-ml flask), cooled at -75°C (see Fig. 5). The tube containing calcium chloride was placed between this receiver and the water pump. After the apparatus had been evacuated (10-20 mmHg), the flask containing the mixture of cumulene and HMPT was gradually heated to 70°C. Pure *tert.*-butylbutatriene (n_D^{21} 1.4898) was obtained in yields varying between 60 and 85%. At 20°C the compound polymerizes within a few hours to a viscous liquid, but it can be stored for at least 24 h at -80°C (under pure nitrogen) without any change.

Note 1. The long neck serves only to prevent the reaction mixture from splashing into the condenser. When a distillation column was used yields of the cumulene were markedly lower, probably because the vapour of the cumulene, escaping from the reaction mixture, dissolved in the refluxing HMPT in the column. Under these conditions it will (partly) polymerize.

Note 2. The sulfinate was prepared from 0.10 mol of $(CH_3)_3C\text{-}CH(OH)\text{-}C{\equiv}C\text{-}CH_2Cl$ as described in Chapter VIII-3, Exp. 2. The alcohol (b.p. 96°C/18 mmHg, n_D^{21} 1.4783, yield 84%) was prepared by lithiating propargyl chloride with ethyllithium (see Chapter II, Exp. 1 and 16) in diethyl ether and subsequently adding pivalyl aldehyde.

Note 3. If more HMPT is used the yield of the cumulene will be lower: the compound can remain longer in the hot reaction mixture, where it can polymerize.

Note 4. It is essential to follow the instructions given. The reaction sometimes starts very soon: if in such a case the desired pressure (10-20 mmHg) has not yet been achieved, part of the cumulene may polymerize in the hot reaction mixture.

(27) *Diacetylene (as a solution in THF or $CH_3OCH_2CH_2OCH_3$)*[167]

$$ClCH_2C{\equiv}CCH_2Cl + 2\ KOH \longrightarrow HC{\equiv}CC{\equiv}CH + 2\ KCl\downarrow$$

Apparatus: See Fig. 7; size of the flask, 2 l; a gas-tight stirrer should be used; the top of the condenser was connected *via* a plastic tube to 2 tubes filled with lumps of $CaCl_2$. These tubes in turn were connected to 2 cold traps (-80°C) each of which contained 40 ml THF or $CH_3OCH_2CH_2OCH_3$ (free from peroxides).

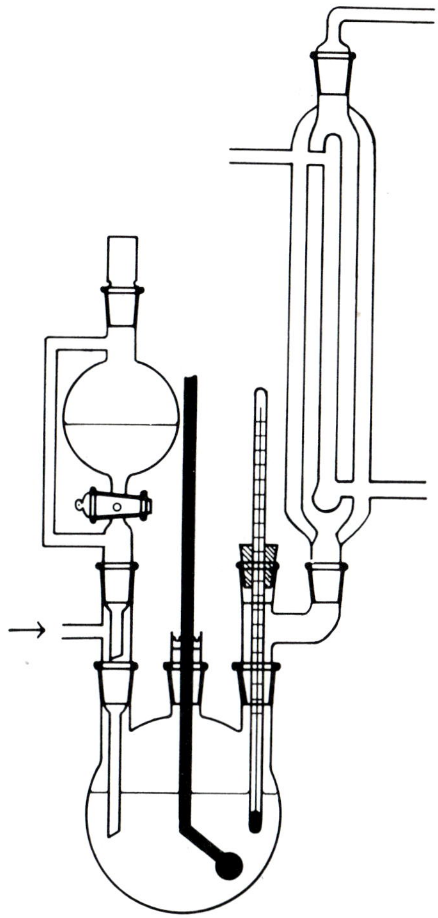

Fig. 7. Preparation of diacetylene.

The flask was charged with a mixture of 250 g of KOH, 400 ml of water and 90 g of DMSO. The air was replaced by passing a vigorous stream of nitrogen during 1-2 min through the flask. The mixture was subsequently brought to a temperature of 65°C. The flow of nitrogen was adjusted to 300 ml/min. Dichlorobutyne (about 10 g, see VIII-2, Exp. 8) was added and stirring was commenced at a rather vigorous rate. The internal temperature rose to about 70°C. The remainder of the 1 mol of dichlorobutyne was added in 5-g portions (note 1), while the temperature was maintained between 67 and 73°C. The addition required about 35 min. After the addition the flask was heated for 45 min in a water-bath of 90-100°C, while nitrogen (200 ml/min) was passed through the flask and stirring continued. The weight increase of the traps amounted to 39-46 g, corresponding to a yield of 78-92% of diacetylene.

If undiluted diacetylene is required, the experiment can be carried out in the same way, but preferably on a smaller scale. 3 Traps should then be used and the stream of nitrogen should be slower. The boiling point of diacetylene is said to be about 10°C. The solutions of diacetylene probably can be stored (under nitrogen in the refrigerator) for at least 24 h without decomposition. Though explosions have never been experienced by the authors, it is advisable to handle diacetylene with the necessary respect.

Instead of THF or 1,2-dimethoxyethane other solvents presumably can be used.

Note 1. Addition of the dichlorobutyne at too fast a rate will give rise to accumulation of the unstable chlorobutatriene $ClCH{=}C{=}C{=}CH_2$, which is partly converted into polymers.

(28) *1,1-Diethoxy-2-propyne*[168]

$$H_2C{=}CH{-}CHO + Br_2 \longrightarrow BrCH_2CHBrCHO$$

$$BrCH_2CHBrCHO + HC(OC_2H_5)_3 \xrightarrow[C_2H_5OH]{ZnCl_2} BrCH_2CHBrCH(OC_2H_5)_2 + HCOOC_2H_5$$

$$BrCH_2CHBrCH(OC_2H_5)_2 + 3\ NaNH_2 \longrightarrow NaC{\equiv}C{-}CH(OC_2H_5)_2 + 2\ NaBr\downarrow$$

$$NaC{\equiv}C{-}CH(OC_2H_5)_2 + H_2O \longrightarrow HC{\equiv}C{-}CH(OC_2H_5)_2 + NaOH$$

Apparatus. 3-l Three-necked round-bottomed flask, provided with dropping funnel, stirrer and thermometer + vent for the bromination and acetalization; 10-l wide-necked flask for the dehydrohalogenation, manual swirling.

To a mixture of 250 ml of ether and 3 moles of freshly distilled acrolein were added about 3 moles of bromine at a rate such that the temperature could easily be maintained between -30 and -90°C (bath of dry-ice-acetone or liquid N_2). After persisting of the brown colour, the temperature was allowed to rise to 0°C. Freshly distilled ethyl orthoformate (3.25 moles) and 96% ethanol (30 ml) were added. The temperature gradually rose to about 40°C (some cooling may be applied). When the temperature had dropped to 30°C, 10 g of anhydrous zinc chloride were added, whereupon again a rise in temperature occurred. (Some cooling should be applied in the case of too rapid a rise.) When the exothermic reaction had ceased and the temperature began to drop, an additional portion of 5 g of zinc chloride was added. Stirring was continued for 1 h. The mixture was poured into 1 l of ice water. After gently shaking the organic layer was immediately separated and (without washing) dried over potassium carbonate. The diethyl ether and ethyl formate were removed in a water-pump vacuum. The residue was heated at 50°C during 1 h in a vacuum of 0.5-0.2 mmHg. A normal distillation apparatus was used. The excess of ethyl orthoformate and other volatile products were condensed in the strongly cooled receiver (see Chapter I, Fig. 5). The residue had almost the theoretically required weight (note 1). It was poured with occasional swirling in 15-20 min into a suspension of 10.5 moles of sodamide (see II, Exp. 11) in 5 l of liquid ammonia which had been prepared at the same time. During the preparation of sodamide the flask was occasionally swirled by hand. The reaction with the dibromoacetal proceeded very vigorously. After rinsing the remaining dibromo acetal into the $NaNH_2$ suspension with ether, the flask was placed in a water-bath at about 60°C. When the stream of evaporating ammonia had become very faint, the bath was removed and nitrogen was blown into the flask during 5 min. Diethyl ether (200 ml) and crushed ice (2.5 kg) were added very quickly, immediately after stopping the flow of nitrogen. The upper walls of the flask were rinsed with some ice water. After the disappearance of the solid material and melting of the ice, the dark upper layer was separated (some solid material can be removed by filtration through a sintered-glass funnel). The aqueous layer was extracted 7 times with 100-ml portions of a 1:1 mixture of diethyl ether and pentane. The unwashed solutions were dried over potassium carbonate. The greater part of the diethyl ether and pentane was distilled off at normal pressure through a 30-cm Widmer column. The residue was distilled in a water-pump vacuum using the same column. The (single) receiver was cooled in ice + ice water (compare Chapter I, Fig. 5). 1,1-Diethoxy-2-propyne, b.p. 38°C/10 mmHg, n_D^{20} 1.4130, was obtained in 80% overall yield.

Note 1. If the dehydrohalogenation is carried out during the next day, the dibromoacetal should be stored in the refrigerator.

(*29*) *Ethoxyacetylene*[169]

$$ClCH_2CH(OC_2H_5)_2 + 3\ NaNH_2 \longrightarrow NaC{\equiv}COC_2H_5 + C_2H_5ONa\downarrow + 2\ NaCl\downarrow$$

$$NaC{\equiv}COC_2H_5 + H_2O \longrightarrow HC{\equiv}COC_2H_5 + NaOH$$

Apparatus: 8-10-l flask, see Chapter I, Fig. 4.

To a suspension of 7.5 mol of sodamide (see Chapter II, Exp. 11) in 4.5-5 l of liquid ammonia were added 2.0 mol of the chloro acetal in 30 equal portions with intervals of 2 min. Twenty minutes after this addition the stirrer was removed, 400 ml of high-boiling light petroleum (b.p. > 180°C) were added and the flask was placed in a water-bath at 40°C (in the case of too strong foaming warming was interrupted). During this evaporation procedure the flask was occasionally swirled by hand. When the flow of ammonia vapour had become considerably fainter, a stopper with a hole was placed on the flask and warming was continued until the stream of ammonia had become very faint. The flask was filled with nitrogen and subsequently finely crushed ice (2.5 kg) was added within 1-2 min. The mass was swirled vigorously until the solid on the glass walls had dissolved. The flask was warmed (with swirling) for some minutes in a water bath until the ice had disappeared. The layers were separated and four extractions with the light petroleum fraction (50-ml portions) were carried out without delay (note 1). The combined extracts were washed with ice-water and subsequently dried over magnesium sulfate. The solution was warmed in a water-pumpvacuum and the ethoxyacetylene condensed in a receiver cooled at -80°C (see Chapter I for the description of this procedure). Redistillation of the contents of the receiver at normal pressure (760 mmHg) through a 40-cm Vigreux column gave ethoxyacetylene, b.p. 52°C, n_D^{20} 1.3820 in 62-70% yield.

Note 1. It is essential to work very quickly. Ethoxyacetylene reacts with aqueous ammonia to give ethyl acetate.

Chapter VI

SYNTHESIS OF ACETYLENES AND CUMULENES BY 1,3-SUBSTITUTION REACTIONS

1. INTRODUCTION

The synthetic methods illustrated in this chapter can be represented schematically as follows (compare Ref. 10):

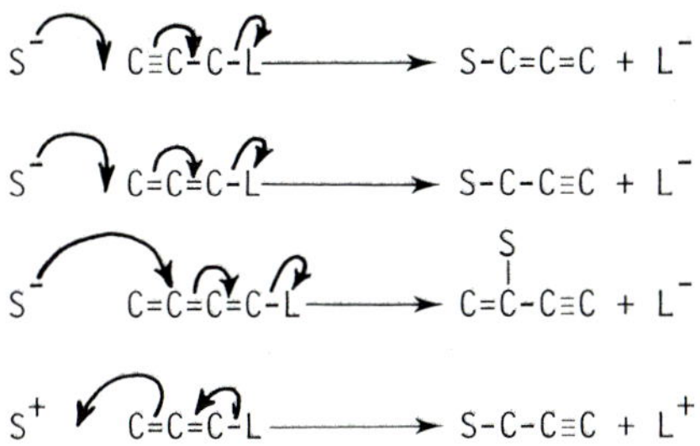

Of course, these schemes indicate only that the overall reactions may be classified as nucleophilic 1,3-substitutions and, in the last case, as electrophilic 1,3-substitutions. The reactions often proceed only in the presence of catalytic or stoichiometric amounts of transition metal salts, while in their absence 1,1--substitutions or other processes are observed. The 1,1-substitutions are also catalyzed by salts of transition metals, and it is not yet well understood, which factors influence the 1,1 to 1,3-ratio. In a number of 1,3-substitutions there is clear evidence for an addition-elimination course[84]. At present the mechanistic aspects of the substitutions are under active investigation (see, for instance, Ref. 85-93). In the next section we give a short survey of 1,3-substitutions that have synthetic value.

2. NUCLEOPHILIC 1,3-SUBSTITUTIONS

2.1. Reaction of acetylenic halides and ethers with Grignard reagents

Although it had been reported several years ago by some research groups[94-97] that allenes are produced in the reaction between certain propargylic halides and Grignard compounds, more recent investigations have led to a better insight into the optimal conditions for allene formation[92]. It was demonstrated that trace amounts of salts of transition metals, e.g. iron salts, cause an enormous increase in the rate of formation of allenes. Yields are generally high. This catalytic method is very attractive in those cases where the starting propargylic halides are either commercially available or can be easily prepared.

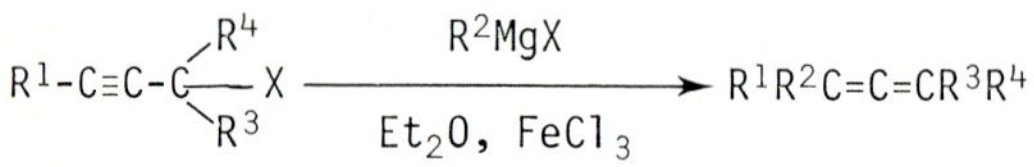

R^1, R^3 and R^4 = H or alkyl
R^2 = alkyl or aryl
X = Cl or Br

The rate of the reaction decreases with increasing number of substituents in the acetylenic halide, and it is higher with acetylenic bromides than with the corresponding chlorides. Methylmagnesium iodide gives equal amounts of 1,1- and 1,3--substitution products, whereas *tert.*-butylmagnesium bromide does not react. However, for some *tert.*-butyl substituted allenes there exists an attractive complementary method: reaction of *tert.*-C_4H_9MgCl with propargyl chlorides[8] or -tosylates in THF[98], using catalytic amounts of copper(I) bromide. Primary alkylmagnesium halides or phenylmagnesium bromide and propargylic halides, if allowed to react in diethyl ether or THF in the presence of copper(I) bromide, give mixtures of the 1,1- and 1,3-substitution products[8]. In a number of cases propargylic ethers or acetals have shown to be suitable starting compounds for the preparation of allenes. The reactions with Grignard compounds proceed in diethyl ether and only a catalytic amount of copper(I) bromide is required[84, 99, 100]:

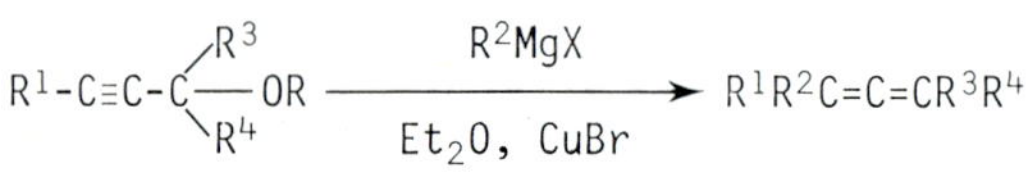

R = alkyl; R^1 = H, alkyl or aryl; R^2 = alkyl or aryl; R^3 = H or alkyl; R^4 = H, alkyl or OR.

Some allenic alcohols can be prepared in analogous ways from acetylenic epoxides or tetrahydropyrans[8, 101, 150]:

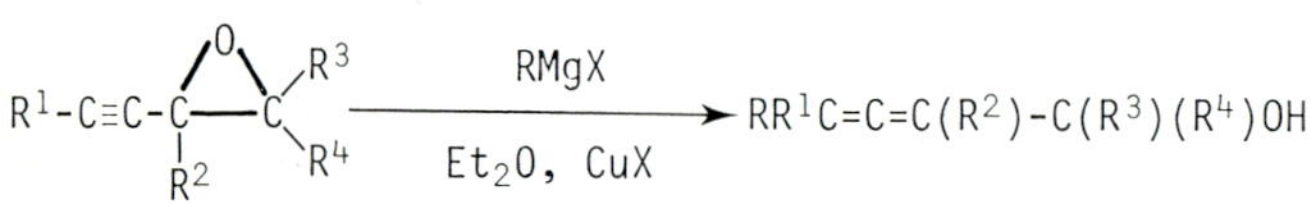

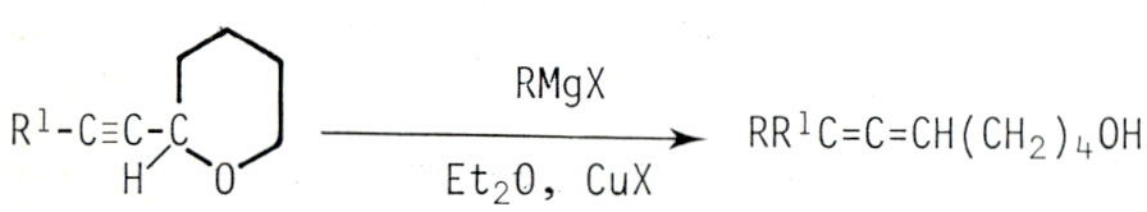

The diethyl ether method sometimes gives unexpected results: the reaction between the amino ethers $R^1R^2C(OCH_3)\text{-}C{\equiv}C\text{-}CH_2NR^3R^4$ and C_4H_9MgBr in the presence of a catalytic amount of CuBr, carried out in THF, affords a mixture of the allenes $R^1R^2C{=}C{=}CH\text{-}CH_2NR^3R^4$ and the acetylenes $R^1R^2CH\text{-}C{\equiv}C\text{-}CH_2NR^3R^4$. These compounds must result from hydrolysis of a metal derivative, $R^1R^2C{=}C{=}C(Mt)CH_2NR^3R^4$. Methylmagnesium bromide, however, affords the normal 1,3-substitution product[102]. *tert.*-Butylmagnesium chloride in general does not react with propargylic ethers in the presence of copper salts[8]; the formation of *tert.*-$C_4H_9CH{=}C{=}CH\text{-}OC_2H_5$ from *tert.*-C_4H_9MgCl and the acetylenic acetal $HC{\equiv}C\text{-}CH(OC_2H_5)_2$ is an exception[100]. Cumulenic ethers react with Grignard compounds in the presence of catalytic amounts of CuBr to yield acetylenes and enynes[103, 104].

$$R^1CH{=}C{=}CH\text{-}OCH_3 \xrightarrow[Et_2O,\ CuBr]{RMgX} RR^1CH\text{-}C{\equiv}CH$$

R^1 = H or alkyl; R = alkyl or aryl

$$R^1R^2CH{=}C{=}C{=}CH\text{-}OR^3 \xrightarrow[Et_2O\ or\ THF,\ CuBr]{RMgX} R^1R^2CH{=}C(R)\text{-}C{\equiv}CH$$

R = alkyl or aryl; R^1 = H or alkyl;
R^2 = H, alkyl or OR^3.

As simple allenic ethers (R^1 = H) and 1,2,3-trienyl ethers ($R^1 = R^2$ = H or alkyl) are readily accessible, the 1,3-substitution method constitutes a good method for the synthesis of some special acetylenes and enynes, e.g. $HC{\equiv}CCH_2Ph$ and $C_2H_5O\text{-}CH{=}C(CH_3)\text{-}C{\equiv}CH$.

2.2. *Reaction of acetylenic esters with organocopper compounds*

The synthesis of allenes from organocopper compounds and acetylenic acetates was reported for the first time more than 10 years ago[105] and this method has been complemented or improved more recently by variants, using sulfonates, sulfinates or other derivatives[86, 98, 106, 110]. An advantage of using sulfinates is that they can be prepared under very mild conditions in excellent yields, even from tertiary acetylenic alcohols; the latter cannot be converted into tosylates.

The organocopper compounds have been prepared from Grignard reagents and also from organolithium compounds and mostly from equivalent amounts of copper(I) halide. One has to use the ester method in those cases where the catalytic method with acetylenic ethers or halides fails because of too low a 1,3-substitution rate and hence decomposition of the intermediate copper compound. The good leaving group character of the ester groups allows the 1,3-substitutions to be carried out in a short time under very mild conditions. This is a fortunate circumstance if the preparation of optically active allenes is intended[86]. The optical rotation is lowered, however, during too long a period of contact between the organocopper compound and the allene[106, 107].

A wide variety of compounds with the allene system has been prepared from the acetylenic esters, including silylallenes with the systems $Me_3Si\text{-}C{=}C{=}C$[108] and $Me_3Si\text{-}C\text{-}C{=}C{=}C$[109], specially substituted alcohols[110] $R^1R^2C{=}C{=}C(R^3)CH_2)_nOH$ (n = 1,2), alkenylallenes[98, 111] C=C-C=C=C, the β-cyanoallene $CH_3CH{=}C{=}CH\text{-}CH_2C{\equiv}N$[8], diallenes[126] C=C=C-C=C=C, β-allenylesters[112] $RCH{=}C{=}CH\text{-}CH_2\text{-}C{\lesssim}^{O}_{OC_2H_5}$.

2.3. *Preparation of haloallenes and cyanoallenes*

Some propargylic halides can be converted into haloallenes by treating them with copper(I) halide and lithium halide, preferably in THF as solvent. A catalytic amount of the copper salt, which forms a soluble complex with lithium halide, is sufficient. In some cases an equilibrium mixture of the allenic and the acetylenic halide is formed[8, 113, 114].

$$HC{\equiv}C\text{-}\underset{\displaystyle R}{\underset{|}{C}H}\text{-}X \xrightarrow[\text{THF}]{\text{LiX.CuX}} X\text{-}CH{=}C{=}CH\text{-}R$$

R = H, alkyl or aryl;
X = Cl, Br or I.

Tertiary acetylenic halides give unsatisfactory results owing to further isomerization of the allenic halide into a conjugated diene system under the influence of the copper salt[10]. Bromo- and iodoallenes with the structures $R^1R^2C{=}C{=}CH\text{-}X$ can also be synthesized by an aqueous procedure, consisting of reaction between a secondary or tertiary acetylenic alcohol and concentrated aqueous HX in the presence of NH_4X and copper(I) halide[115]:

$$R^1R^2\text{-}\underset{\displaystyle \diagdown OH}{C}\text{-}C{\equiv}CH \xrightarrow[NH_4X,\ H_2O]{HX,\ CuX} R^1R^2C{=}C{=}CH\text{-}X$$

R^1 = H, alkyl or aryl;
R^2 = alkyl;
X = Br or I.

Iodoallenes can also be prepared in reasonable yields by treatment of secondary acetylenic alcohols with triphenylphosphite-methyl iodide[116], using DMF as a solvent. One of the -OPh groups is probably first replaced by the propargyloxy group. This intermediate subsequently undergoes attack by iodide on the terminal acetylenic carbon atom, affording the iodoallene in a 1,3-substitution:

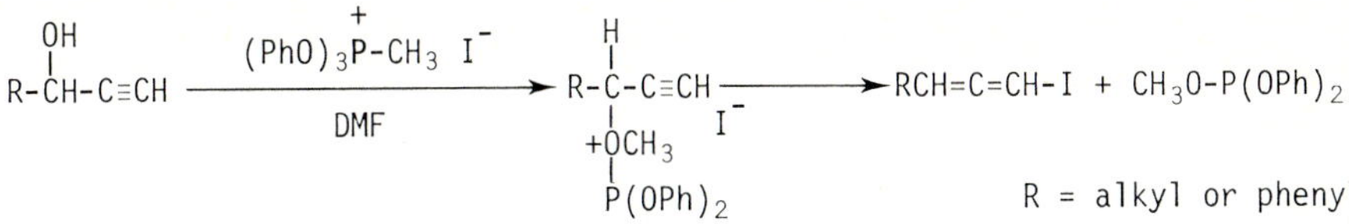

R = alkyl or phenyl.

Propargyl alcohol gives a 70:30 mixture of iodoallene and propargyl iodide. For cyanoallenes we developed two satisfactory methods[8] based on existing procedures (see Ref. 117, 118):

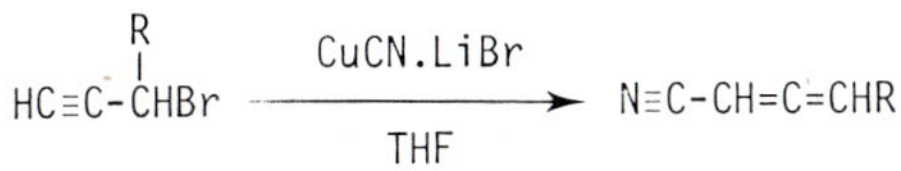

(R = alkyl)

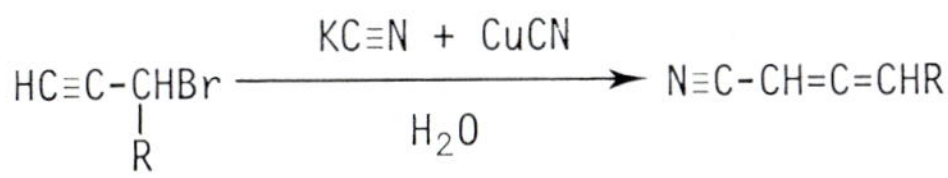

(R = H or alkyl)

In the first method a secondary acetylenic bromide is warmed in THF with an equivalent amount of copper(I) cyanide. We found that a small amount of anhydrous lithium bromide is necessary to effect solubilization of the copper cyanide. Primary acetylenic bromides, $RC{\equiv}CCH_2Br$, under these conditions afford mainly the acetylenic nitriles, $RC{\equiv}CCH_2C{\equiv}N$ (see Chapter VIII). The aqueous procedure for the allenic nitriles is more attractive, in our opinion, because only a catalytic amount of copper cyanide is required; the reaction of the acetylenic bromide with the KCN.CuCN complex is faster than the reaction with KCN. Excellent yields of allenic nitriles can be obtained if the potassium cyanide is added at a moderate rate during the reaction. Excess of KCN has to be avoided, as it causes resinification of the allenic nitrile. In the case of propargyl bromide 1,1-substitution may also occur, but the propargyl cyanide immediately isomerizes under the influence of the potassium cyanide.

2.4. *Synthesis of allenic alcohols by 1,3-substitution of chlorine or ether groups by hydride*

A well known, synthetically useful reaction of propargylic alcohols is their reduction to *trans*-allylic alcohols with lithium alanate in THF[2]; the reduction involves an intramolecular transfer of hydride to the triple bond:

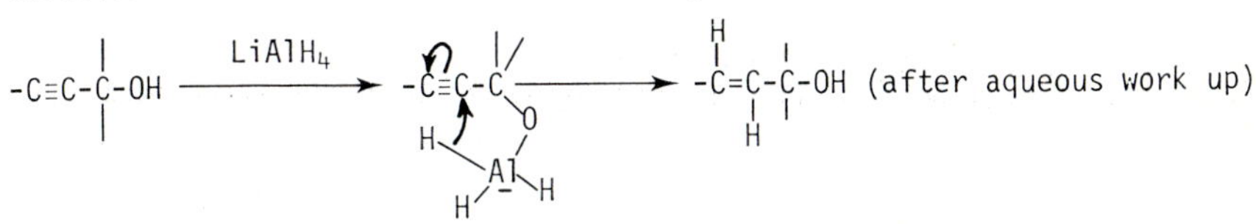

1,3-Substutution can occur by a similar mechanism, if a leaving group (OR or Cl) is present in a suitable position of the triple bond[119, 120]:

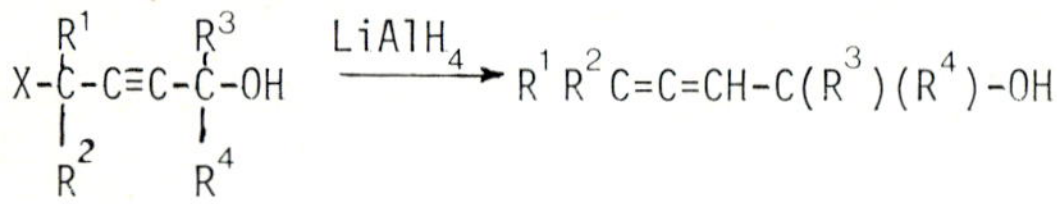

X = Cl, O-tetrahydropyranyl, O-CH(CH_3)OC_2H_5

The starting compounds can be easily synthesized by standard procedures in acetylene chemistry.

3. ELECTROPHILIC 1,3-SUBSTITUTIONS

It was discovered in 1940[75] that some acetylenic chlorides are converted into allenes by treatment with a zinc-copper couple in ethanol. This method appears to be very suitable for the preparation of allenic hydrocarbons. An organozinc compound is thought to be the intermediate[121-123]. If this is the case, then the reaction can be considered as an electrophilic 1,3-substitution (SE'):

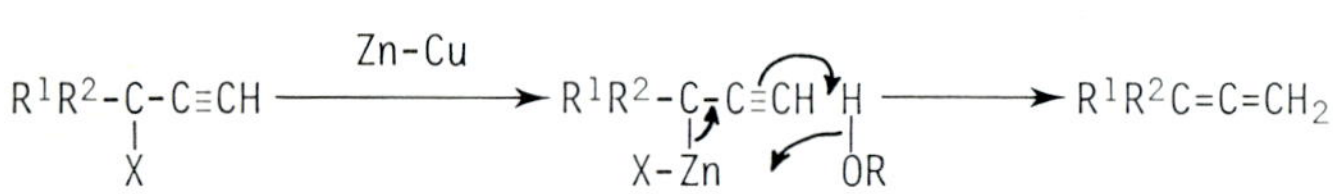

One or both groups R can be alkyl, or R^1R^2C = cycloalkyl. 1,2,4-Trienes are formed from enyne halides, -C≡C-C=C-C-X, and the zinc-copper couple[75]. We found that the results of the reaction in ethanol are very variable; sometimes the triene was obtained in low yields, together with considerable amounts of high-boiling products, possibly "dimers" of the enyne radical. Furthermore, reduction products, C=C-C-C=C, may be present, especially if the zinc is strongly activated. If the reaction with the couple is carried out in hexanol and the volatile 1,2,4-triene is removed during the reaction by evacuation, excellent yields can be obtained[8].

Tributylallenyltin reacts smoothly with acetyl chloride in the presence of a small amount of zinc chloride to give methyl propargyl ketone in a reasonable yield[8]:

$$Bu_3Sn\text{-}CH{=}C{=}CH_2 + CH_3C({=}O)Cl \xrightarrow{ZnCl_2} HC{\equiv}C\text{-}CH_2\text{-}C({=}O)\text{-}CH_3 + Bu_3SnCl$$

This reaction is related to the formation of iodoallene from tributylpropargyltin and iodine[124] and of a propargylic sulfone from the system C=C=C-Sn(CH_3)$_3$ and Aryl-SO_2Cl[125]. At present only a few data concerning electrophilic 1,3-substitution with allenic and acetylenic tin or silicon compounds are available.

4. EXPERIMENTS

(1) *Butylallene*[99]

$$HC{\equiv}CCH_2OCH_3 + C_4H_9MgCl \xrightarrow[Et_2O]{CuBr} C_4H_9CH{=}C{=}CH_2$$

Apparatus: 2-l flask, see Fig. 1.

To a mixture of 1.00 mol of freshly distilled methyl propargyl ether (VIII-6, Exp. 7) and 150 ml of dry diethyl ether was added 1 g of copper(I) bromide. A solution of butylmagnesium chloride (note 1) in 600 ml of diethyl ether, prepared from 1.20 mol of butyl chloride (see Chapter II, Exp. 3) was added with vigorous stirring and efficient cooling, so that the temperature of the reaction mixture could easily be kept between 0 and -10°C. The addition took 30 min. The cooling bath was then removed and a small amount (*ca.* 0.5 g) of CuBr was added and stirring was continued for a further 30 min. The greyish suspension was cautiously poured with manual swirling on to a mixture of 400 g of finely crushed ice, 50 g of ammonium chloride and 100 ml of 36% hydrochloric acid in a 4-l conical flask. The remaining salt mass in the reaction flask was rinsed with dilute (2 N) hydrochloric acid. After separation of the layers the aqueous layer was extracted four times with small portions of diethyl ether and the combined ethereal solutions were dried over magnesium sulfate. Careful distillation through a 50-cm Widmer column gave butylallene, b.p. 105°C/760 mmHg, n_D^{20} 1.4332. The remaining liquid was distilled in a partial vacuum (60-100 mmHg, b.p. 40-70°C) and the distillate was redistilled at normal pressure to give an additional amount of butylallene, bringing the yield to 72-78%.

Note 1. Butylmagnesium *bromide* was also used but the yield was lower (about 65%) owing to the inactivation of the CuBr in a later stage of the reaction by the CH_3OMgBr slurry, which was deposited on the walls of the flask. Moreover, this made stirring, and therefore efficient mixing of the reagents, difficult.

(2) *tert.-Butylallene*[8]

$$HC{\equiv}C\text{-}CH_2Cl + \textit{tert.}\text{-}C_4H_9MgCl \xrightarrow[CuBr]{THF} \textit{tert.}\text{-}C_4H_9CH{=}C{=}CH_2$$

Apparatus: 2-l flask, see Fig. 1.

A mixture of 100 ml of dry THF, 1 mol of propargyl chloride and 2 g of copper(I) bromide was cooled to about -40°C. A solution of about 1.1 mol of *tert.*-butylmagnesium chloride (obtained from magnesium and 1.85 mol of *tert.*-butyl chloride) in about 600 ml of THF (see Chapter II, Exp. 4) was added from the dropping funnel over a period of 60 min. The temperature of the reaction mixture was initially kept between -15 and -20°C, but precipitation of large amounts of salt made it necessary to increase the temperature gradually to 0-10°C. Stirring then became more efficient (note 1). After the addition of the Grignard solution stirring was continued for an additional 30 min, then the mixture was poured into 1 l of ice-cold 3 N hydrochloric acid. High-boiling light petroleum (b.p. > 170°C at normal pressure) (250 ml) was added and, after vigorous shaking, the layers were separated. The organic layer was shaken ten times with 300-ml portions of 3 N HCl in order to remove the THF. The combined aqueous layers were extracted once with 100 ml of light petroleum and the upper layer was freed from THF by shaking it five times with 50-ml portions of 3 N HCl. The combined light petroleum layers were dried over a small amount of magnesium sulfate, then the solution was decanted from the magnesium sulfate and poured into a 1-l round-bottomed flask. After adding some boiling stones the flask was connected to a 40-cm Vigreux column, condenser and receiver cooled at -70°C and the system was evacuated (10-20 mmHg) whilst the flask was heated in a water-bath. The volatile allene condensed in the receiver. The "distillation" was stopped when the temperature in the top of the column had reached 55-60°C. Redistillation of the contents of the receiver through a 40-cm Widmer column gave *tert.*-butylallene, b.p. 79-82°C/760 mmHg, n_D^{26} 1.4196, in 77-84% yield (note 2).

Note 1. If the reaction mixture becomes too thick, one can add some more THF.

Note 2. In order to minimize the hold-up, a partial vacuum (*ca.* 100 mmHg) may be applied during the last stage of the distillation; the fraction obtained in this way can be redistilled at normal pressure in a small distillation apparatus.

(3) *Cyclohexylallene*[99]

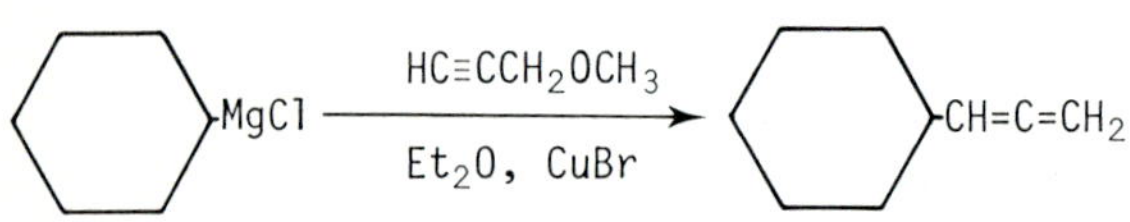

Apparatus: 2-l flask, see Fig. 1.

In the flask were placed 0.50 mol of freshly distilled methyl propargyl ether (VIII-6, Exp. 7), 100 ml of dry diethyl ether and 1.5 g of finely powdered copper(I) bromide. A solution of cyclohexylmagnesium chloride in 400 ml of diethyl ether, prepared from 0.75 mol of cyclohexyl chloride (see Chapter II, Exp. 2), was added from the dropping funnel over a period of 30-40 min. During this addiiton the temperature of the reaction mixture was kept between -10 and +5°C (a cooling bath of dry-ice and acetone should be used). A greyish brown precipitate or slurry was formed. After the addition, stirring was continued for a further 30 min without external cooling. Water (200 ml) was then added cautiously with vigorous stirring and cooling in an ice-bath. After the disappearance of the solid material, enough dilute (2-4 N) hydrochloric acid was added just to give clear layers. The aqueous layer was extracted three times with small portions of diethyl ether and the combined solutions were washed with concentrated ammonium chloride solution and subsequently dried over magnesium sulfate. The diethyl ether was removed by evaporation in a water-pump vacuum. Careful distillation of the remaining liquid afforded cyclohexylallene, b.p. 55°C/17 mmHg, n_D^{23} 1.4801, in 71-77% yield.

(4) *Phenylallene*[99]

$$HC{\equiv}CCH_2OCH_3 + PhMgBr \xrightarrow[CuBr]{Et_2O} PhCH{=}C{-}CH_2$$

Apparatus: 1-l flask, see Fig. 1.

A solution of phenylmagnesium bromide, prepared in the usual way (see Chapter II, Exp. 5) from 0.35 mol of bromobenzene, a 100% excess of magnesium and 300 ml of diethyl ether (note 1), was cooled to 0°C (internal temperature). Copper(I) bromide (1 g, note 2) was added and 0.20 mol of methyl propargyl ether (VIII-6, Exp. 7), dissolved in 100 ml of dry diethyl ether, was added in 10 min. The temperature of the mixture was kept between 0 and 5°C. After the addition, stirring was continued for 20 min at 5-10°C, a dark slurry being formed. Working up was carried out by slow addition of 200 ml of an aqueous solution of 3 g of KCN and 30 g of ammonium chloride to the vigorously stirred mixture. During this addition the flask was cooled in ice-water. After separation of the layers the aqueous layer was extracted with diethyl ether and the combined ethereal solutions were dried over magnesium sulfate and concentrated in a water-pump vacuum. Distillation of the residue through a 30-cm Vigreux column gave phenylallene, b.p. 68°C/15 mmHg, n_D^{20} 1.5815, in 80% yield. The residue was mainly biphenyl.

Note 1. The preparation was carried out in another flask. The Grignard solution was decanted from the excess of magnesium.

Note 2. When more CuBr was used the yields of phenylallene were lower and the residue contained a large proportion of $PhCH{=}CHCH_2OCH_3$, formed by hydrolysis of the adduct from $PhCu.MgBr_2$ and $HC{\equiv}CCH_2OCH_3$.

(*5*) *2,3-Pentadiene*[8]

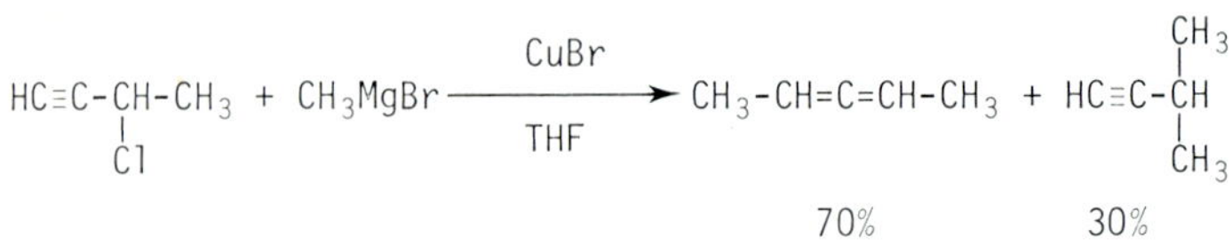

Apparatus: 2-l flask, see Fig. 1.

To a mixture of 1.00 mol of the acetylenic chloride (commercially available, see also VIII-2, Exp. 3) and 200 ml of THF were added 4 g of copper(I) bromide. A solution of methylmagnesium bromide in 600 ml of THF, prepared from 1.30 mol of methyl bromide (see Chapter II, Exp. 7) was added dropwise over a period of 45 min. During the addition the temperature of the mixture was at first kept between -20 and -10°C and towards the end between 0 and +5°C (note 1). Ten minutes after addition of the Grignard solution, 250 ml of high-boiling light petroleum (b.p. > 170°C) were added, then the salt slurry was cautiously poured into 500 ml of 2 N hydrochloric acid, pre-cooled to about -10°C. The remaining salt in the flask was dissolved by cautious addition of 2 N HCl. The upper layer was shaken twelve times with 250-ml portions of 2 N HCl and the combined aqueous layers were extracted once with 100 ml of light petroleum, the organic layer being washed eight times with 100-ml portions of 2 N HCl. The light petroleum solutions were combined and dried over magnesium sulfate. The dried extract was gradually warmed in a distillation apparatus (consisting of a 500-ml round-bottomed flask, 40-cm Vigreux column, condenser and receiver, cooled at -75°C, see Fig. 5) which was evacuated by means of the water pump (10-20 mmHg). Between the water pump and the receiver was placed a tube filled with KOH pellets. Evacuation and heating were stopped as soon as the light petroleum began to reflux in the top of the column. The receiver contained a 70:30 mixture of 2,3-pentadiene and isopropylacetylene, yield about 65%. Extremely careful fractionation (preferably through a spinning band column) gave 2,3-pentadiene, b.p. 48°C/760 mmHg, n_D^{20} 1.4278, in 44% yield (note 2). The first fraction obtained (b.p. 28-47°C) consisted mainly of isopropylacetylene.

Note 1. This increase in temperature was necessary to make stirring more easily.

Note 2. The reaction between CH_3Cu and $HC{\equiv}C{-}CH(CH_3)OTs$[98] gave pure 2,3-pentadiene in about 70% yield, using the isolation procedure described above. This means that for the preparation of about 4 g of the allene about 16 g of CuBr and 250 ml of THF are required!

(6) *3-Phenyl-1,2-pentadiene*[99]

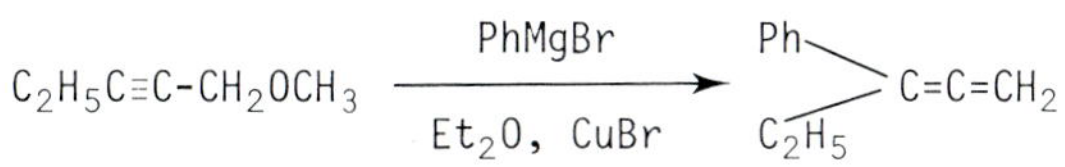

Apparatus: 1-l flask, provided with a gas inlet, a mechanical stirrer and a reflux condenser.

To a solution of phenylmagnesium bromide in 350 ml of diethyl ether, prepared from 0.27 mol of bromobenzene, was added 1.5 g of copper(I) bromide at 0°C. The temperature was increased to 15-20°C and 0.20 mol of 1-methoxy-2-pentyne (prepared by alkylation of $HC{\equiv}CCH_2OCH_3$, note 1) was added in 15 min through the reflux condenser. A vigorous reaction took place and the diethyl ether started to reflux. Heating under reflux was continued for 40 min after the addition, a black slurry being formed. After cooling in an ice-bath, 200 ml of an aqueous solution of 30 g of ammonium chloride and 4 g KCN or NaCN were added with vigorous stirring. The upper layer was separated and combined with the layers from three extractions. After drying the solutions over magnesium sulfate the diethyl ether was removed in a water-pump vacuum. Distillation of the residue through a 30-cm Vigreux column gave the allene, b.p. 95°C/20 mmHg, n_D^{20} 1.5633, in 74% yield.

Note 1. To a suspension of 0.60 mol of $LiNH_2$ (see Chapter II, Exp. 11) in 600 ml of liquid NH_3 was added in 10 min 0.50 mol of $HC{\equiv}CCH_2OCH_3$ (see Chapter VIII-6, Exp. 7). Subsequently ethyl bromide (0.65 mol) was added dropwise in 25 min. After an additional 1 h stirring was stopped and the ammonia was allowed to evaporate. The remaining salt was dissolved by addition of 250 ml of water and the product was extracted with diethyl ether. The extract was dried and the diethyl ether distilled off at atmospheric pressure. Distillation of the remaining liquid in a partial vacuum gave $C_2H_5C{\equiv}CCH_2OCH_3$ b.p. 60°C/90 mmHg, n_D^{20} 1.4247 in 87% yield.

(7) *1,1-Di-tert.-butylallene*[98]

$$(CH_3)_3C\text{-}C{\equiv}CH \xrightarrow[(2)\ (CH_2O)_n]{(1)\ C_2H_5Li/Et_2O} (CH_3)_3C\text{-}C{\equiv}C\text{-}CH_2OH \xrightarrow[\text{tosyl chloride}]{\text{KOH}}$$

$$\longrightarrow (CH_3)_3C\text{-}C{\equiv}C\text{-}CH_2OTs$$

$$(CH_3)_3C\text{-}C{\equiv}C\text{-}CH_2OTs + (CH_3)_3C\text{-}MgCl \xrightarrow[\text{THF}]{\text{CuBr}} [(CH_3)_3C]_2C{=}C{=}CH_2$$

Apparatus: 1-l flask see Fig. 1 for the last reaction.

tert.-Butylacetylene (see Chapter V, Exp. 1) was hydroxymethylated as described in Ref. 1: $(CH_3)_3C\text{-}C{\equiv}C\text{-}CH_2OH$ has b.p. 66°C/17 mmHg and n_D^{20} 1.4424. The alcohol was subsequently converted into the tosylate by adding 200 g of machine-powdered KOH at 0°C to a stirred solution of 0.20 mol of the alcohol and 0.14 mol of tosyl chloride in 350 ml of diethyl ether (see Chapter VIII-3, Exp. 3 and also Ref. 1). The crude tosylate obtained after evaporation of the diethyl ether was dissolved in 160 ml of THF. After addition of 1 g of CuBr the solution was cooled to -10°C and a solution of *tert*.-butylmagnesium chloride in 250 ml of THF, prepared from 0.40 mol of *tert*.-butyl chloride and magnesium (see Chapter II, Exp. 4) was added in 20 min while keeping the temperature between -5 and 0°C. The cooling bath was then removed and stirring was continued for a further 30 min. The solution was cautiously poured into 300 ml of 2 N hydrochloric acid, pre-cooled to -5°C. After shaking, the upper layer was separated off and the aqueous layer was extracted twice with 100-ml portions of distilled pentane. The THF and pentane solutions were combined and then washed ten times with 100-ml portions of 2 N HCl. The combined washings were extracted once with 50 ml of pentane, then the pentane layer was washed five times with 2 N HCl. The combined pentane solutions were dried over magnesium sulfate. The greater part of the solvent was distilled off at normal pressure through a 40-cm Vigreux column. Distillation of the residue afforded the desired allene, b.p. 45°C/20 mmHg, n_D^{22} 1.4538, in 87% yield.

(8) Butenylidenecyclohexane[99]

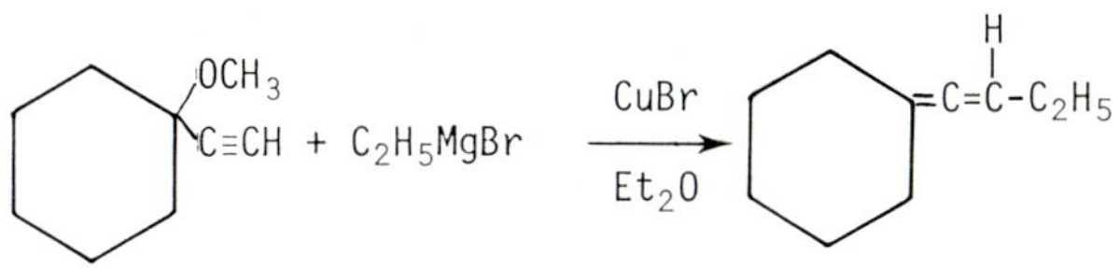

Apparatus: 1-l flask, see Fig. 1.

To a mixture of 0.10 mol of the acetylenic ether (see Chapter VIII-6, Exp. 3), 100 ml of dry diethyl ether and 1 g of powdered copper(I) bromide was added in 20 min at 5°C a solution of about 0.15 mol of ethylmagnesium bromide in 150 ml of diethyl ether (prepared from 0.18 mol of ethyl bromide, see Chapter II, Exp. 5). A black slurry was precipitated on the walls of the flask. After the addition, the cooling bath was removed and stirring was continued for an additional 1 h. The mixture was hydrolysed by gradually adding (with vigorous stirring) a solution of 20 g of ammonium chloride and 3 g of NaCN or KCN in 150 ml of water. After separation of the layers, two extractions with diethyl ether were carried out. The combined ethereal solutions were dried over magnesium sulfate and concentrated in a water-pump vacuum. Careful distillation of the residue through a 40-cm Vigreux column gave the allene, b.p. 62°C/15 mmHg, n_D^{19} 1.4872, in 75% yield.

(9) 1-Butenylidenecyclohexane[86]

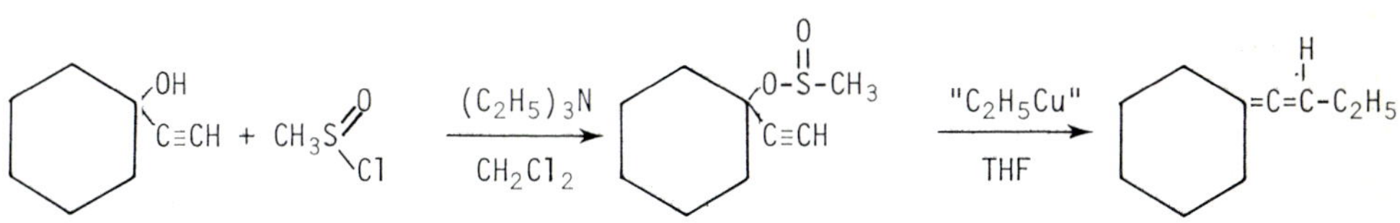

Apparatus: 1-l flask, see Fig. 1.

To a solution of about 0.07 mol of ethylmagnesium bromide in 120 ml of THF (prepared from 0.08 mol of ethyl bromide, see Chapter II, Exp. 6) was added in 15 min at -50°C a solution of 10.0 g of copper(I) bromide and 6 g of anhydrous lithium bromide in 30 ml of THF. Fifteen minutes later a solution of 0.05 mol of the sulfinate from 1-ethynylcyclohexanol (see Chapter VIII-3, Exp. 2) in 30 ml of THF was introduced at -40°C over a period of 15 min. The cooling bath was then removed and when the temperature had risen to -10°C, the mixture was hydrolyzed by adding slowly a solution of 25 g of ammonium chloride and 15 g of NaCN or KCN in 150 ml of water with vigorous stirring. The layers were separated and three extractions with diethyl ether were carried out. The combined organic solutions were dried over magnesium sulfate and subsequently concentrated in a water-pump vacuum. Distillation of the residue in a small apparatus gave the allene, b.p. 67°C/18 mmHg, n_D^{23} 1.4860, in 70% yield.

(*10*) *3-Phenyl-1,2-butadiene*[98]

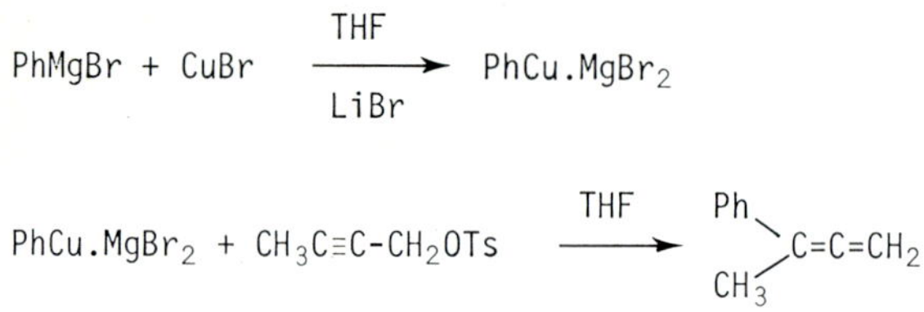

Apparatus: 1-l flask, see Fig. 1.

A solution of PhMgBr in 200 ml THF was prepared from 21 g of bromobenzene and 5 g of magnesium (see Chapter II, Exp. 6). This solution was transferred into the reaction flask and a solution of 17 g of copper(I) bromide (prepared according to Ref. 128) and 8.5 g of anhydrous lithium bromide in 50 ml of THF was added in 10 min at -30°C. Fifteen minutes later a solution of 0.10 mol 2-butynyl tosylate (VIII-3, Exp. 3) in 30 ml of THF was introduced in 20 min at -35°C. The cooling bath was then removed and the temperature was allowed to rise to -10°C. The dark reaction mixture was poured into a solution of 25 g of NaC≡N and 25 g of ammonium chloride in 250 ml of water. After vigorous shaking the layers were separated and the aqueous layer was extracted three times with diethyl ether. The combined solutions were dried over magnesium sulfate and subsequently concentrated in a water-pump vacuum. The remaining liquid was distilled through a 25-cm Vigreux column to give the allene, b.p. 75°C/15 mmHg, n_D^{20} 1.5736, in 80% yield.

(*11*) *1,3-Diphenylallene*[86]

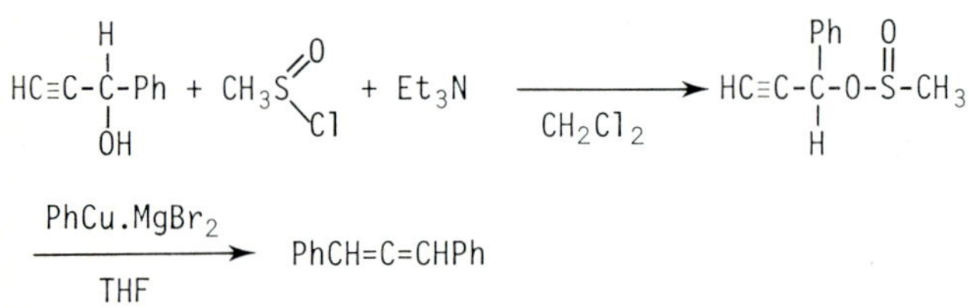

Apparatus: 500-ml flask, see Fig. 1.

To a solution of PhMgBr in 120 ml of THF, prepared from 0.06 mol of bromobenzene (see Chapter II, Exp. 6) was added with cooling at -20^{0}C a solution of 0.06 mol of dry copper(I) bromide and 0.06 mol of anhydrous lithium bromide (note 1) in 40 ml of dry THF. After this addition which took about 10 min the cooling bath was removed and the temperature was allowed to rise to 0^{0}C. The mixture was then again cooled to -40^{0}C and a solution of 0.05 mol of the sulfinate of the acetylenic alcohol (Exp. 27 andIII, Exp. 59) (for the preparation of sulfinates see Chapter VIII-3, Exp. 2) in 40 ml of THF was added in 10 min. After stirring for an additional 15 min at -40^{0}C the cooling bath was removed and, when the temperature had reached 0^{0}C, the reaction mixture was poured into 150 ml of an aqueous solution of 30 g of ammonium chloride and 15 g of KCN or NaCN. After vigorous shaking, the product was extracted four times with diethyl ether and the combined extracts were dried over magnesium sulfate and subsequently concentrated in a water-pump vacuum. The residue, a yellow oil, solidified upon standing at room temperature. It was dissolved in 75 ml of pentane and the solution was cooled to about -30^{0}C with swirling. Some oil precipitated: NMR spectroscopy showed the presence of only a very small amount of the allene. The supernatant yellow liquid was decanted from the oil and, after some pentane had been removed by evacuation, the solution was cooled below -40^{0}C. The pale yellow crystals (m.p. 52^{0}C) were filtered off on a sintered-glass funnel (note 2). From the mother liquor an additional small amount of product was obtained, bringing the yield of NMR-pure material to 62%.

Note 1. Prepared by heating the commercial anhydrous product for 45 min at 150-160^{0}C in a vacuum of 10-20 mmHg.

Note 2. The compound is very air-sensitive.

(*12*) *3-Ethyl-4-methyl-1,2,4-pentatriene*[99]

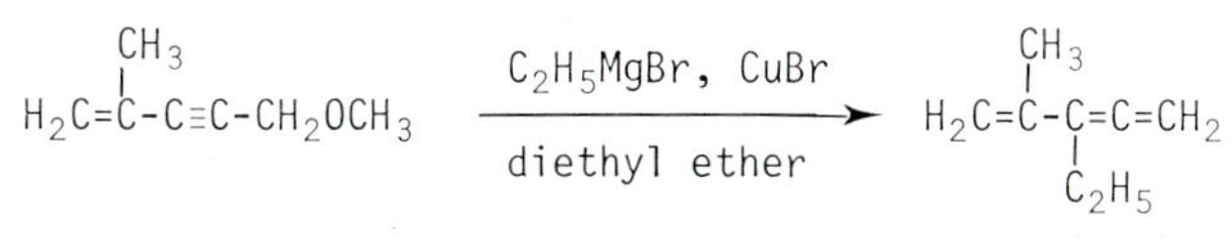

Apparatus: 1-l flask, see Fig. 1.

To a mixture of 0.20 mol of the enyne ether (see Chapter IV, Exp. 11), 100 ml of dry diethyl ether and 1.5 g of copper(I) bromide was added with cooling at about 0^{0}C a solution of about 0.25 mol of ethylmagnesium bromide in 200 ml of diethyl ether (see Chapter II, Exp. 5) in 15 min. After stirring for an additional 30 min at 10-20°C the dark mixture was hydrolysed by careful addition of a cold (0°C) solution of 3 g of NaCN or KCN and 25 g of ammonium chloride in 250 ml of water. During this operation the mixture was vigorously stirred. After separation of the layers and two extractions with diethyl ether, the organic solutions were dried over magnesium sulfate. The greater part of the diethyl ether was distilled off at normal pressure through a 40-cm Vigreux column under nitrogen, keeping the bath temperature below 100°C. The remaining liquid was distilled in a partial vacuum to give the 1,2,4-triene, b.p. 55°C/50 mmHg, n_D^{23} 1.4747, in 76% yield.

(13) 1-Ethoxy-1,2-heptadiene[100]

$$HC{\equiv}C\text{-}CH(OC_2H_5)_2 \xrightarrow[Et_2O,\ CuBr]{C_4H_9MgCl} C_4H_9CH{=}C{=}CHOC_2H_5$$

Apparatus: 1-l flask, see Fig. 1.

To a mixture of 0.40 mol of 1,1-diethoxy-2-propyne (Ref. 1 and V, Exp. 28) and 150 ml of dry diethyl ether was added 1 g of powdered copper(I) bromide. The mixture was cooled to -30°C and from the dropping funnel was added in 20 min a solution of butylmagnesium chloride in 300 ml of diethyl ether, prepared from 0.50 mol of butyl chloride (see Chapter II, Exp. 3). During the first 10 min the temperature was kept at about -30°C, but the remainder of the Grignard solution was added at a somewhat higher temperature (-10 to -20°C) since stirring became more difficult at -30°C. A thick suspension of ethoxymagnesium chloride was formed. The conversion was terminated by stirring for an additional 20 min at about 0°C. The mixture was hydrolysed by cautious addition of a solution of 3 g of KCN and 30 g of ammonium chloride in 200 ml of ice-water: during this operation, carried out with vigorous stirring, the flask was immersed in a bath of ice-water. After separating the layers, three extractions with diethyl ether were carried out. The combined etherea solutions were washed with a saturated ammonium chloride solution and dried over potassium carbonate. The diethyl ether was removed by evaporation in a water-pump vacuum. Careful distillation of the remaining liquid through a 40-cm Widmer column gave the allenic ether, b.p. 63°C/15 mmHg, n_D^{20} 1.4471, in 82% yield.

(*14*) *3-Methyl-1-ethoxy-1,2-pentadiene*[84]

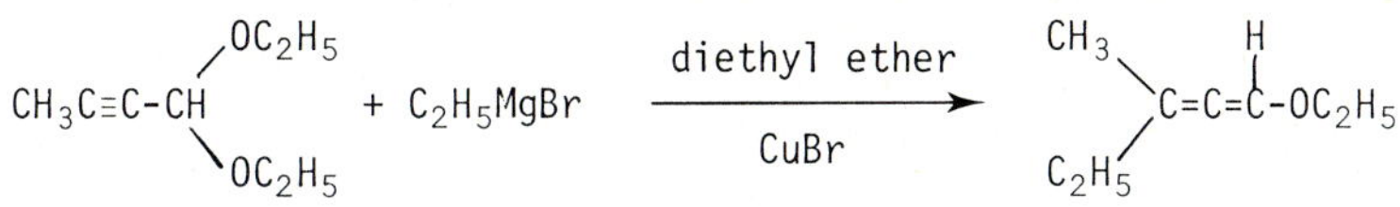

Apparatus: 1-l flask, see Fig. 1.

A solution of ethylmagnesium bromide, prepared from 0.30 mol of ethyl bromide and 170 ml of diethyl ether (see Chapter II, Exp. 5), was added in 30 min to a mixture of 0.20 mol of the acetylenic acetal (prepared from $CH_3C{\equiv}CMgBr$ and $HC(OC_2H_5)_3$, see Ref. 1), 100 ml of dry diethyl ether and 1.5 g of copper(I) bromide. During the addition the temperature of the reaction mixture was kept at about 15^oC. Stirring was continued for an additional 1.5 h without external cooling and the temperature rose to about 30^oC. The black reaction mixture was cautiously poured into 300 ml of an aqueous solution of 20 g of ammonium chloride and 3 g of NaCN. After vigorous shaking the layers were separated and the aqueous layer was extracted three times with diethyl ether. After drying of the combined solutions over magnesium sulfate, most of the diethyl ether was distilled off at normal pressure through a 40-cm Vigreux column. Careful distillation of the remaining liquid through a 40-cm Widmer column gave the allenic ether, b.p. $40^oC/17$ mmHg, n_D^{20} 1.4447, in 76% yield.

(*15*) 1-*Ethoxy-4,4-dimethyl-1,2-pentadiene*[100]

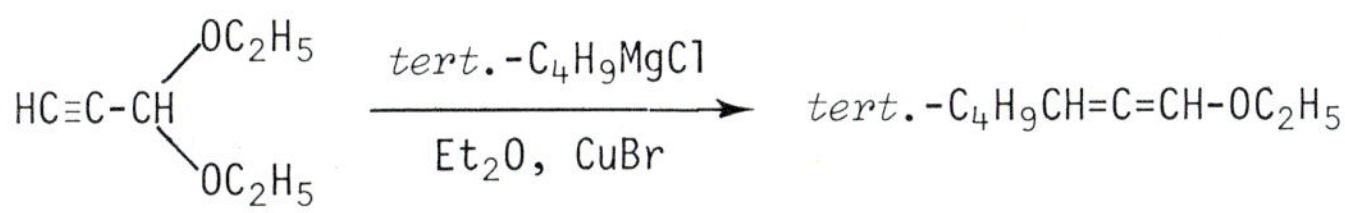

Apparatus: 1-l flask, see Fig. 1.

An ethereal solution of *tert.*-C_4H_9MgCl (300 ml), prepared from 0.50 mol of *tert.*-butyl chloride (note 1) as described in Chapter II, Exp. 2, was added in 20 min at 20^oC to a mixture of 0.15 mol of the acetylenic acetal (V, Exp. 28), 150 ml of diethyl ether and 1.5 g of CuBr. After the addition was completed, the reaction mixture was heated under reflux for 40 min. Working-up was carried out by cautiously pouring the mixture into 200 ml of saturated ammonium chloride solution, to which 3 g of KCN had been added (note 2), in a 3-l conical flask. After swirling until the layers had become clear, the ethereal layer was separated off.

The aqueous layer was extracted four times with small portions of diethyl ether and the combined ethereal solutions were washed with saturated ammonium chloride solution and dried over magnesium sulfate. The greater part of the solvent was distilled off at normal pressure through a 40-cm Vigreux column. Distillation of the residue afforded the allenic ether, b.p.47°C/20 mmHg, n_D^{23} 1.4378, in 80% yield.

Note 1. Although the "yield" of *tert.*-C_4H_9MgCl is only 60-70%, this amount corresponds to a larger excess. This is necessary to ensure that all $HC{\equiv}CCH(OC_2H_5)_2$ is converted; the separation of this compound and the allenic ether by distillation would be impossible.

Note 2. Since an excess of *tert.*-C_4H_9MgCl has been used, the hydrolysis may proceed vigorously.

(*16*) *3,4-Di-tert.-butyl-1,2,4,5-hexatetraene*[8]

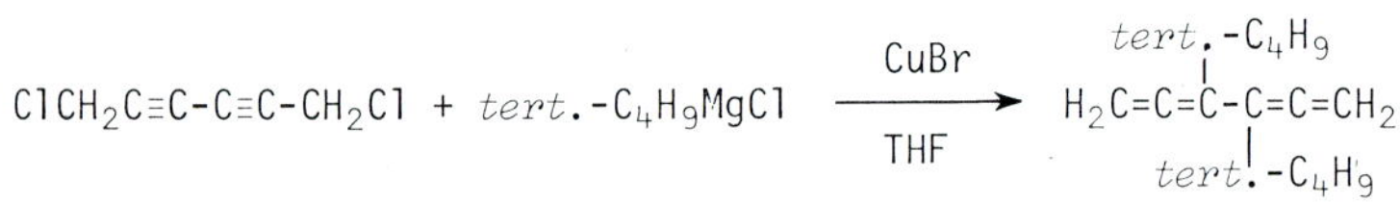

Apparatus: 500-ml flask, see Fig. 1.

To a solution of 0.05 mol of 1,6-dichloro-2,4-hexadiyne (VIII-2, Exp. 9) in 50 ml of dry THF was added 1.3 g of powdered copper(I) bromide. A solution of *tert.*-butylmagnesium chloride in 150 ml of THF, prepared from 0.25 mol of *tert.*-butylchloride (see Chapter II, Exp. 4) was added in 15 min with cooling at -40°C. After the addition the temperature was allowed to rise to 0°C. The mixture was cautiously poured into a solution of 20 g of ammonium chloride and 5 g of NaCN or KCN in 150 ml of water. After vigorous shaking, the upper layer was separated off and the aqueous layer was extracted three times with diethyl ether. The combined solutions were washed with saturated ammonium chloride solution, dried over magnesium sulfate and concentrated in a water-pump vacuum. During the last operation the temperature of the warming bath was kept below 30°C. Pure bis-allene, n_D^{22} 1.4882, was obtained in almost quantitative yield. Upon standing at room temperature or during distillation in a high vacuum (b.p. *ca.* 55°C/0.1 mmHg) the compound underwent partial cyclization to the corresponding dimethylidenecyclobutene.

(*17*) *3-Trimethylsilyl-1,2-heptadiene*[108]

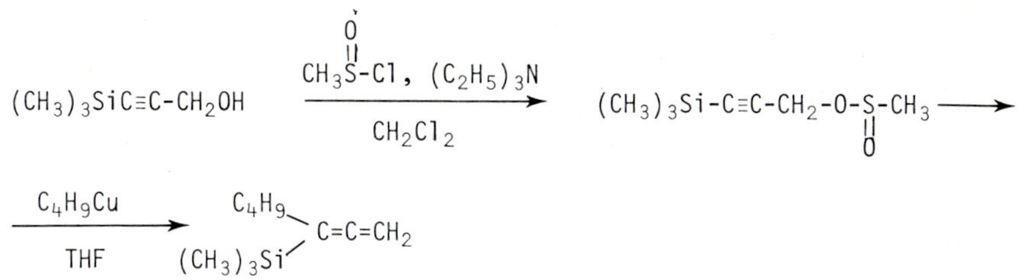

Apparatus: 1-l flask, see Fig. 1.

To a solution butylmagnesium bromide in 200 ml of THF, prepared from 0.12 mol of butyl bromide (see Chapter II, Exp. 6) was added with cooling between -50 and $-60^{\circ}C$ a solution of 0.12 mol of dry copper(I) bromide and 0.12 mol of anhydrous lithium bromide (note 1) in 50 ml of dry THF. After an additional 30 min at $-50^{\circ}C$ a solution of 0.10 mol of the sulfinate [prepared from 0.10 mol of $(CH_3)_3SiC{\equiv}CCH_2OH$ (see Chapter III, Exp. 34) and $CH_3S(=O)Cl$, as described in Chapter VIII-3, Exp. 2) (note 2)] in 40 ml of dry THF was added in 15 min with cooling at about $-50^{\circ}C$. The cooling bath was removed after this addition and the temperature was allowed to rise to $0^{\circ}C$. The greyish solution was poured into 250 ml of an aqueous solution of 40 g of ammonium chloride and 25 g of KCN or NaCN. After vigorous shaking the layers were separated. The aqueous layer was extracted three times with redistilled pentane and the combined solutions were dried over magnesium sulfate. The greater part of the solvents was distilled off at normal pressure through a 40-cm Vigreux column (bath temperature not higher than $110^{\circ}C$). The remaining liquid was carefully distilled through the same column to afford the allene, b.p. $55^{\circ}C/15$ mmHg, n_D^{21} 1.4516, in 74% yield.

Note 1. See Exp. 11, note 1.

Note 2. A large number of experimental examples are given in Ref. 108. In some cases methylsulfonates can be successfully applied when the use of the sulfinic esters leads to mixtures of 1,1- and 1,3-substitution products.

(*18*) *2-Methyl-1-NN-diethylamino-2,3-butadiene*[102*]

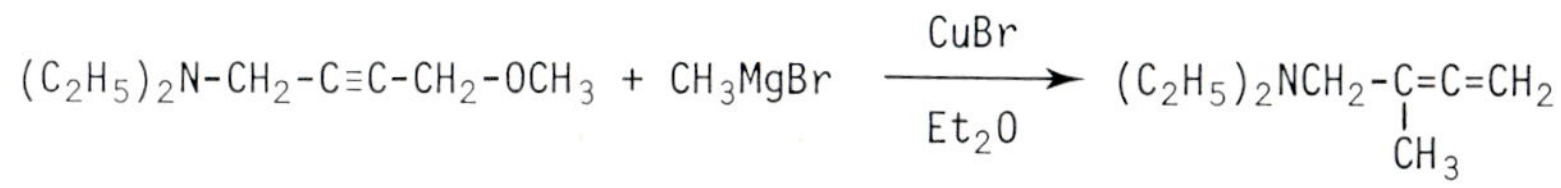

Apparatus: 1-l flask, see Fig. 1.

To a mixture of 0.30 mol of 1-diethylamino-4-methoxy-2-butyne (see Chapter V, Exp. 15) and 175 ml of dry diethyl ether were added 2.5 g of copper(I) bromide. A solution of methylmagnesium bromide in 150 ml of diethyl ether, prepared from 0.5 mol of methyl bromide (see Chapter II, Exp. 5) was subsequently added in 20 min with cooling at about 20°C. After the addition the mixture was warmed for 2 h under reflux (the thermometer and gas outlet were replaced with a reflux condenser), a black slurry being formed on the bottom of the flask. The mixture was cooled in a bath of dry-ice and acetone and a solution of 30 g of ammonium chloride in 200 ml of water was added with vigorous stirring. The organic layer and four ethereal extracts were combined, dried over potassium carbonate and subsequently concentrated in a water-pump vacuum. Careful distillation of the residue through a 40-cm Widmer column gave the allenic amine, b.p. 48°C/15 mmHg, n_D^{21} 1.4532, in 70% yield.

*We are indebted to Prof. A. Claesson (University of Uppsala, Sweden) for submitting detailed instructions.

(*19*) *tert.-Butyl-3,4-hexadienoate*[112]

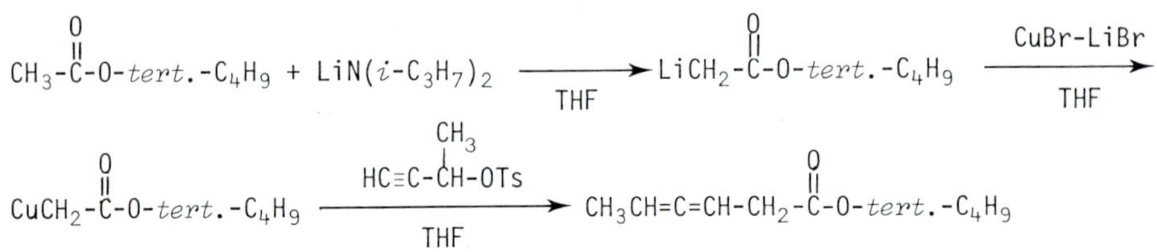

Apparatus: 1-l flask, see Fig. 1.

The performance of this experiment was very similar to the preparation of 3,4-hexadienenitrile (see this Chapter, Exp. 20). The same molar amounts of reagents and the same amounts of solvents were used. The work-up was also similar to that in Exp. 20. The allenic ester, b.p. 75°C/15 mmHg, n_D^{21} 1.4502, was obtained in 60% yield. For the preparation of acetylenic tosylates see VIII-3, Exp. 3.

(*20*) *3,4-Hexadienenitrile*[112]

$$CH_3C{\equiv}N + LiN(i\text{-}C_3H_7)_2 \xrightarrow[THF]{} LiCH_2C{\equiv}N \xrightarrow[THF]{CuBr\text{-}LiBr} CuCH_2C{\equiv}N$$

$$CuCH_2C{\equiv}N + HC{\equiv}C\text{-}\underset{}{\overset{CH_3}{\overset{|}{C}}}H\text{-}OTs \xrightarrow[THF]{} CH_3\text{-}CH{=}C{=}CH\text{-}CH_2\text{-}C{\equiv}N$$

Apparatus: 1-l flask, see Fig. 1.

To a solution of 0.12 mol of dry diisopropylamine in 200 ml of dry THF was added at -20°C a solution of 0.10 mol of butyllithium in about 70 ml of hexane. After cooling the solution to -40°C, 0.13 mol of dry acetonitrile (dried over phosphorus pentoxide and subsequently distilled) was added in 10 min. A white suspension was formed. Ten minutes after this addition a solution of 18.0 g of CuBr[128] and 12.0 g of anhydrous LiBr (see Exp. 11, note 1) in 50 ml of dry THF was added at -25°C in 10 min. After an additional 15 min (at -20°C) the mixture was cooled to -35°C and a solution of 0.12 mol of the acetylenic tosylate (see Chapter VIII-3, Exp. 3) in 30 ml of THF was added in 15 min. During this addition the temperature was kept between -25 and -35°C. A two-layer system was formed. The temperature was allowed to rise to -10°C, then the mixture was poured into 300 ml of a solution of 50 g of ammonium chloride to which 40 ml of 36% HCl (note 1) had been added. Seven extractions with diethyl ether were carried out, the combined extracts were washed with 100 ml of a saturated solution of ammonium chloride to which 10 ml of 20% ammonia solution had been added (to remove traces of copper salts) and were subsequently dried over magnesium sulfate. After the greater part of the solvents had been distilled off at 760 mmHg through a 30-cm Vigreux column, the remaining liquid was distilled to give the nitrile, b.p. 56°C/15 mmHg, n_D^{22} 1.4682, in 76% yield.

Note 1. No cyanide was used to remove the copper salts, since the nitrile is probably very base-sensitive (isomerization to a conjugated diene).

(*21*) *2-Methyl-4-phenyl-2,3-pentadien-1-ol*[101]

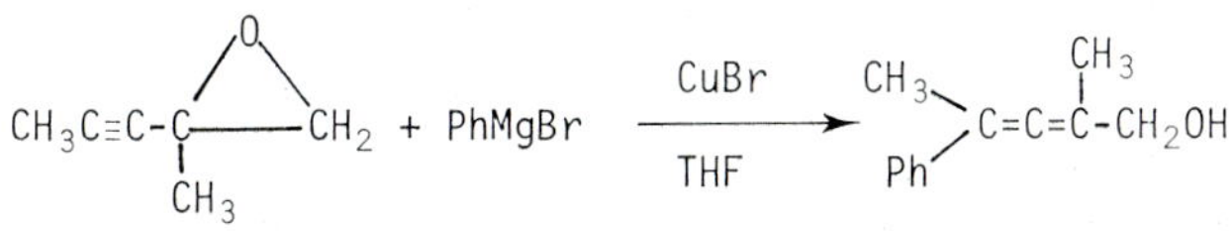

Apparatus: 500-ml flask, see Fig. 1.

To a mixture of 100 ml of THF and 0.10 mol of the epoxide (note 1) was added 0.5 g of copper(I) bromide. A solution of phenylmagnesium bromide (prepared from 0.18 mol of bromobenzene, see Chapter II, Exp. 6) in 130 ml of THF was added dropwise in 20 min at 20-30°C. After an additional 30 min the black reaction mixture was hydrolysed with a solution of 2 g of NaCN or KCN and 20 g of ammonium chloride in 150 ml of water. The aqueous layer was extracted three times with diethyl ether. The combined organic solutions were washed with water and dried over magnesium sulfate. The residue obtained after concentration of the solution in a water-pump vacuum was distilled through a short column, giving the allenic alcohol, b.p. 100°C/0.2 mmHg, n_D^{20} 1.5705, in 75% yield.

Note 1. Prepared by treating the chlorohydrine, obtained from $CH_3C{\equiv}CMgBr$ and $ClCH_2COCH_3$, with powdered KOH in ether at 10-20°C[170].

(22) *2-Ethynyltetrahydropyran and 5,6-nonadien-1-ol*[150]

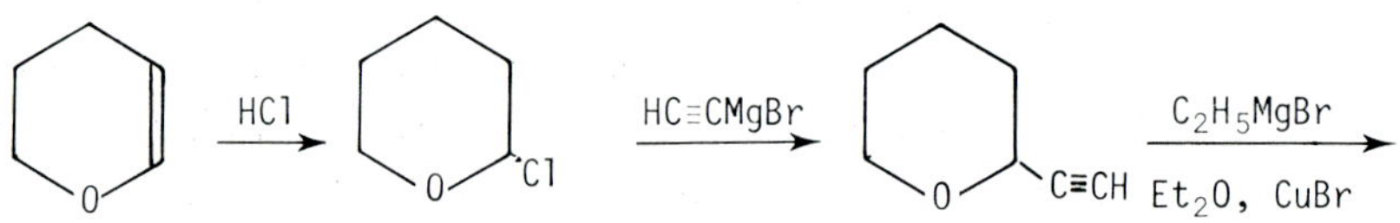

→ $C_2H_5CH{=}C{=}CH(CH_2)_4OH$

Apparatus: 500-ml round-bottomed, three-necked flask with a gas inlet tube, thermometer and a gas outlet for the preparation of chlorotetrahydropyran; 1-l four-necked, round-bottomed flask with a gas inlet tube, a dropping funnel, a mechanical stirrer and a thermometer, combined with a gas outlet for the preparation of HC≡CMgBr and its reaction with chlorotetrahydropyran; 1-l three-necked, round-bottomed flask with a dropping funnel, combined with a gas inlet, a mechanical stirrer and a thermometer, combined with a gas outlet for the conversion into the allenic alcohol.

Dry gaseous hydrogen chloride (17.5 g, weight increase) was introduced with manual swirling into a mixture of 0.50 mol of dihydropyran and 50 ml of dry diethyl ether. During the introduction of the gas the temperature of the mixture was maintained at about -20°C. After the required amount of HCl had been passed in, the flask was closed with well greased stoppers and kept in a refrigerator. Ethynylmagnesium bromide, about 0.6 mol, was prepared by addition at 25°C of a solution of ethylmagnesium bromide in 250 ml of THF (prepared from 0.70 mol of ethyl bromide) to 500 ml of dry THF, through which acetylene was bubbled (for a description of this procedure see Organic Synthesis, Collect. Vol. 4, 792). When the conversion into ethynylmagnesium bromide was completed, nitrogen was passed through the flask and the solution was cooled to -20°C. The ethereal solution of chlorotetrahydropyran was added in 30 min with vigorous stirring, while keeping the temperature of the mixture between -20 and 0°C. After the addition the cooling bath was removed and the temperature of the reaction mixture was allowed to rise to 15 or 20°C and stirring was continued for an additional 30 min at that temperature. The mixture was then poured into an aqueous solution of 150 g of ammonium chloride. Redistilled pentane (200 ml) was added and after vigorous shaking, the upper layer was separated off. The aqueous layer was extracted three times with small portions of pentane and the combined organic layers were dried over magnesium sulfate. The drying agent was filtered off and rinsed with pentane. The solvents were distilled off at normal pressure as thoroughly as possible through a 40-cm Vigreux column. Careful distillation of the remaining liquid through a 40-cm Widmer column afforded 2-ethynyltetrahydropyran, b.p. 58°C/35-40 mmHg, n_D^{20} 1.4572, in 75% yield.

A solution of about 0.5 mol of ethylmagnesium bromide in 300 ml of diethyl ether, prepared from 0.6 mol of ethyl bromide, was added dropwise to a mixture of 0.35 mol of ethynyltetrahydropyran, 200 ml of dry diethyl ether and 1 g of copper(I) bromide. During this addition, which was carried out in 30 min, the temperature was kept between -5 and +5°C. The cooling bath was then removed and stirring was continued for a further 30 min. The dark reaction mixture was poured into 500 ml of an aqueous solution of 50 g of ammonium chloride and 5 g of KCN or NaCN. The black copper suspension disappeared after vigorous shaking. The aqueous layer was extracted with diethyl ether. The ethereal solutions were dried over magnesium sulfate and then concentrated in a water-pump vacuum. Distillation of the residue through a 40-cm Vigreux column gave the allenic alcohol, b.p. 110°C/24 mmHg, n_D^{20} 1.4768, in 82% yield.

(23) *Cyanoallene*[117] *(note 1)*

$$HC{\equiv}CCH_2Br \xrightarrow[KCN,\ H_2O]{CuCN} [HC{\equiv}CCH_2C{\equiv}N + H_2C{=}C{=}CHC{\equiv}N] \xrightarrow{isomer.} H_2C{=}C{=}CH{-}C{\equiv}N$$

Apparatus: 1-l three-necked flask with a dropping funnel, a mechanical stirrer and a thermometer, combined with a vent.

In the flask were placed 1.0 mol of propargyl bromide, 10 g of copper(I) cyanide and 50 ml of water. The mixture was heated to 50°C and a solution of 70 g of potassium cyanide in 115 ml of water was added dropwise (note 2) in 50 min with vigorous agitation, whilst keeping the temperature between 55 and 60°C (occasional cooling or heating). Stirring and heating at about 55°C were continued for an additional 30 min. After cooling to room temperature, seven extractions with small portions of diethyl ether were carried out. The combined extracts were dried over magnesium sulfate (washing was omitted). The greater part of the diethyl ether was distilled off through a 40-cm Vigreux column, keeping the bath temperature below 90°C. The remaining liquid was subsequently distilled in a water-pump vacuum, using a single receiver, cooled at -15°C. Practically no residue remained. The contents of the receiver were distilled once more to remove traces of diethyl ether that had been swept with during the first distillation. Cyanoallene was collected between 20 and 35°C/15 mmHg, n_D^{20} 1.4637 in a yield of 90% or higher (note 3).

Note 1. Caution: Cyanoallene is an *extremely strong* lachrymator. It also irritates the skin, causing blisters.

Note 2. The reaction probably proceeds via the complex of CuCN and KCN, affording a mixture of propargyl cyanide and cyanoallene. The acetylene isomerizes under the influence of the slightly alkaline KCN. If the latter is added at too fast a rate, resinification of the allene may occur. Care should therefore be taken that some CuCN is always present.

Note 3. The ^1H NMR spectrum of the product is misleadingly simple, with just one peak at δ 5.5 ppm!

(24) 2,3-Pentadienenitrile (note 1)

$$HC{\equiv}C\text{-}CH(CH_3)Br \xrightarrow[H_2O,\ C_2H_5OH]{KCN,\ CuCN} CH_3CH{=}C{=}CH\text{-}C{\equiv}N$$

Apparatus: 1-l three-necked, round-bottomed flask with a dropping funnel, a stirrer and a thermometer, combined with a vent.

In the flask were placed 40 ml of ethanol, 10 ml of water, 12 g of finely powdered CuCN and 0.40 mol of 3-bromo-1-butyne (compare VIII-2, Exp. 3). The mixture was warmed to 55°C and a solution of 26 g of KCN in 60 ml of water was added dropwise or in small portions: care was taken that complete dissolution of the copper cyanide did not occur (note 2). The temperature of the mixture was maintained close to 60°C throughout the period of addition. The conversion was terminated by heating the mixture for a further 30 min at 65-70°C whilst stirring vigorously. After cooling to 20°C 300 ml of ice-water were added and the product was extracted seven times with small portions of diethyl ether. The ethereal extracts were combined and washed twice with saturated ammonium chloride solution. After drying over magnesium sulfate, most of the diethyl ether was distilled off at normal pressure through a 30-40-cm Vigreux column. Distillation of the remaining liquid gave the allenic compound, b.p. 39°C/15 mmHg, n_D^{17} 1.4713, in 90% yield.

Note 1. The compound has lachrymatory properties and may cause blisters on the skin.

Note 2. If the addition is performed at too fast a rate, all of the copper cyanide may dissolve temporarily. The free KCN, present in the solution may cause partial resinification of the allenic nitrile.

(*25*) *2,3-Decadienenitrile (note 1)*

$$n\text{-}C_6H_{13}CH{=}C{=}CH\text{-}Br \xrightarrow[\text{THF}]{\text{CuCN, LiBr}} n\text{-}C_6H_{13}CH{=}C{=}CH\text{-}C{\equiv}N$$

Apparatus: 100-ml three-necked flask, provided with a mechanical stirrer, thermometer (dipping into the liquid) and reflux condenser.

A mixture of 0.15 mol of dry CuCN (commercially available), 6 g of anhydrous LiBr (see Exp. 11, note 1), 0.10 mol of 1-bromo-1,2-nonadiene (see Exp. 32) and 25 ml of dry THF was heated to 90°C and kept at that temperature for 30 min. The reaction mixture gradually became homogeneous. After cooling to about 50°C the brown solution was poured into a vigorously stirred solution of 30 g of sodium cyanide and 30 g of ammonium chloride in 200 ml of water, to which 100 ml of diethyl ether had been added. The reaction flask was rinsed with warm (40°C) THF and the solution obtained was also hydrolysed. After separation of the layers the aqueous layer was extracted three times with diethyl ether. The combined ethereal solutions were washed with concentrated ammonium chloride solution, dried over magnesium sulfate and subsequently concentrated in a water-pump vacuum. Rapid distillation of the residue through a 25-cm Vigreux column gave the allenic nitrile, b.p. 110°C/15 mmHg, n_D^{18} 1.4705, in 62% yield (note 2).

Note 1. A simpler procedure is the conversion of $n\text{-}C_6H_{13}\text{-}CH(Br)\text{-}C{\equiv}CH$ (prepared in Chapter VIII-2, Exp. 4) with a mixture of KCN and CuCN in water-ethanol at 65-70°C (compare the preparation of $CH_3CH{=}C{=}CH\text{-}C{\equiv}N$, this chapter, Exp. 24).

Note 2. The nitrile polymerizes (dimer ?) upon prolonged heating.

(26) *1-Chloro-1,2-butadiene*

$$HC{\equiv}C\text{-}CH(CH_3)Cl \xrightarrow[\text{THF}]{\text{CuCl-LiCl}} ClCH{=}C{=}CH\text{-}CH_3$$

Apparatus: 250-ml flask, provided with a gas inlet, a thermometer (dipping in the liquid) and a reflux condenser.

A mixture of 10 g of finely powdered anhydrous LiCl, 20 g of powdered anhydrous CuCl (note 1) and 40 g of dry tetrahydrofuran was heated under reflux for 15 min. The mixture was then cooled to about 40°C, 50 g of dry 3-chloro-1-butyne* (see VIII-2, Exp. 3) were added and the flask was placed in a heating bath. The mixture was heated under reflux for 3 h, during which period the boiling temperature of the liquid rose from about 80 to 87°C, owing to the dissolution of salts and conversion into the chloroallene, which has a higher boiling point than that of the acetylenic chloride. The reaction can be followed by taking samples and determining the integration ratio of the allenic hydrogens and the tertiary H of the starting compound by ^{1}H NMR spectroscopy. After 3 h of heating no further change occurred. The mixture was cooled to 0°C and 175 ml of a high-boiling light petroleum fraction (b.p. > 170°C) at 760 mmHg) were added. The mixture was shaken fifteer times with 40-ml portions of a cold (0°C) solution, prepared by addition of 100 g of ammonium chloride and 150 ml of 36% HCl to 500 ml of ice-water. The combined aqueous layers were extracted once with 50 ml of light petroleum and the upper layer was washed five times with 40-ml portions of the mixture. The combined light petroleum solutions were washed with water and dried over magnesium sulfate. The volatile components were isolated by gradually heating the extract at 15-20 mmHg in a normal distillation apparatus, consisting of a 500-ml flask, 40-cm Vigreux column, condenser and receiver, cooled at -70°C (see Fig. 5). A tube filled with calcium chloride was placed between the receiver and the water pump. The distillation was stopped when the extraction solvent began to distil (b.p. *ca.* 60°C/15 mmHg). The contents of the receiver were carefully redistilled at normal pressure. Two fractions were collected: 4.5 g, b.p. 60-80°C/760 mmHg, n_D^{20} 1.4586, and 30.5 g, b.p. 80-84 ^{o}C/760 mmHg, n_D^{20} 1.4682. The main fraction was a 92:8 mixture of chlorobutadiene and starting compound (note 2), the initial fraction contained about 20% of the acetylenic chloride and a trace of THF.

Note 1. Prepared from the commercial products by heating the powdered materials in round-bottomed flasks at 160 ^{o}C/0.5-0.1 mmHg.

Note 2. The same procedure with $HC{\equiv}C\text{-}CH_2Cl$ gave only 18% conversion into $ClCH{=}C{=}CH_2$.

*We are indebted to BASF (Arnhem) for providing a sample of the acetylenic chloride.

(27) *3-Phenyl-3-chloro-1-propyne and 1-chloro-3-phenylallene*[114]

$$PhCH(OH)C{\equiv}CH \xrightarrow[Et_2O]{SOCl_2} Ph\text{-}CH(Cl)C{\equiv}CH \xrightarrow[CuBr,\ THF]{LiCl} PhCH{=}C{=}CHCl$$

Apparatus: For the first reaction a 500-ml round-bottomed flask with a thermometer and a gas outlet, connected with a tube filled with calcium chloride; for the second reaction a 100-ml round-bottomed flask with a thermometer.

a. 3-Phenyl-3-chloro-1-propyne

To a mixture of 0.50 mol of $PhCH(OH)C{\equiv}CH$ (note 1), 5 ml of pyridine and 250 ml of dry diethyl ether were added 66 g of redistilled thionyl chloride with cooling between -5 and +5 ^{o}C. The addition was carried out in 5-10 equal portions at intervals of about 3 min. The temperature of the mixture was allowed to rise to 15-20 ^{o}C in 2 h. After standing for 12 h (overnight) at room temperature the mixture was added with vigorous stirring to a solution of 20 g of sodium hydrogen carbonate in 200 ml of water in a 2-l round-bottomed flask. When the evolution of gas had stopped the upper layer was separated and the aqueous layer was extracted three times with diethyl ether. The combined solutions were dried over magnesium sulfate and were concentrated in a water-pump vacuum. The residue was distilled through a 30-cm Vigreux column to give the acetylenic chloride, b.p. 88 ^{o}C/18 mmHg, n_D^{20} 1.5559 in 78-85% yield.

b. 1-Chloro-3-phenylallene

To a solution (note 2) of 2 g of copper(I) chloride (commercial product) and 4 g of dry lithium chloride (note 3) in 15 g of dry THF was added with swirling 10 g of the acetylenic chloride. The refractive index n_D^{20} (note 4) of the solution rose during 40 min of warming at 40°C from the initial value of 1.487 to a maximum of 1.505. The solution was poured into 100 ml of 4N hydrochloric acid and, after vigorous shaking, the product was extracted with a 1:1 mixture of diethyl ether and pentane. The extracts were washed with water and dried over magnesium sulfate. Concentration in a water-pump vacuum gave the chloroallene, n_D^{20} 1.5980, in more than 90% yield. The NMR spectrum showed that no starting compound was present and the purity was satisfactory. Attempts to distil the allene led to extensive polymerization.

Note 1. The procedure described in Ref. 1 was modified. To a solution of 2.0 mol of lithium acetylide in 1.2 l of liquid ammonia in a 4-l round-bottomed, three-necked flask (see Fig. 2) was added 1.5 mol of freshly distilled benzaldehyde with cooling at about -45°C. After an additional 30 min finely powdered ammonium chloride (2 mol) was introduced in 15 min. The ammonia was allowed to evaporate, then water (1.1 l) was added and the product was extracted with diethyl ether. After drying over magnesium sulfate the extract was concentrated in a water-pump vacuum. High-vacuum distillation, ($p < 0.5$ mmHg), followed by careful redistillation at 10-20 mmHg afforded the desired carbinol in 55-60% yield. For another procedure see III, Exp. 59.

Note 2. Obtained by briefly heating the mixture under reflux.

Note 3. Commercial lithium chloride can be dried by heating at about 200°C/15 mmHg for 30 min.

Note 4. After placing the liquid on the prism, the apparatus should be closed immediately because evaporation of THF may give too high values.

(*28*) *Bromoallene*[113]

$$HC{\equiv}C\text{-}CH_2Br \xrightarrow[\text{THF}]{\text{CuBr-LiBr}} BrCH{=}C{=}CH_2$$

Apparatus: Same as in Exp. 26.

A mixture of 10 g of copper(I) bromide[128], 35 ml of dry THF and 20 g of anhydrous lithium bromide (see Exp. 11, note 1) was heated gently until a clear solution had formed, then 1 mol of freshly distilled propargyl bromide was added. The mixture was heated under reflux for 3 h. The temperature of the boiling solution, initially about 87°C, had then dropped to the minimum value of about 82.5°C. NMR spectroscopy indicated that the ratio of bromoallene and propargyl bromide was about 70:30 (note 1). After cooling to 20°C the mixture was shaken vigorously with a cold (0°C) solution of 20 g of NaCN in 500 ml of water. The heavy lower layer was separated off as sharply as possible. The aqueous layer was extracted twice with 30-ml portions of high-boiling light petroleum (b.p. > 170°C). The extracts and the undiluted liquid were combined and transferred (without drying) into a 250-ml three-necked flask provided with a dropping funnel, a mechanical stirrer and a thermometer, combined with a vent. The solution was cooled to 10°C and a mixture of 0.6 mol of diethylamine and 40 ml of water was added in 10 min with vigorous stirring, while keeping the temperature between 10 and 5°C (note 2). Stirring was continued for an additional 15 min at +5°C. The mixture was then poured into 500 ml of cold (0°C) 2 N hydrochloric acid and, after vigorous shaking, the organic layer was separated off. The aqueous layer was extracted twice with 70-ml portions of light petroleum and the combined solutions were washed (note 3) seven times with 100-ml portions of 2 N HCl, saturated with ammonium chloride and then dried over magnesium sulfate, and transferred into a 1-l distillation flask, equipped for distillation at water-pump pressure (see Fig. 5). By gradually heating the solution at 10-15 mmHg the bromoallene condensed in the receiver which was cooled at -75°C. The evacuation was terminated as soon as light petroleum began to distil (b.p. > 50°C/15 mmHg). The contents of the receiver were freed from traces of light petroleum by repeating the procedure in the same apparatus, but keeping the temperature of the heating bath below 40°C so that the small amount of light petroleum remained in the distillation flask. The receiver now contained pure bromoallene, n_D^{20} 1.5220, the yield being 56-62%.

Note 1. This corresponds to the equilibrium value.

Note 2. This operation is necessary to remove the propargyl bromide. At higher temperatures bromoallene also reacts with the amine.

Note 3. This washing procedure is necessary to remove the dissolved THF.

(*29*) *Iodoallene*[113]

$$HC{\equiv}CCH_2Cl + NaI \xrightarrow{\text{acetone}} H_2C{=}C{=}CHI + HC{\equiv}CCH_2I + NaCl\downarrow$$

equilibration with CuI-LiI

Apparatus: 1-l three-necked flask with a mechanical stirrer and a reflux condenser for the exchange reaction.

To a solution of 95 g of sodium iodide in 350 ml of dry acetone was added 0.50 mol of propargyl chloride. The separation of salt started immediately. The mixture was heated for 1 h under reflux, then poured into 1.5 l of ice-water. After standing for about 30 min (with occasional gentle swirling of the separating funnel) the brown lower layer was separated off as sharply as possible and dried over magnesium sulfate. The drying agent was filtered off through a sintered-glass funnel and rinsed with 50 ml of dry acetone. To the filtrate (a mixture of acetone and of propargyl iodide and iodoallene ratio $\sim$3:2) was added a solution of 5 g of copper(I) iodide and 6 g of anhydrous lithium iodide in 25 ml of acetone. The mixture was transferred into a 250-ml two-necked, round-bottomed flask, equipped with a gas inlet (for nitrogen) and a reflux condenser. The mixture was heated under reflux until the NMR spectrum of a sample drawn from the liquid showed a ratio of about 3:2 for the allene and the acetylene (about 3 h). The gas inlet was then replaced with a stopper and the reflux condenser with a 30-cm Vigreux column. The acetone and reaction products were distilled in a water-pump vacuum (10-15 mmHg) and collected in a receiver cooled at -75°C (see Fig. 5). The distillate was shaken five times with 100 ml portions of cold 3 N HCl in order to remove the acetone. The lower layer was transferred into a 250-ml three-necked flask, equipped with a dropping funnel, a mechanical stirrer and a thermometer, combined with a vent. A mixture of 0.35 mol of diethylamine and 10 ml of water was added in 10 min with vigorous stirring and cooling at 5°C (note 1), followed by stirring at 5°C for 20 min. A mixture of 15 ml of 36% HCl and 100 ml of water was then added with vigorous stirring at 10-15°C. The lower layer was separated off in a small separating funnel, washed with water and dried over a small amount of magnesium sulfate. The practically pure product, n_D^{20} 1.6060, was obtained in 53% yield.

Note 1. See note 2, Exp. 28.

(*30*) *1-Bromo-1,2-butadiene*[115]

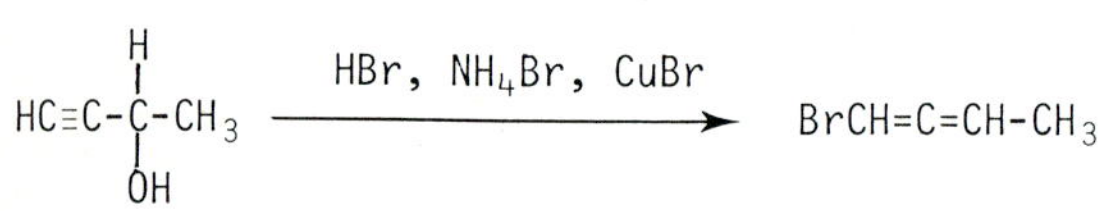

Apparatus: 1-l three-necked, round-bottomed flask, provided with a gas inlet, a mechanical stirrer and a gas outlet, combined with a thermometer.

To 200 ml of 48% hydrobromic acid was added 0.40 mol of phosphorus tribromide (note 1). The mixture was agitated vigorously, while the temperature was kept between 20 and 30°C by cooling in a water-bath at 10-15°C. After about 1 h the lower layer had disappeared completely. The solution was cooled to 0°C, then 0.40 mol of ammonium bromide, 0.10 mol (note 2) of copper(I) bromide (commercial product), 2 g of copper bronze, 140 ml of redistilled pentane and 0.40 mol of the acetylenic alcohol (note 3) were added successively. The mixture was stirred for 5 h at about 0°C and subsequently for 18 h at 15-20°C. After separation of the pentane layer two extractions with 50-ml portions of pentane were carried out. The combined solutions were washed with water and dried over magnesium sulfate. Most of the pentane was distilled off at normal pressure through a 40-cm Vigreux column, keeping the bath temperature below 100°C. The remaining liquid was carefully distilled through a 40-cm Widmer column, giving the bromoallene, b.p. about 60°C/160-170 mmHg, n_D^{20} 1.5230, in 62-74% yield. The product contained a small amount (< 5%) of $HC{\equiv}C{-}CH(Br)CH_3$.

Note 1. The concentration of the aqueous HBr solution is increased by conversion of phosphorus tribromide with water. If available, a corresponding amount of gaseous HBr can be introduced into the 48% solution at 0°C.

Note 2. In Ref. 115 an *equivalent* amount of CuBr is used.

Note 3. This compound is commercially available as a 55% aqueous solution. It can be isolated by saturating the solution with anhydrous potassium carbonate The upper layer is dried over a small amount of potassium carbonate and subsequently distilled: b.p. ~ 40°C/35 mmHg.

(*31*) *1-Bromo-3-methyl-1,2-butadiene*[115]

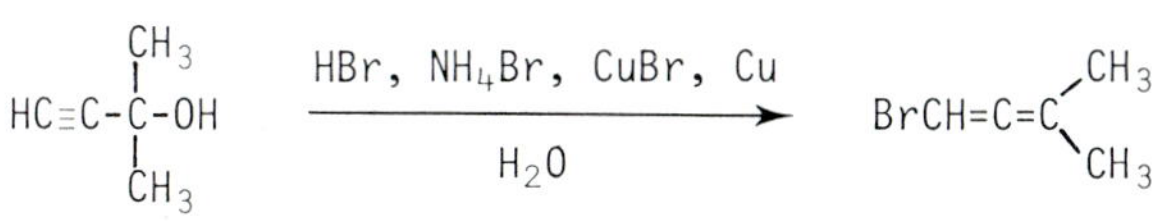

Apparatus: 250-ml three-necked, round-bottomed flask with a mechanical stirrer, a thermometer and a vent.

A mixture of 50 ml of 48% HBr, 10 g of commercial CuBr, 8 g of ammonium bromide, 0.5 g of copper bronze (note 1) and 0.20 mol of the acetylenic alcohol (note 2) was stirred for 15 min at 40°C. After cooling to room temperature the upper layer was separated off as sharply as possible and transferred into a 250-ml flask containing 10 g of sodium hydrogen carbonate. After shaking, the flask was fitted with a short column connected with a condenser and a receiver, cooled at -75°C. By evacuating with the water-pump (10-15 mmHg), and heating the flask at about 50°C the bromoallene condensed in the receiver. The yield of pure product, n_D^{20} 1.5184, was 78%.

Note 1. Copper bronze reduces (traces of) Cu^{II} to Cu^{I}.
Note 2. This compound is commercially available. It can be prepared in high yields by adding acetone to a solution of NaC≡CH in liquid NH_3 and hydrolyzing the residue, remaining after evaporation of the ammonia, with ice-water[1].

(*32*) *1-Bromo-1,2-nonadiene*[8]

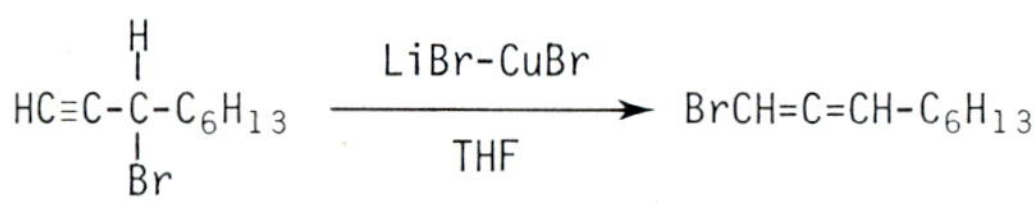

Apparatus: 100-ml two-necked, round-bottomed flask, provided with a gas inlet and a reflux condenser.

A mixture of 40 ml of dry THF, 6.0 g of anhydrous LiBr (note 1), 2.5 g of CuBr and 0.10 mol of the acetylenic bromide (see Chapter VIII-2, Exp. 4) was heated under reflux. The solid material disappeared after a short time. The refractive index (n_D^{20}) of a sample, taken from the liquid after cooling to about 50°C, which was about 1.460 after dissolution of the solids, rose to the maximum value of about n_D^{20} 1.466 after refluxing for 2.5 h (note 2). After 3 h the mixture was cooled to room temperature and poured into a solution of 10 g of ammonium chloride and 6 g of NaCN (or KCN) in 100 ml of water. The mixture was shaken vigorously, then five extractions with diethyl ether were carried out. The combined extracts were dried over magnesium sulfate and subsequently concentrated in a water-pump vacuum. Careful distillation of the residue afforded the bromoallene, b.p. 90°C/ /15 mmHg, n_D^{20} 1.4920, in 78% yield. The small first fraction contained some starting compound.

Note 1. See Exp. 11, note 1.

Note 2. In order to obtain reproducible values, the determination should be carried out very quickly: a few drops are placed on the prism and the apparatus is closed *immediately*, otherwise the THF will evaporate and too high a value is measured.

(33) *1-Bromo-3-phenylpropadiene*[115]

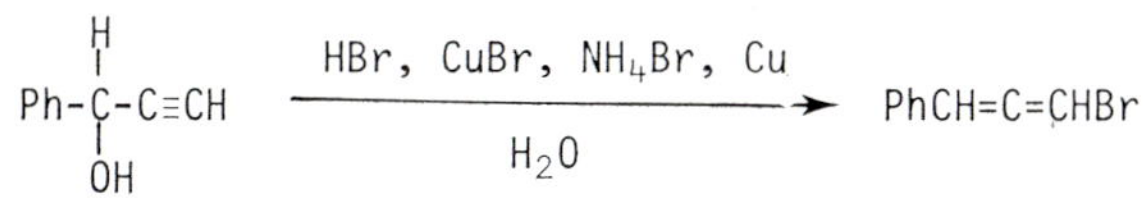

Apparatus: 250-ml flask, see Fig. 1.

To a mixture of 50 ml of 47% hydrobromic acid, 0.03 mol of copper(I) bromide, 0.1 mol of ammonium bromide and 0.1 g of copper bronze was added 0.10 mol of the acetylenic carbinol (Exp. 27 and II, Exp. 59) dissolved in 20 ml of pentane at $0^{o}C$ in 3 min. After stirring for 2 h at $0^{o}C$ 150 ml of pentane and 200 ml of ice--water were added successively. The pentane layer was shaken with 25-ml portions of 47% HBr until the aqueous layer remained colourless. The pentane solution was dried over magnesium sulfate and then concentrated in a water-pump vacuum. High--vacuum distillation afforded the bromoallene, b.p. *ca.* $50^{o}C$/0.01 mmHg, n_D^{20} 1.6540, in 80% yield. In view of the thermolability of the compound too high bath temperatures have to be avoided.

(34) *1-Iodo-1,2-butadiene*[116]

$$CH_3CH(OH)C{\equiv}CH + (PhO)_3\overset{+}{P}\text{-}CH_3I^- \xrightarrow[DMF]{} CH_3CH{=}C{=}CHI$$

Apparatus: 250-ml flask, see Fig. 1.

A solution of 45 g of triphenyl phosphite methiodide[116] in 100 ml of dry dimethylformamide was heated at $100^{o}C$ and 0.08 mol of the acetylenic alcohol (commercially available) was added in 2 min. After stirring for 30 min at $100^{o}C$ the mixture was cooled. The flask was equipped for vacuum distillation. On the flask were placed a 25-cm Vigreux column and two stoppers and the column was connected with a condenser and a receiver. The DMF and the iodoallene distilled between

40 and 50°C/15 mmHg and after addition of 200 ml of water to the distillate four extractions with small amounts of diethyl ether or pentane were carried out. The extracts were washed with water and dried over magnesium sulfate. The residue, remaining after evaporation of the solvent in a water-pump vacuum, was distilled through a short column, affording the iodoallene, b.p. 40°C/15 mmHg, n_D^{20} 1.5879, in 74% yield.

(35) *1-Iodo-3-phenyl propadiene*[115]

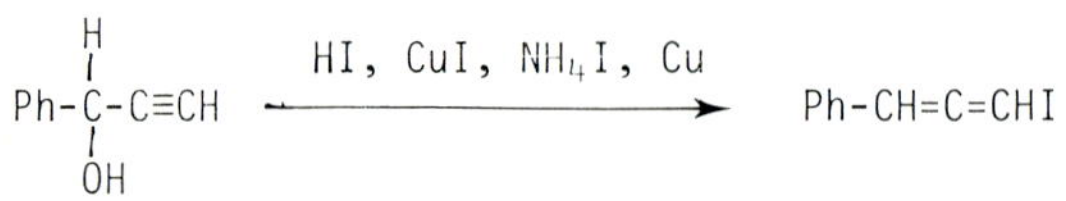

Apparatus: 250-ml flask, see Fig. 1.

A mixture of 40 ml of 50% hydroiodic acid, 0.03 mol copper(I) iodide, 0.1 mol of ammonium iodide, 0.2 g of copper bronze, 0.10 mol of the acetylenic alcohol (Exp. 27 and III, Exp. 59) and 20 ml of pentane was stirred for 1 h at 0-5°C. Ice-water (200 ml) was then added and the product was extracted three times with 50-ml portions of pentane. The combined pentane solutions were washed with water, dried over magnesium sulfate and subsequently concentrated in a water-pump vacuum. The weight of the residue corresponded to 90% of the iodoallene. High-vacuum distillation gave the pure product, b.p. 70°C/0.01 mmHg, n_D^{20} 1.6850 in 60% yield. Owing to its low thermal stability much of the product polymerized during the distillation.

(36) *1-Methylthio-3-iodo-1-propyne*[8]

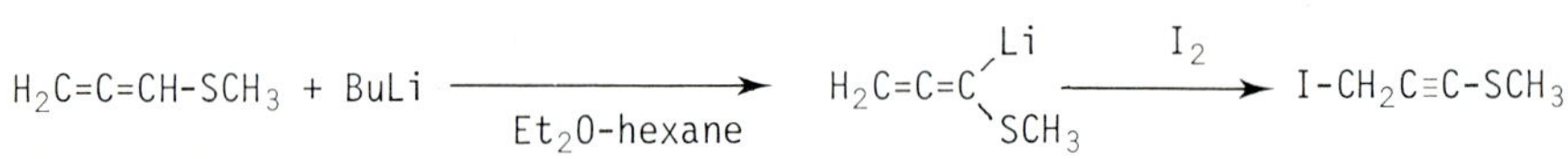

Apparatus: 1-l three-necked flask, see Fig. 1.

To a solution of 0.12 mol of butyllithium in about 75 ml of hexane were added 130 ml of dry diethyl ether with cooling below 0°C. The mixture was cooled to -40°C and 0.12 mol of methylthioallene (see Chapter IV, Exp. 23) was added in one portion. The temperature was kept between -20 and -40°C. After an additional 10 min a solution of 0.10 mol of iodine in 200 ml of diethyl ether was added dropwise, while maintaining the temperature between -45 and -35°C. The addition, which took about 15 min, was stopped as soon as the brown colour of the iodine persisted; usually more than 90% of the iodine can be added before this will occur (note 1). To the light-brown solution was added a solution of 5 g of sodium thiosulfate in 200 ml of water. After vigorous stirring the layers were separated. The organic layer was dried over magnesium sulfate and then concentrated in a water-pump vacuum. Distillation of the residue in a high vacuum gave 1-methylthio-3-iodo-1--propyne, b.p. *ca.* 70°C/0.5 mmHg, n_D^{20} 1.6645, in 72% yield.

Note 1. Addition of more iodine gives rise to vigorous decomposition during the distillation.

(*37*) *3-Cyclopentyl-1-propyne*[103]

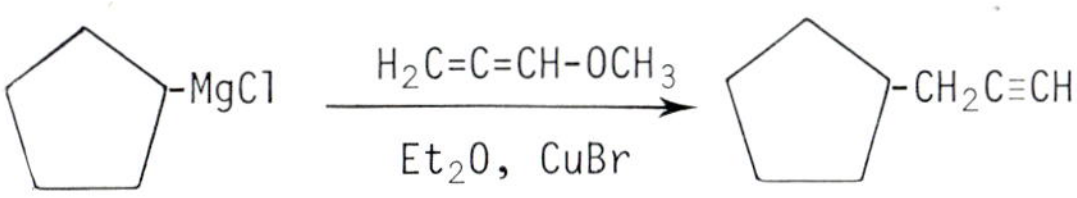

Apparatus: 1-l three-necked, round-bottomed flask with dropping funnel, combined with a gas inlet, mechanical stirrer and a thermometer, combined with a gas outlet.

A solution of cyclopentylmagnesium chloride in 400 ml of diethyl ether, prepared from 0.62 mol of cyclopentyl chloride (see Chapter II, Exp. 2) was added in 30 min to a mixture of 0.40 mol of freshly distilled methoxyallene (see Chapter IV, Exp. 4), 150 ml of diethyl ether and 1 g of finely powdered copper(I) bromide. During this addition the temperature of the reaction mixture was kept between -5 and +5°C by cooling in a bath of dry-ice and acetone. A white suspension was formed. After the addition the cooling bath was removed and stirring was continued for a further 45 min, then the reaction mixture was poured cautiously into 400 ml of ice-water (some cooling may be necessary). After dissolution of the solid material a small amount of 4 N HCl was added, so that the layers became clear. The aqueous layer was extracted three times with small portions of diethyl ether. The combined extracts were washed with concentrated ammonium chloride solution and subsequently dried over magnesium sulfate. The greater part of the diethyl ether was distilled off at normal pressure through a 40-cm Vigreux column. The remaining liquid was distilled and collected in a single receiver, cooled at 0°C. The desired product, b.p. 30°C/20 mmHg, n_D^{20} 1.4502, was obtained in 76-82% yield.

(38) Benzylacetylene[103]

$$PhMgBr + H_2C{=}C{=}CHOCH_3 \xrightarrow[CuBr]{Et_2O} PhCH_2C{\equiv}CH$$

This preparation was very similar to that of phenylallene (see this chapter, Exp. 4). For the preparation of methoxyallene see Chapter IV, Exp. 4. The yield of benzylacetylene, b.p. 65°C/17 mmHg, $n_D^{20.5}$ 1.5278, was 80%.

(39) 3-Butyl-1-heptyne[8]

$$C_4H_9CH{=}C{=}CH\text{-}OC_2H_5 + C_4H_9MgBr \xrightarrow[Et_2O]{CuBr} C_4H_9\text{-}C(C_4H_9)(H)\text{-}C{\equiv}CH$$

Apparatus: 1-l flask, see Fig. 1.

To a mixture of 0.10 mol of 1-ethoxy-1,2-heptadiene (see this chapter, Exp. 13) and 120 ml of diethyl ether was added 1 g of copper(I) bromide. A solution of butylmagnesium bromide in about 200 ml of diethyl ether, prepared from 0.25 mol of butyl bromide (see Chapter II, Exp. 5) was added in 15 min. The reaction was weakly exothermic and the temperature rose slowly to about 32°C. The mixture was held for an additional 40 min at that temperature, then the black reaction mixture was cautiously poured on to a mixture of 300 g of finely crushed ice, 3 g of NaCN (or KCN) and 30 g of ammonium chloride. After vigorous shaking and melting of the ice, the upper layer (almost colourless) was separated off. The aqueous layer was extracted four times with diethyl ether. The combined solutions were dried over magnesium sulfate and the diethyl ether was removed by evaporation in a water--pump vacuum. Distillation of the residue through a 40-cm Vigreux column afforded the acetylene, b.p. 60°C/17 mmHg, n_D^{27} 1.4268, in 85% yield.

(40) Conjugated enynes from cumulenic ethers[104]

$$R^1R^2C{=}C{=}C{=}CHOCH_3 + R^3MgCl \xrightarrow[CuBr]{Et_2O} R^1R^2C{=}C(R^3)\text{-}C{\equiv}CH$$

Apparatus: 1-l flask, see Fig. 1.

After the air in the flask had been replaced completely with nitrogen, 100 ml of dry diethyl ether, 0.20 mol of the cumulenic ether (see Chapter V, Exps. 7, 8 and 11) and 1 g (note 1) of copper(I) bromide were placed in it. A solution of the Grignard-reagent, prepared from 0.50 mol of the chloride (see Chapter II, Exp. 3, note 2) in 250 ml of diethyl ether was added in 30 min, while keeping the temperature between 5 and 10°C, then the cooling bath was removed. The temperature was allowed to rise, but was kept below 30°C by occasional cooling. Stirring at 25-30°C was continued for an additional 15 min. The reaction mixture was cautiously poured into a stirred solution of 50 g of ammonium chloride in 300 ml of 1 N hydrochloric acid, cooled at -10°C. The black salt slurry which remained in the flask was dissolved by addition of 50 ml of this mixture. After separation of the layers the aqueous layer was extracted three times with small portions of diethyl ether. The combined organic solutions were washed twice with 50 ml of saturated ammonium chloride solution and subsequently dried over magnesium sulfate. In the case of volatile enynes (b.p. $<$ 140°C) the greater part of the diethyl ether was distilled off at normal pressure through a 40-cm Widmer column, in other cases the diethyl ether was removed by evaporation in a water-pump vacuum. The remaining liquids were distilled through a 30-40-cm Vigreux column, either at normal pressure (under nitrogen) or in a water-pump vacuum.

The physical constants of the products were as follows:

$R^1 = R^2 = H$, $R^3 = n\text{-}C_4H_9$: b.p. 40°C/40-45 mmHg, n_D^{20} 1.4322, yield 76%;

$R^1 = R^2 = H$, R^3 = cyclohexyl: b.p. 52°C/15 mmHg, n_D^{20} 1.4800, yield 84%;

$R^1 = R^2 = CH_3$, $R^3 = C_2H_5$: b.p. 93°C/760 mmHg, n_D^{20} 1.4572, yield 71%.

More examples can be found in Ref. 104.

Note 1. If more CuBr is used, the yields of the enyne are lower and more residue [consisting of the adducts $R^1R^2C{=}C(R^3)CH{=}CHOCH_3$] remains after the distillations.

Note 2. A large excess of R^3MgCl is required, as the subsequent metallation of the enyne is rapid. In the case of $R^3 = C_2H_5$ the Grignard reagent was prepared from ethyl *bromide*.

(41) 2-Methyl-1-ethoxy-1-buten-3-yne[104]

$$C_2H_5O\text{-}CH_2C{\equiv}C\text{-}CH(OC_2H_5)_2 \xrightarrow[(2)\ H_2O]{(1)\ 2\ C_2H_5Li} C_2H_5O\text{-}CH{=}C{=}C{=}CHOC_2H_5 \longrightarrow$$

$$\xrightarrow[Et_2O,\ CuBr]{CH_3MgBr} C_2H_5O\text{-}CH{=}\overset{\overset{CH_3}{|}}{C}\text{-}C{\equiv}CH$$

Apparatus: 1-l flask, see Fig. 1.

Starting from 0.25 mol of the acetal, an ethereal solution of 1,4-diethoxy--1,2,3-butatriene was prepared as described in Chapter V, Exp. 9. The solution (volume 400-500 ml) was decanted from the $MgSO_4$ after drying and transferred into the reaction flask. Copper(I) bromide (1 g) was added and the solution was cooled to about $0^{o}C$. A solution of about 0.5 mol of methylmagnesium bromide (prepared from 0.56 mol of methyl bromide and a 100% excess of magnesium at $15^{o}C$, see Chapter II, Exp. 5) in 300 ml of diethyl ether was added in 20 min from the dropping funnel, while keeping the temperature of the reaction mixture between 5 and $10^{o}C$. The cooling bath was removed and stirring was continued for a further 1.5 h. The reaction mixture was then poured into a solution of 50 g of ammonium chloride and 4 g of NaCN or KCN in 200 ml of water. After vigorous shaking, the layers were separated. The aqueous layer was extracted three times with small amounts of diethyl ether, then the combined ethereal solutions were dried over magnesium sulfate and the greater part of the diethyl ether was distilled off through a 40-cm Vigreux column at normal pressure. The residue was distilled through the same column, giving the enyne ether (one isomer), b.p. $48^{o}C/15$ mmHg, n_D^{20} 1.4648, in 60-67% yields.

(42) 2,3-Butadien-1-ol[119]

$$ClCH_2C{\equiv}C\text{-}CH_2OH \xrightarrow[Et_2O]{LiAlH_4} H_2C{=}C{=}CH\text{-}CH_2OH$$

Apparatus: 1-l flask, provided with a dropping funnel, combined with a gas inlet, a mechanical stirrer and an efficient reflux condenser.

To a solution of 0.20 mol of 4-chloro-2-butyn-1-ol (see Chapter III, Exp. 42) in 150 ml of dry diethyl ether was added a solution of 0.22 mol of lithium alanate in 250 ml of diethyl ether. The addition was performed at a rate such that the diethyl ether gently refluxed. A thick white suspension was formed. The mixture was warmed for an additional 30 min under reflux and was subsequently cooled by complete immersion of the flask in a bath of ice and ice-water. Ice-water (about 100 ml) was added dropwise with vigorous stirring until the refluxing of the diethyl ether (in spite of cooling) had ceased. The ethereal layer was decanted and the white slurry was extracted ten times with small portions of diethylether, which were decanted after each extraction. The combined extracts were well dried over magnesium sulfate, after which the greater part of the diethyl ether was distilled off at normal pressure through a 40-cm Vigreux column. Distillation of the residue (using a single receiver, cooled at $0^{o}C$) gave the allenic alcohol, b.p. $38^{o}C$/12 mmHg, n_D^{20} 1.4750, in 78% yield.

In a similar way were prepared: $H_2C{=}C{=}CH{-}CH(CH_3)OH$, b.p. $65^{o}C$/50 mmHg, n_D^{20} 1.4620, in 70% yield from $ClCH_2C{\equiv}C{-}CH(CH_3)OH$ (for the preparation of this compound from $LiC{\equiv}CCH_2Cl$ and acetaldehyde see Ref. 1) and $H_2C{=}C{=}CH{-}CH(Ph)OH$, b.p. $100^{o}C$/2 mmHg, n_D^{20} 1.5680, in 85% yield from $ClCH_2C{\equiv}C{-}CH(Ph)OH$ (see also Ref. 1).

(*43*) *4-Methyl-2,3-pentadien-1-ol*[120]

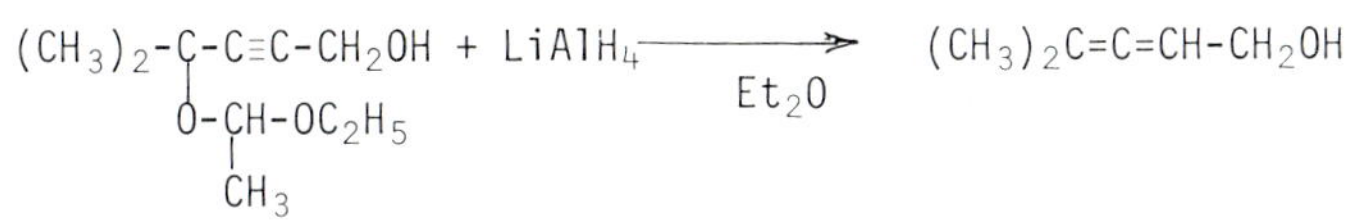

Apparatus: 1-l round-bottomed flask, provided with a dropping funnel, combined with a gas inlet, a mechanical stirrer and a reflux condenser.

To a solution of 0.24 mol of lithium alanate in 500 ml of diethyl ether was added 0.20 mol of the acetylenic alcohol (note 1) at a rate such that gentle refluxing of the diethyl ether was maintained. After the addition the mixture was warmed under reflux for an additional 1 h. It was then cooled to $0^{o}C$ and subsequently poured on to 400 g of finely crushed ice. After the remaining ice had melted the layers were separated (note 2). The aqueous layer was extracted several times with diethyl ether. The combined ethereal solutions were dried over magnesium sulfate and subsequently concentrated in a water-pump vacuum. Distillation of the residue through a 30-cm Vigreux column gave the allenic alcohol, b.p. $76^{o}C$/32 mmHg, n_D^{20} 1.4767, in 77% yield.

In a similar way were prepared: $(CH_3)_2C{=}C{=}CH{-}CH(CH_3)OH$, b.p. 70°C/25 mmHg, n_D^{20} 1.4610, in 72% yield and $=C{=}CH{-}CH_2OH$, b.p. 110°C/18 mmHg, n_D^{20} 1.5135, in 70% yield.

Note 1. Prepared by converting $HC{\equiv}C{-}C(CH_3)_2O{-}CH(CH_3)OC_2H_5$ into its lithium derivative by reaction with ethyllithium in diethyl ether (prepared from C_2H_5Br and Li: see Chapter II, Exp. 1), subsequently adding the required amount of dry paraformaldehyde and heating the mixture for 2 h under reflux[1]. The acetal was prepared from $HC{\equiv}C{-}C(CH_3)_2OH$ and excess of $H_2C{=}CHOC_2H_5$ in the presence of a small amount of *p*-toluenesulfonicacid[1].

Note 2. The separation of the layers may give some difficulties, owing to the presence of aluminium hydroxide. Too vigorous shaking during the extraction procedure should be avoided. The best way to separate the layers is to run off as much of the aqueous layer as possible and subsequently decant the ethereal layer.

(44) 1,2-Butadiene[75]

$$HC{\equiv}C{-}CH(CH_3)Cl \xrightarrow[C_2H_5OH]{Zn/Cu} CH_3{-}CH{=}C{=}CH_2$$

Apparatus: 1-l three-necked, round-bottomed flask, provided with a dropping funnel, combined with a thermometer, reaching into the liquid, a mechanical stirrer and a 40-cm Widmer column, connected with a condenser and a receiver, cooled at -20°C or lower.

Freshly prepared Zn/Cu couple from 70 g of zinc, and 130 ml of 100% ethanol were placed in the flask, stirring was started and 0.2 mol of the acetylenic chloride* (see VIII-2, Exp. 3) was added. When the Zn/Cu couple was of good quality (see Chapter II, Exp. 18) the temperature began to rise after a few minutes, and methylallene began to distil. The remaining 0.5 mol of the chloride was added dropwise over a period of about 30 min. The temperature of the reaction mixture was kept between 65 and 70°C, and the temperature in the head of the distillation column below 45°C by occasional cooling or heating. After the addition the temperature of the reaction mixture was increased gradually. Heating and stirring were stopped when ethanol began to distil. The distillate was redistilled carefully through the same column, using a cold receiver. The allene distilled between 20 and 35°C (b.p. *ca.* 20°C/760 mmHg) and was obtained in 82-86% yield (note 1).

Note 1. Some batches of zinc powder gave yields of only about 60%. The reaction then proceeded much more slowly, and heating (at 65-70°C) had to be continued for 2-3 h in order to achieve complete conversion.

In a similar way $(CH_3)_2C=C=CH_2$, b.p. 41°C, n_D^{20} 1.4143, in 72% yield, was prepared from $HC\equiv C-C(CH_3)_2Cl$ (see Chapter VIII-2, Exp. 5).

* We are indebted to BASF (Arnhem, The Netherlands) for kindly supplying this compound.

(45) *Vinylidenecyclohexane*[75]

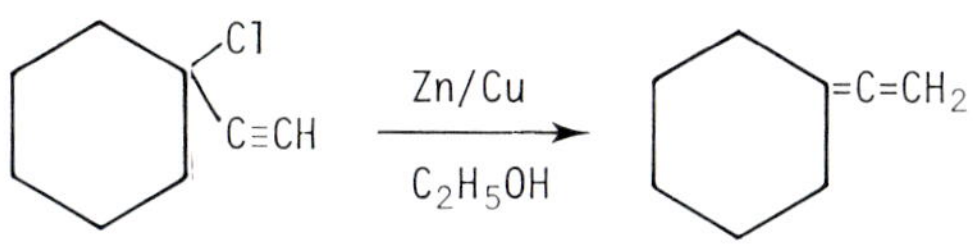

Apparatus: 1-l flask, see Fig. 1.

To a suspension of a zinc-copper couple in 150 ml of 100% ethanol, prepared from 80 g of zinc powder (see Chapter II, Exp. 18), was added at room temperature 0.10 mol of the acetylenic chloride (see Chapter VIII-2, Exp. 7). After a few minutes an exothermic reaction started and the temperature rose to 45-50°C (note 1). When this reaction had subsided, the mixture was cooled to 35-40°C and 0.40 mol of the chloride was added over a period of 15 min, while maintaining the temperature around 40°C (occasional cooling). After the addition stirring was continued for 30 min at 65°C, then the mixture was cooled to room temperature and the upper layer was decanted off. The black slurry of zinc was rinsed five times with 50-ml portions of diethyl ether. The alcoholic solution and the extracts were combined and washed three times with 100-ml portions of 2 N HCl, saturated with ammonium chloride. After drying over magnesium sulfate the greater part of the diethyl ether was distilled off at normal pressure through a 40-cm Vigreux column. The remaining liquid was distilled at 15 mmHg through the same column. The (single) receiver was cooled in an ice-bath. The allene, b.p. 32°C/15 mmHg, n_D^{20} 1.4900, was obtained in 86% yield

Note 1. A prompt start occurs when the Zn/Cu couple is of good quality. If the reaction does not start at room temperature, the mixture should be warmed stepwise (first to 25°C, then to 30°C, etc.) until a further rise of the temperature is observed.

(*46*) *1,2,4-Hexatriene*[75]

$$\mathrm{HC{\equiv}C\text{-}CH{=}CH\text{-}\underset{}{\overset{CH_3}{\overset{|}{C}}}H\text{-}Br \xrightarrow[\text{hexanol}]{\text{Zn/Cu}} H_2C{=}C{=}CH\text{-}CH{=}CH\text{-}CH_3}$$

Apparatus: 1-l three-necked flask, provided with an evacuatable dropping funnel, a gas-tight mechanical stirrer and a 45-cm Vigreuxcolumn. The column was connected to a condenser and a single receiver, cooled at -75^{0}C. Between the receiver and the water pump was placed a tube filled with KOH pellets.

A Zn/Cu couple was prepared from 70 g of zinc powder (see Chapter II, Exp. 18). The black slurry was transferred into the reaction flask . After the greater part of the absolute ethanol had been poured off from the zinc, the zinc was rinsed at least ten times with small portions of dry diethyl ether. The diethyl ether was then decanted off, 100 ml of hexanol were added and the flask was connected to the other parts of the distillation apparatus. The diethyl ether and traces of ethanol were subsequently removed by evacuating the apparatus (the receiver being cooled at -75^{0}C) and heating the reaction flask. This operation was stopped when about 10 ml of hexanol had passed over. The receiver and condenser were cleaned and the apparatus again evacuated (10-15 mmHg). Stirring was started and the flask heated until the hexanol started to reflux in the lower part of the column. From the dropping funnel was added in 20 min 0.20 mol of the bromide (see Exp. 48). The reaction was very vigorous and external heating was not necessary. A mixture of hexatriene and hexanol condensed in the receiver (note 1). The conversion was terminated by external heating, so that 5-10 ml of hexanol distilled at 55-60^{0}C/ /15 mmHg. The contents of the receiver were "redistilled", using the apparatus shown in Fig. 5, collecting the vapour of the hexatriene in a receiver cooled at -75^{0}C. The yield of this hydrocarbon was 78-84% (n_D^{20} 1.4870) (with comparable amounts of the *cis*- and *trans*-isomer). The compound can be distilled at normal pressure (b.p. *ca.* 78^{0}C/760 mmHg) but some polymerization occurs (note 2).

Note 1. If the temperature in the top of the column rises above 50^{0}C, the addition should be interrupted.

Note 2. We have also carried out this synthesis in ethanol as a solvent but the results were not reproducible. Although a series of experiments with zinc powder from one flask gave reasonable results (60-78% yields), a new flask with the same batch number gave low yields of impure products. The main impurity was probably the non-conjugated diene, $H_2C{=}CH\text{-}CH_2\text{-}CH{=}CH\text{-}CH_3$, possibly resulting from reduction of the 1,2,4-triene by the zinc. The advantage of using hexanol is that the triene can be removed directly from the reaction mixture, so that no further reduction can occur.

(47) *Methyl propargyl ketone*[8]

$$(C_4H_9)_3Sn\text{-}CH{=}C{=}CH_2 + CH_3\text{-}\overset{O}{\overset{\|}{C}}\text{-}Cl \xrightarrow{ZnCl_2} HC{\equiv}C\text{-}CH_2\text{-}\overset{O}{\overset{\|}{C}}\text{-}CH_3 + (C_4H_9)_3SnCl$$

Apparatus: 250-ml flask, see Fig. 1.

In the flask were placed 0.20 mol of tributylallenyltin (see Chapter III, Exp. 4) and 0.20 mol of freshly distilled acetyl chloride. The mixture was cooled to -15°C and 300 mg of powdered anhydrous zinc chloride were added. After stirring for 30 min at -10 to -15°C, the cooling bath was removed and the temperature was allowed to rise gradually in 1.5 h to 20°C (occasional cooling may be necessary). The flask was then equipped for vacuum distillation: two stoppers were placed on the outer necks and a 30-cm Vigreux column on the middle neck. This column was connected with a condenser and a receiver, cooled at -15°C. Between the receiver and the water pump was placed a tube filled with anhydrous calcium chloride. The apparatus was evacuated at 10-15 mmHg and the flask gradually heated until the temperature in the top of the column had risen to about 60°C. A dark residue (tributyltin chloride) remained in the distillation flask. The contents of the receiver were redistilled in a partial vacuum of 40 mmHg. Methyl propargyl ketone, b.p. *ca.* 50°C, n_D^{22} 1.4317, was obtained in 71% yield.

(48) *5-Chloro-3-hexen-1-yne and 5-bromo-3-hexen-1-yne*[171]

$$HC{\equiv}C\text{-}\underset{OH}{\underset{|}{\overset{H}{\overset{|}{C}}}}\text{-}CH{=}CHCH_3 + HX \longrightarrow HC{\equiv}C\text{-}CH{=}CH\text{-}\underset{X}{\underset{|}{\overset{H}{\overset{|}{C}}}}\text{-}CH_3 + H_2O$$

(X = Cl, Br)

Apparatus: 1-l three-necked flask with dropping funnel, stirrer and thermometer.

To 250 ml of concentrated HCl (36%) or HBr (note 1) was added in 30 min with vigorous stirring 0.5 mol of the acetylenic alcohol (see III, Exp. 58). In the case of HCl the temperature was maintained at about 20°C; the reaction with HBr was carried out at 5-10°C. The reactions were slightly exothermic. After an additional 30 min (maintaining the temperature indicated) 250 ml of ice water were added. Six extractions with diethyl ether were carried out (note 2). The extracts were washed with water and dried over magnesium sulfate . The greater part of the diethyl ether was then distilled off through a 40-cm Widmer column, keeping the bath temperature below 80°C. The remaining liquid was distilled through the same column. $HC{\equiv}CCH{=}CH{-}CH(Cl)CH_3$ (*cis-trans* ~ 30:70), b.p. 55-60°C/50 mmHg, n_D^{20} 1.4829, was obtained in 72% yield. $HC{\equiv}CCH{=}CH{-}CH(Br)CH_3$ (*cis-trans* ~30:70), b.p. 44-50°C/10 mmHg, n_D^{20} 1.5242, was obtained in 96% yield.

Note 1. The required strength of the HBr solution was obtained as follows: 250 ml of 48% HBr were vigorously stirred with 50 g of PBr_3 at 40°C, until the underlayer had disappeared.

Note 2. A sufficient amount of diethyl ether should be used for the first extraction of the bromide in order to obtain an upper layer.

This method gives poor results with $H_2C{=}CH{-}CH(OH){-}C{\equiv}CH$: a 30:70 mixture of $HC{\equiv}C{-}CH(Br)CH{=}CH_2$ and $HC{\equiv}C{-}CH{=}CH{-}CH_2Br$ was obtained in about 60% yield.

(49) *4,5-Hexadien-2-yn-1-ol*[172]

$$HC{\equiv}C{-}CH_2OH + HC{\equiv}C{-}CH_2Cl \xrightarrow[NH_4OH]{CuCl} H_2C{=}C{=}CH{-}C{\equiv}C{-}CH_2OH + HCl$$

Apparatus: 500-ml four-necked flask with dropping funnel, gas inlet tube (reaching beneath the surface of the liquid), stirrer and thermometer and vent.
Caution: Use rubber gloves during the work-up, the allenyne alcohol probably is a skin irritant.

In the flask were placed 70 ml of methanol, 50 ml of 25% aqueous ammonia, 0.5 mol of propargyl alcohol (distilled in a partial vacuum before use), 1.5 g of CuCl and 2 g of hydroxylamine HCl. The air in the flask was completely replaced by nitrogen by passing a vigorous stream of this gas through the flask for 1-2 min. A mixture of 0.25 mol of propargyl chloride and 40 ml of methanol was added dropwise over a period of 1 h. Nitrogen (50 ml/min) was introduced continuously. The temperature of the mixture was kept between 25 and 30°C by occasional cooling. Fortyfive minutes after the propargyl chloride had been added, a solution of 5 g of NaCN or KCN in 150 ml of water was added. The mixture was extracted ten times with diethyl ether. The combined extracts were washed with a saturated solution of ammonium chloride and dried over magnesium sulfate. The diethyl ether was thoroughly removed in a water-pump vacuum. After termination of this operation nitrogen was admitted. The residue weighed 18.3 g (~77%) and was almost pure $H_2C{=}C{=}CHC{\equiv}CCH_2OH$ (by NMR), n_D^{20} 1.5405. The liquid was mixed with 40 ml of paraffin oil and some boiling stones were added after which the compound was distilled in a high vacuum. The receiver was cooled at -20°C. The yield of distilled product was only 61%. The compound should be stored in the refrigerator.

(*50*) *1-Phenylthio-3-methyl-1,2-butadiene*[180]

$$\mathrm{PhSH} \xrightarrow[\text{hexane-THF}]{\text{BuLi}} \mathrm{PhSLi}$$

$$\mathrm{HC{\equiv}C\text{-}C(CH_3)_2Cl} \xrightarrow[\text{CuBr.LiBr-THF}]{\text{PhSLi}} \mathrm{PhS\text{-}CH{=}C{=}C(CH_3)_2}$$

Apparatus: 1-l flask, see Fig. 1.

To a solution of 0.12 mol of butyllithium in about 75 ml of hexane were added 100 ml of dry THF with cooling below -20°C. Subsequently 0.12 mol of freshly distilled thiophenol (note 1) and a solution of 1.5 g of copper(I) bromide and 3 g of anhydrous lithium bromide in 30 ml of THF were successively introduced, keeping the temperature below 0°C. The brown solution was cooled down to -15°C and 0.10 mol of freshly distilled 3-chloro-3-methyl-1-butyne (Chapter VIII-2, Exp. 5) was added in 10 min. The cooling bath was then removed and the temperature of the solution was allowed to rise. After 30 min the flask was placed in a heating bath and the light-brown solution heated under reflux (the thermometer-gas-outlet combination being replaced with a reflux condenser) for 15 min (note 1). Then the solution was cooled down to 20°C and subsequently poured into a solution of 5 g of potassium cyanide, 5 g of sodium hydroxide and 20 g of ammonium chloride in 150 ml of water. After vigorous shaking the upper layer was separated off and the aqueous layer was extracted twice with diethyl ether. The combined organic solutions were dried over potassium carbonate and then concentrated in a water-pump vacuum. Distillation under about 0.5 mmHg gave the allenic sulfide (b.p. 75-80°C), n_D^{20} 1.592-1.596 in 70% yield. During the distillation a small part of the allenic sulfide dimerised. The product contained 2-6% of the acetylenic isomer $HC{\equiv}C\text{-}C(CH_3)_2SPh$ (note 1).

Note 1. Heating under reflux during a longer period will probably result in a decrease of the amount of the acetylenic isomer, due to isomerisation of this compound into the allenic sulfide under the influence of PhSLi.CuBr-LiBr (compare also Ref. 180).

With propargyl chloride a mixture consisting of about 85% of PhS-CH=C=CH_2 and 15% of HC≡C-CH_2-SPh was obtained, when using *catalytic* amounts of CuBr.

Chapter VII

MISCELLANEOUS SYNTHESES

1. 2,3-SIGMATROPIC REARRANGEMENTS

(1) *Allenyl methyl sulfoxide (methylsulfinyl propadiene)*[129-131]

$$HC{\equiv}CCH_2OH + CH_3SCl \xrightarrow[CH_2Cl_2]{(C_2H_5)_3N} [HC{\equiv}C{-}CH_2{-}O{-}S{-}CH_3] \xrightarrow{[2,3]} H_2C{=}C{=}CH{-}\overset{O}{\overset{\|}{S}}{-}CH_3$$

Apparatus: 1-l flask, see Fig. 1.

To a mixture of 200 ml of dry dichloromethane, 0.22 mol of dry propargyl alcohol and 0.22 mol of dry triethylamine was added with cooling between -95 and -100°C (occasional cooling with liquid nitrogen) a solution of about 0.1 mol of methanesulfenyl chloride (note 1) in 60 ml of dichloromethane. After the addition, which took about 10 min, 10 g of methyl iodide (note 2) were added and the cooling bath was removed. When the temperature had reached -20°C, the salt was filtered off on a sintered-glass funnel (G-2) and was rinsed well with dichloromethane. To the residue, remaining after evaporation of the solvent in a water-pump vacuum were added 100 ml of dry diethyl ether in order to precipitate some remaining salt. Filtration and subsequent removal of the diethyl ether in a water-pump vacuum gave pure allenyl methyl sulfoxide in about 100% yield as an oily liquid n_D^{25} 1.5508.

Note 1. Prepared by adding 0.10 mol of dimethyl disulfide at -30°C to a solution of 0.10 mol of chlorine in dichloromethane and subsequently raising the temperature to 0°C.

Note 2. Traces of unreacted triethylamine might cause partial isomerization of the allenyl sulfoxide into the propargyl sulfoxide. The methyl iodide is added to ensure that no triethylamine remains.

(2) *Allenyl phenyl sulfoxides*[129-131]

$$HC{\equiv}C{-}CH(R)OH + PhSCl \xrightarrow[CH_2Cl_2]{(C_2H_5)_3N} HC{\equiv}C{-}CH(R)O{-}S{-}Ph \xrightarrow{[2,3]} RCH{=}C{=}CH{-}\overset{O}{\overset{\|}{S}}{-}Ph$$

$R = H, CH_3$

Apparatus: 500-ml flask, see Fig. 1.

In the flask were placed 0.22 mol of the acetylenic alcohol, 0.22 mol of triethylamine and 200 ml of dichloromethane (note 1). To this mixture was added in 10 min with cooling between -90 and -100°C (occasional cooling with liquid nitrogen) a solution of 0.20 mol of PhSCl in 75 ml of dichloromethane (note 2). Five minutes after this addition 10 g of methyl iodide were added (note 3) and the cooling bath was removed. When the temperature had reached 10-15°C, the mixture was poured into 200 ml of water to which 2 ml of 36% HCl had been added. After vigorous shaking the lower layer was separated off. The aqueous layer was extracted with 50 ml of dichloromethane. The combined solutions were washed with water and dried over magnesium sulfate. The oily residues remaining after evaporation of the solvent in a water-pump vacuum, appeared to be reasonably pure sulfoxides; yields were almost quantitative. Small amounts of diphenyl disulfide, which were sometimes present, could be removed by shaking the oils with 50 ml of pentane. The disulfide was extracted in this way. The n_D^{20} of PhS(=O)CH=C=CH_2, obtained after this purification was 1.6180, the n_D^{20} of PhS(=O)CH=C=CH-CH_3 was 1.5993.

Note 1. All reagents and solvents must be thoroughly dry.

Note 2. Prepared by adding at -20°C a cold (-30°C) solution of 7.2 g of chlorine in 20 ml of CCl_4 to 0.10 mol of diphenyl disulfide in CH_2Cl_2. After the addition, which was carried out in 15 min, the temperature was allowed to rise to 0°C.

Note 3. See Exp. 1, note 2.

(3) *3-Methyl-1,2-butadienyl methyl sulfone*[129-131]

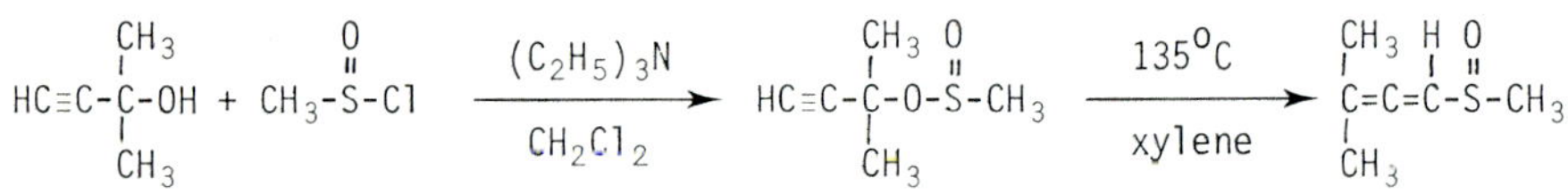

Apparatus: 500-ml flask (see Fig. 1) for the preparation of the sulfinate; 250-ml two-necked, round-bottomed flask with a reflux condenser and a thermometer for the 2,3-sigmatropic rearrangement.

To a mixture of 0.10 mol of the dry acetylenic alcohol (commercially available, see also Ref. 1), 0.12 mol of dry triethylamine and 150 ml of dry dichloromethane 0.12 mol of methanesulfinyl chloride[73] was added in 10 min with cooling at -50°C. The cooling bath was removed and after 15 min a mixture of 3 ml of 36% HCl and

200 ml of water was added with vigorous stirring. The lower layer was separated off, washed with water and dried over magnesium sulfate. The residue remaining after evaporation of the solvent in a water-pump vacuum was dissolved in 80 ml of xylene. The solution was heated under reflux for 15 min, then the xylene was distilled off at 10-20 mmHg. The residue was warmed for a few minutes with 75 ml of dry diethyl ether. After cooling to 20°C, the brown solution was decanted from the small amount of insoluble material. Evaporation of the diethyl ether in a water-pump vacuum gave the reasonably pure sulfone as a brown oil, in a yield of about 85%.

Note 1. The sulfinate of propargyl alcohol did not rearrange upon heating at 130-140°C in xylene.

(*4*) *Allenyldiphenylphosphine oxide*[132]

$$HC{\equiv}CCH_2OH + Ph_2PCl \xrightarrow[CH_2Cl_2]{(C_2H_5)_3N} HC{\equiv}CCH_2O\text{-}P(Ph)_2 \xrightarrow{[2,3]} H_2C{=}C{=}CH\text{-}\overset{O}{\overset{\|}{P}}Ph_2$$

Apparatus: 1-l flask, see Fig. 1.

To a mixture of 100 ml of dry dichloromethane, 0.10 mol of propargyl alcohol and 0.11 mol of triethylamine was added a solution of 0.05 mol of Ph_2PCl in 75 ml of dichloromethane in 3 min between -80 and -90°C. The cooling bath was removed, and when the temperature had reached 10°C, the reaction mixture was poured into a solution of 2.5 ml of 36% HCl in 100 ml of water. After vigorous shaking the lower layer was separated and the aqueous layer was extracted twice with 25-ml portions of dichloromethane. The combined solutions were washed twice with water, dried over magnesium sulfate and then concentrated in a water-pump vacuum, giving almost pure allenyl phosphine oxide as a white solid, m.p. 98-100°C, in almost 100% yield.

2. 3,3-SIGMATROPIC REARRANGEMENTS

(1) *2-Methyl-3,4-pentadiene-NN-diethylamide*[133*]

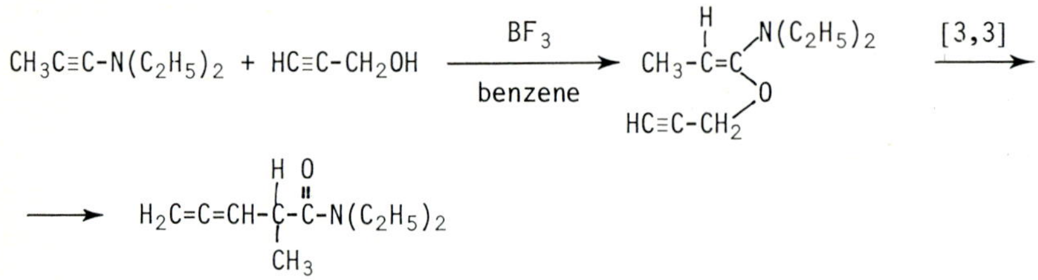

Apparatus: 250-ml one-necked, round-bottomed flask and thermometer.

To a mixture of 65 ml of dry benzene and 0.10 mol of freshly distilled NN-diethylamino-1-propyne[1] were added 3 drops of BF_3.ether and 0.12 mol of dry propargyl alcohol was added to the reddish solution in 5 min. The temperature rose in 5-10 min to about 45°C, remained at this level for about 10 min and then began to drop. The mixture was warmed to 60°C, whereupon the exothermic reaction made the temperature rise in a few minutes to 85°C. This level was maintained by occasional cooling. After the exothermic reaction (3,3-sigmatropic rearrangement) had subsided, the mixture was heated for an additional 10 min at 80°C and the benzene was then removed in a water-pump vacuum. The red residue was practically pure acid amide (NMR). Distillation through a 10-cm Vigreux column gave the pure compound, b.p. *ca.* 70°C/0.5 mmHg, n_D^{20} 1.4807, in 84% yield. There remained 2 g of brown residue.

*We are indebted to Prof. J. Ficini (Paris, France) for detailed instructions.

(2) *6-Methylhepta-4,5-dien-2-one*[134]

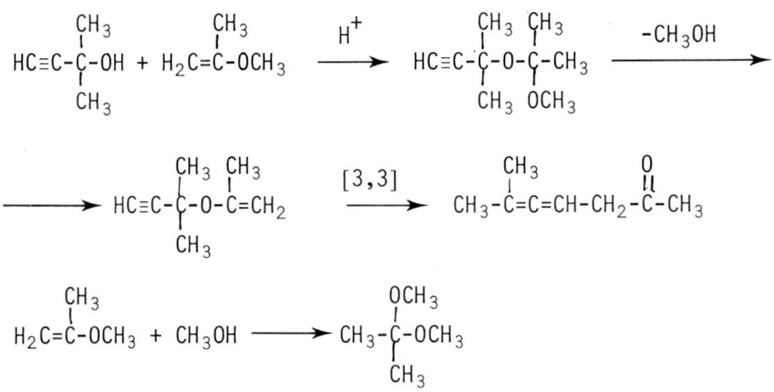

Apparatus: 500-ml three-necked round-bottomed flask, with a dropping funnel, a mechanical stirrer and a thermometer, combined with a vent, for the addition to the isopropenyl ether; for the further reaction the dropping funnel was replaced with a thermometer and the combination of thermometer and vent with a reflux condenser.

In the flask were placed 1.50 mol of the isopropenyl ether (commercially available) and 0.10 mol of the (dry) acetylenic alcohol[1] (also commercially available), and in the dropping funnel 0.4 mol of the latter compound. The mixture was cooled to $0^{\circ}C$ and 100 mg of anhydrous *p*-toluenesulfonic acid were added with stirring. An exothermic reaction started immediately and the temperature rose by several degrees. When this reaction had subsided, the mixture was cooled to about $10^{\circ}C$ and the acetylenic alcohol was added from the dropping funnel in 15 min, whilst keeping the temperature between 10 and $20^{\circ}C$. The apparatus was then modified as indicated above and the solution was heated under reflux. In order to maintain refluxing the bath temperature had to be increased gradually. After 2 h the temperature in the boiling liquid had become constant (about $95^{\circ}C$). The brown solution was cooled, then the volatile components (mainly the dimethylacetal of acetone) were removed in a water-pump vacuum by means of a rotary evaporator. Subsequent distillation of the remaining liquid through a 30-cm Vigreux column gave the allenic ketone, b.p. $65^{\circ}C/15$ mmHg, n_D^{20} 1.4695, in 74% yield (note 1). There remained about 8 g of viscous residue.

Note 1. This yield is lower than that reported in the literature. In our procedure no low-boiling light petroleum is used as co-solvent, so that the temperature of the boiling reaction mixture can become considerably higher, which may give rise to the formation of polymeric products and tars. Our reaction time is much shorter than that in the literature. The reaction with $HC{\equiv}CCH_2OH$ and $HC{\equiv}C\text{-}CH(CH_3)OH$ failed.

(3) *Ethyl 3,4-pentadienedithioate*[135]

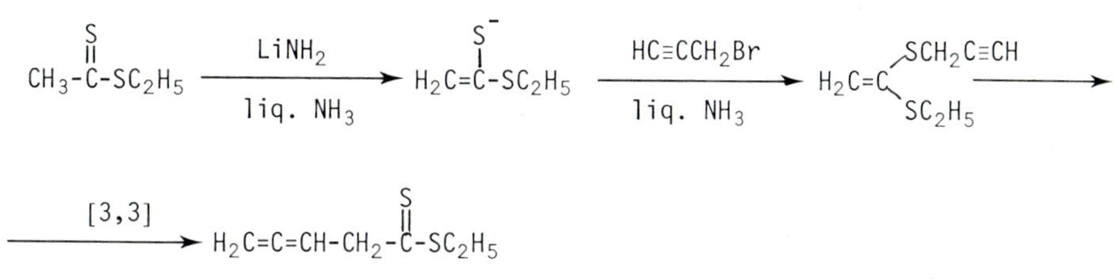

Apparatus: 2-l round-bottomed, three-necked flask with a mechanical stirrer, a stopper and a gas outlet.

Anhydrous liquid ammonia (water content < 0.1%) (400 ml) was transferred into a 2-l wide-necked, round-bottomed flask. A suspension of lithium amide was prepared starting from 0.30 mol of lithium (see Chapter II, Exp. 11). To the white suspension was added 0.31 mol of ethyl dithioacetate (note 1) in 3 min with swirling. Care was taken that the dithioester did not flow along the inner walls, but came directly into the suspension. Prior to preparing this solution of the enethiolate 200 ml of liquid ammonia were placed in the three-necked flask. Directly after the addition of the dithioester the stopper and gas outlet were removed from the three-necked flask and 0.35 mol of propargyl bromide was run in during 5-10 s. Stirring was started and the solution of the enethiolate was cautiously poured into the ammoniacal solution of propargyl bromide (note 2). During this addition, which was carried out in 5 min, the outer necks of the reaction flask were open. The reaction was very vigorous. Two minutes after the addition 5 g of solid ammonium chloride were introduced and stirring was continued for an additional 1 h (note 3). The reaction mixture was then cautiously poured on to 300 g of finely crushed ice in a 3-l beaker and the salt in the flask was rinsed into the beaker with ice-water. The product was immediately extracted 5-7 times with redistilled pentane (note 4). The extracts were kept below $0^{O}C$ (note 5). The combined extracts were washed with 5% acetic acid and subsequently dried over magnesium sulfate (note 6). The extract was concentrated in a water-pump vacuum to about 60 ml by means of the rotary evaporator, care being taken that the bath temperature remained below $25^{O}C$. The remaining pale yellow solution was warmed to about $35^{O}C$ (internal temperature). The temperature rose gradually but was kept at about $45^{O}C$ by occasional cooling. When after about 45 min the exothermic reaction had subsided, the flask was placed in a water-bath at $55^{O}C$. After 30 min the remaining pentane was removed in a water--pump vacuum. The orange residue, n_D^{22} 1.5878, yield *ca.* 92% was almost pure allenic dithioester.

Note 1. Prepared by adding CS_2 to CH_3MgBr in THF at $0^{O}C$, preferably in the presence of a small amount of CuBr (2 mol %), and subsequently alkylating with ethyl iodide[136].

Note 2. Propargyl bromide itself also reacts with ammonia and therefore the interval between the addition of this compound and its conversion with the enethiolate should be kept as short as possible. *Inverse* addition was applied: if propargyl bromide is added to the enthiolate solution, the primary product partly isomerizes into $H_2C{=}C(SC_2H_5)SCH{=}C{=}CH_2$ under the catalytic influence of the enethiolate.

Note 3. During this period the excess of propargyl bromide is converted into propargylamines.

Note 4. In principle diethyl ether can be used, but with pentane a faster separation of the layers is effected.

Note 5. At room temperature the 3,3-sigmatropic rearrangement begins. The ammonia, still present during the work up, will cause ring closure of the allenic dithioester to a 2-*H*-thiopyran derivative.

Note 6. The glassware used for drying and evaporation of the solvent should be rinsed with a dilute solution of acetic acid in diethyl ether or acetone, in order to ensure that no traces of alkali are present or the glass walls, which might cause cyclization of the allenic dithioester.

(*4*) *Ethyl 3,4-pentadienoate*[137]

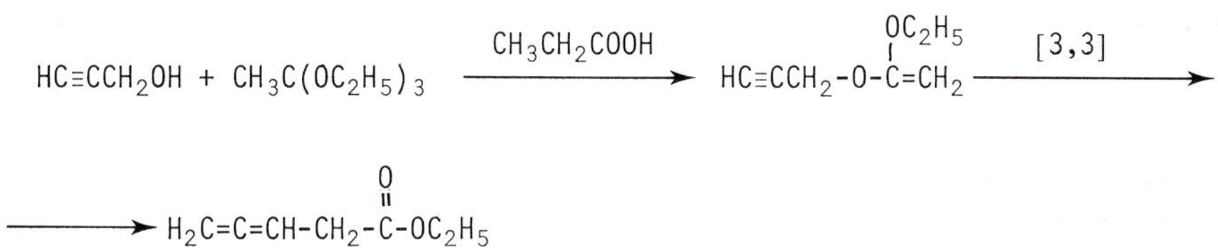

Apparatus: 500-ml three-necked, round-bottomed flask with a thermometer, dipping well into the liquid, a 40-cm Widmer column and a dropping funnel; the column was connected to a condenser and receiver.

A mixture of 0.50 mol of ethyl orthoacetate, 3 g of propionic acid and 0.22 mol of propargyl alcohol was gradually heated in an oil-bath. When about 5 ml of ethanol had passed over (between 75 and 85°C), 0.30 mol of propargyl alcohol was added dropwise over a period of 15 min. When the temperature of the liquid in the distillation flask had reached 150°C, the heating bath was removed and the liquid was allowed to cool to 120°C. The alcohol which had been collected was placed in the dropping funnel and mixed with 2 g of propionic acid, and the flask was heated again. The mixture of alcohol and acid was added gradually in 20 min. After the greater part of the alcohol had been distilled off (about 50 g) the internal temperature could rise to about 155°C. This temperature was maintained for 30 min. The contents of the flask were allowed to cool to 20-25°C. Very careful fractionation gave the allenic ester, b.p. 54°C/17 mmHg, n_D^{20} 1.4492 in 64% yield.

(*5*) *Ethyl 3,4-hexadienoate*[137]

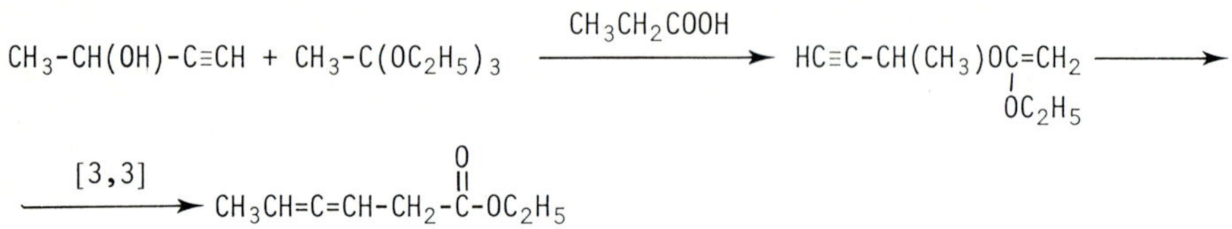

Apparatus: Same as in Exp. 4.

In the flask were placed 100 g of ethyl orthoacetate and 2 ml of propionic acid. From the dropping funnel, which contained 0.45 mol of $HC{\equiv}C{-}CH(CH_3)OH$ (commercially available), were added 10 g of this acetylenic alcohol. The flask was heated in an oil bath. The temperature of the boiling liquid, which initially was 113°C, rose to about 130°C in 20 min, while ethanol distilled off at 76-80°C. The remainder of the acetylenic alcohol was added dropwise in 30 min. After heating for 40 min about 50 g of ethanol had passed over and the temperature of the liquid in the distillation flask had risen to about 150°C. Heating at this temperature was continued for an additional 30 min. After cooling to 30°C the remaining liquid was carefully distilled. After the first fraction, consisting mainly of ethyl orthoacetate had passed over, the allenic ester distilled at 65°C/15 mmHg, n_D^{22} 1.4522, and was obtained in 69% yield.

(*6*) *2,2-Dimethyl-3,4-pentadienal*[159]

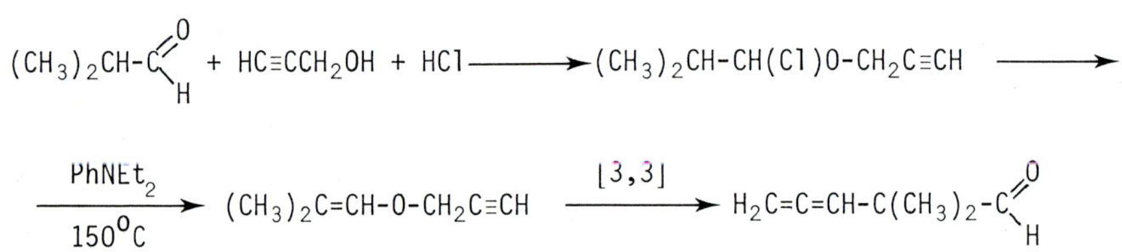

Apparatus: 500-ml round-bottomed, three-necked flask, provided with a gas inlet tube, a mechanical stirrer and a thermometer-vent combination for the preparation of the α-chloroether; 500-ml round-bottomed flask, provided with a gas inlet, a reflux condenser and a thermometer, dipping in the liquid for the elimination of HCl and the 3,3-sigmatropic rearrangement.

Gaseous hydrogen chloride was introduced into a mixture of 0.52 mol of propargyl alcohol and 0.50 mol of freshly distilled isobutyraldehyde, while keeping the temperature of the mixture between -20 and 0°C. The introduction of gas was stopped when copious fumes of HCl escaped from the vent. After standing for 1-2 min at 0°C the clear upper layer was decanted from the small turbid aqueous layer. The latter was extracted (shaking followed by decantation) four times with 20-ml portions of pentane. The main portion and the extracts were dried over a small amount of magnesium sulfate and subsequently the pentane was removed in a water-pump vacuum (rotary evaporator), keeping the bath temperature at about 25°C. The residue (weight about 70 g), crude α-chloroisobutyl propargyl ether, was mixed with 120 g of *NN*-diethylaniline and the mixture was heated at 100°C (internal). A weakly exothermic reaction started, during which the temperature rose gradually. Some cooling (water-bath) was applied in order to keep the temperature between 100 and 105°C. Twenty minutes after the exothermic reaction had ceased, the mixture was heated. The diethylaniline salt of HCl, which had precipitated on the glass wall, dissolved. The reaction mixture was heated under reflux for 10 min (temperature of the reaction mixture 165-170°C, but soon dropping to 150-155°C). The mixture was cooled to 80°C and 300 ml of ice-water were added. After shaking (until the salt had dissolved) and separation of the upper layer, the aqueous layer was extracted once with 30 ml of pentane. The extract and the upper layer were combined and washed five times with water. After drying over magnesium sulfate, the mixture was distilled in a slow stream of nitrogen through a 30-cm Vigreux column. The fraction passing over between 110 and 175°C/760 mmHg was redistilled through the same column, giving the allenic aldehyde, b.p. 132°C, n_D^{24} 1.4532 in 68-76% yield (towards the end of the distillation and after cooling, a partial vacuum was applied to minimize the hold up).

Starting from propionaldehyde the aldehyde $H_2C{=}CH{-}CH{=}C(CH_3)CH{=}O$ was obtained in an impure state and in moderate yield. Its precursor $H_2C{=}C{=}CH{-}CH(CH_3)CH{=}O$ was present in traces only.

(7) *3-Methyl-1,2-butadienyl acetate (1-acetoxy-3-methyl-1,2-butadiene)*[158]

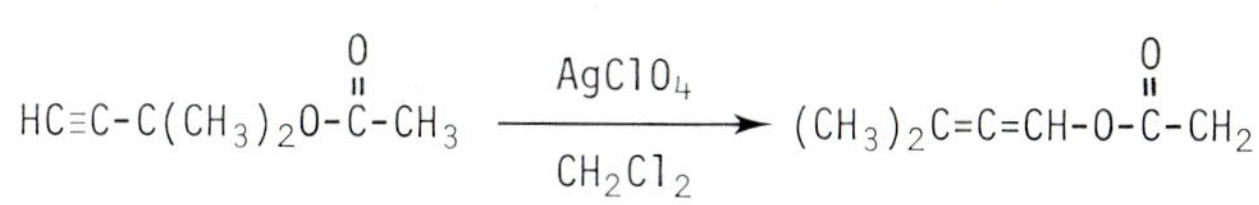

Apparatus: 100-ml round-bottomed flask and thermometer.

To a mixture of 15 g of the acetylenic acetate (see Chapter VIII-3, Exp. 1) and 20 g of dry dichloromethane was added 0.75 g of silver perchlorate at 20°C. The mixture was swirled for 2-3 min; the perchlorate dissolved completely, turbidity appeared, the temperature rose gradually to about 30°C and the solution turned brown to black. After standing for 2 h (note 1) the solution was poured into 100 ml of water to which 4 ml of concentrated ammonia solution had been added. After vigorous shaking the organic layer was separated and combined with two dichloromethane extracts. The combined solutions were dried over magnesium sulfate and subsequently concentrated in a water-pump vacuum. Distillation of the remaining brown liquid through a short Vigreux column gave the almost pure allenic acetate, b.p. 60°C/20 mmHg, n_D^{23} 1.4597 in 57% yield. A considerable amount of viscous residue remained.

Note 1. Longer reaction times gave rise to lower yields and more polymer.

With 1-ethynylcyclohexyl acetate (1-acetoxy-1-ethynylcyclohexane) only about 50% conversion was effected after 2 h. Addition of more $AgClO_4$ gave rise to a vigorous decomposition.

3. ADDITIONS TO UNSATURATED SYSTEMS

(*1*) *3-Pentyn-2-one*[8]

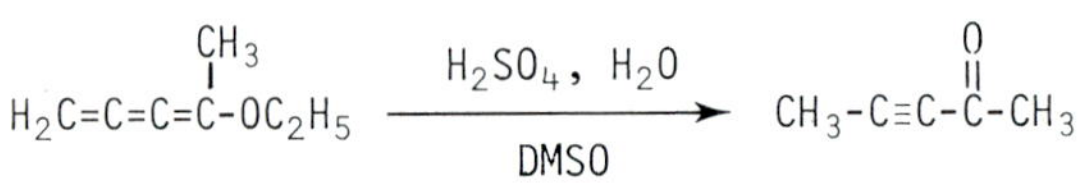

Apparatus: 500-ml flask, see Fig. 1.

A mixture of 12 g of water and 5 g of 96% sulfuric acid was added to 150 ml of DMSO. The solution was cooled to 25°C and 0.20 mol of the cumulenic ether (see Chapter III, Exp. 20) was immediately added. The temperature rose slowly to about 40°C in about 20 min. After stirring for an additional 30 min at 45°C, the mixture was poured into 300 ml of a solution of 100 g of ammonium chloride and 8 g of sodium hydrogen carbonate. After swirling and subsequent shaking, the reaction product was extracted at least 15 times with small portions of diethyl ether. The extracts were dried (without washing) over magnesium sulfate. The greater part of the diethyl ether was distilled off at normal pressure through a 40-cm Vigreux column. Distillation of the remaining liquid at 10-15 mmHg, using a single receiver cooled in an ice-bath, gave the acetylenic ketone, b.p. *ca.* 20-22°C/15 mmHg, in 68% yield.

(2) *Ethyl 4-hexynoate*[138]

$$CH_3C{\equiv}CCH_2C{\equiv}C\text{-}OC_2H_5 \xrightarrow[\text{DMSO}]{H_2O,\ H^+} CH_3C{\equiv}C\text{-}CH_2CH_2\text{-}\overset{O}{\overset{\|}{C}}\text{-}OC_2H_5$$

Apparatus: 250-ml round-bottomed three-necked flask, provided with a dropping funnel, a mechanical stirrer and a thermometer, combined with a vent.

To a mixture of 25 ml of water and 3 ml of 96% sulfuric acid were added 40 ml of DMSO. The mixture was cooled to $10^{o}C$ and 0.20 mol of 1-ethoxy-1,4-hexadiyne (see Chapter III, Exp. 51) was added with vigorous stirring in 15 min. During this addition, which was exothermic, the temperature of the mixture was kept between 20 and $25^{o}C$. After the addition stirring was continued for 30 min at $35^{o}C$, then 150 ml of water were added and six extractions with diethyl ether were carried out. The combined extracts were washed with water and dried over magnesium sulfate. Evaporation of the solvent in a water-pump vacuum, followed by distillation through a 25-cm Vigreux column gave the acetylenic ester, b.p. $72^{o}C/15$ mmHg, n_D^{20} 1.4416, in 89% yield.

(3) *2-Ethynylselenophen*[8]

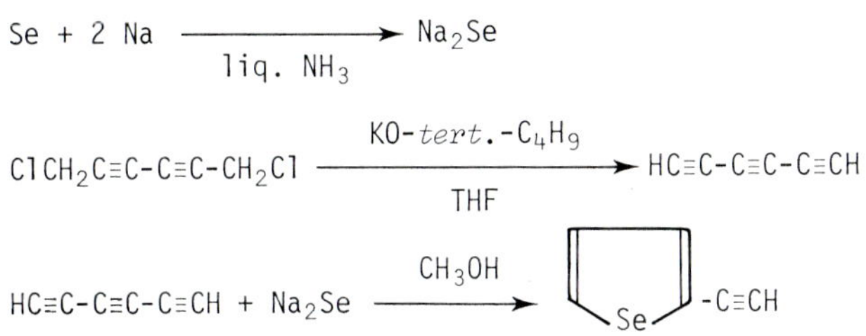

Apparatus: 1-l flask (see Fig. 2) for the preparation of Na_2Se and its addition to triacetylene; 1-l flask (see Fig. 1) for the preparation of HC≡C-C≡C-C≡CH.

A solution of 4.8 g of sodium in 300 ml of anhydrous liquid ammonia (note 1) was prepared by introducing the sodium, cut into 1-g pieces, into the ammonia. When the sodium had dissolved (after 10 min), 0.12 mol of selenium was added in 0.5-g portions over a period of 10 min with vigorous stirring. A dropping funnel, containing 250 ml of methanol, combined with a gas inlet was then placed on the neck through which the selenium powder had been added. The methanol was run in over a period of 10 min, while a stream of nitrogen was passed through the flask. A solution of hexatriyne in THF was prepared, starting from 0.10 mol of 1,6-dichloro-2,4-hexadiyne (VIII-2, Exp. 9) as described in Chapter V, Exp. 17 (note 2). This solution, together with the precipitated salt, was poured into the flask containing the solution of the sodium selenide. The slurry on the wall was rinsed into the second reaction flask with 50 ml of methanol and the mixture was stirred for 2 h. It was then poured into 1.5 l of water, after which eight extractions with small portions of redistilled pentane (note 3) were carried out. The combined pentane extracts were washed four times with 200-ml portions of 2 N HCl and once with water and then dried over magnesium sulfate. The greater part of the pentane was distilled off at normal pressure (bath temperature not higher than 75°C) through a 30-cm Vigreux column. Distillation of the remaining liquid gave 2-ethynylselenophen, b.p. 58°C/17 mmHg, n_D^{22} 1.6212, in 56-63% yield (calculated on 1,6-dichloro-2,4-hexadiyne).

Note 1. Water content of the ammonia not more than 0.1%.

Note 2. No high-boiling light petroleum and water should be added! The solution has to be kept below -30°C during the period between its preparation and its addition to the sodium selenide.

Note 3. Commercial pentane may contain small amounts of higher boiling hydrocarbons.

(4) *3,4-Octadien-1-ol*[139]

$$H_2C{=}CH{-}C{\equiv}C{-}CH_2CH_2OH + 2\ C_2H_5Li \xrightarrow[\text{ether}]{\text{diethyl}} C_2H_5CH_2CH{=}C{=}\overset{Li}{\overset{|}{C}}{-}CH_2CH_2OLi \xrightarrow{H_2O}$$

$$\longrightarrow C_3H_7CH{=}C{=}CHCH_2CH_2OH$$

Apparatus: 500-ml flask, see Fig. 1.

To a solution of 0.10 mol of the enyne alcohol (see Chapter III, Exp. 41) in 140 ml of dry diethyl ether was added in 15 min with cooling at about -30°C a solution of 0.22 mol of ethyllithium in about 180 ml of diethyl ether (see Chapter II, Exp. 1). The clear solution was then warmed to 25°C and after standing for 10 min at that temperature, it was cautiously poured into 200 ml of an aqueous solution of 30 g of NH_4Cl. The layers formed after standing were separated, and the aqueous layer was extracted three times with diethyl ether. The combined ethereal solutions were dried over magnesium sulfate and concentrated in a water-pump vacuum. Distillation of the remaining liquid through a 30-cm Vigreux column gave the allenic alcohol, b.p. 87°C/15 mmHg, n_D^{20} 1.4744, in 93% yield.

(*5*) *1-NN-diethylamino-2,3-heptadiene*[139]

$$C_2H_5Li + H_2C{=}CH{-}C{\equiv}C{-}CH_2N(C_2H_5)_2 \longrightarrow C_2H_5{-}CH_2{-}CH{=}C{=}\overset{\overset{\displaystyle Li}{|}}{C}{-}CH_2N(C_2H_5)_2 \longrightarrow$$

$$\xrightarrow{H_2O} C_2H_5CH_2CH{=}C{=}CH{-}CH_2{-}N(C_2H_5)_2$$

Apparatus: 500-ml flask, see Fig. 1.

To a solution of 0.10 mol of the enyne amine[1] in 100 ml of dry diethyl ether was added with cooling at -30°C during 15 min a solution of 0.13 mol of ethyllithium in about 130 ml of diethyl ether (see Chapter II, Exp. 1). After completion of the addition the cooling bath was removed, the temperature was allowed to rise to 10°C and the solution was cautiously poured into 100 ml of saturated NH_4Cl solution. After shaking, the upper layer was separated off and three extractions with diethyl ether were carried out. The combined ethereal solutions were dried over potassium carbonate and then concentrated in a water-pump vacuum. Distillation of the residue (note 1) through a 30-cm Vigreux column gave the allenic amine, b.p. 84°C/15 mmHg, n_D^{18} 1.4648, in about 90% yield.

Note 1. A 250-ml distillation flask should be used, as strong foaming may occur during the distillation.

(*6*) *1-Methylthio-1,2-hexadiene*[139]

$$H_2C{=}CHC{\equiv}C{-}SCH_3 + C_2H_5Li \xrightarrow[\text{ether}]{\text{diethyl}} C_2H_5CH_2CH{=}C{=}\overset{\overset{\displaystyle Li}{|}}{C}{-}SCH_3 \xrightarrow{H_2O}$$

$$\longrightarrow C_2H_5CH_2CH{=}C{=}CH{-}SCH_3$$

Apparatus: 1-l flask, see Fig. 1.

A solution of 0.10 mol of 1-methylthio-3-buten-1-yne (see Chapter V, Exp. 16) in 100 ml of dry diethyl ether was cooled to -40^{0}C. A solution of 0.12 mol of ethyllithium in about 120 ml of diethyl ether (see Chapter II, Exp. 1) was added at this temperature in 10 min. The cooling bath was then removed, the temperature was allowed to rise to -10^{0}C and the yellowish solution was poured into 200 ml of ice-water. After shaking, the layers were separated and the aqueous layer was extracted three times with diethyl ether. The extracts were dried over magnesium sulfate, the diethyl ether removed by evaporation in a water-pump vacuum and the residue distilled through a 30-cm Vigreux column to give the product, b.p. 60^{0}C/ /15 mmHg, n_D^{21} 1.5153 in a yield of 79%.

(7) *3,4-Pentadien-1-ol*[140]

$$HC{\equiv}C\text{-}CH{=}CHCH_2OH \xrightarrow[Et_2O]{LiAlH_4} H_2C{=}C{=}CH\text{-}CH_2CH_2OH$$

Apparatus: 1-l three-necked, round-bottomed flask with a dropping funnel, a mechanical stirrer and a reflux condenser.

To a solution of 8 g of lithium alanate in 250 ml of diethyl ether was added in 15 min 24 g (0.3 mol) of 2-penten-4-yn-1-ol (III, Exp. 57). The diethyl ether began to reflux and a rubber-like greyish precipitate was formed. After heating for 1 h under reflux the flask was placed in an ice + ice-water bath and water (150 ml) was added dropwise with vigorous stirring. After this hydrolysis procedure the ethereal solution was decanted and the aqueous jelly layer was extracted ten times with diethyl ether. The ethereal extracts were dried (without washing) over magnesium sulfate and subsequently concentrated in a water-pump vacuum. Careful distillation of the residue afforded the allenic alcohol, b.p. 55^{0}C/15 mmHg, n_D^{20} 1.4763, in 76-82% yield. The first fraction (2-3.5 g), n_D^{20} 1.4662, contained some water.

(8) *2,4-Dimethyl-2,3-pentadien-1,5-bis(diethylamide)*[*141]

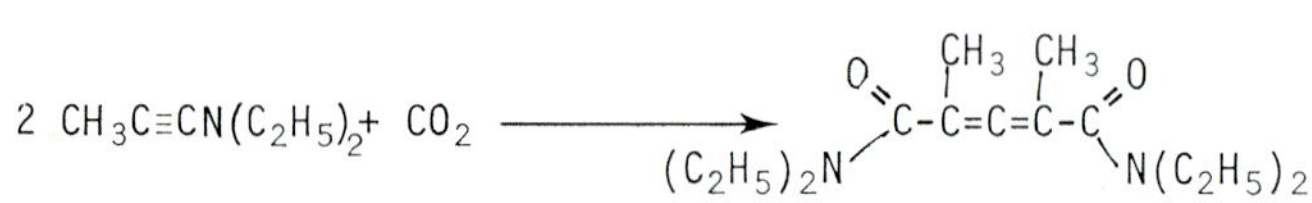

Apparatus: 500-ml three-necked, round-bottomed flask, provided with a gas inlet tube, a mechanical stirrer and a thermometer, combined with a gas outlet. The apparatus should be perfectly dry.

A solution of 0.10 mol of freshly distilled diethylaminopropyne[1] in 80 ml of dry (distilled from phosphorus pentoxide) acetonitrile was cooled to $5^{0}C$ and dry carbon dioxide was introduced into the vigorously agitated solution at a rate of about 0.3 l/min. The temperature rose above $20^{0}C$ within a few minutes, but was kept at about $30^{0}C$ by occasionally immersing the flask in a bath of ice-water. The introduction of CO_2 was continued until the temperature had dropped to $25^{0}C$ and the typical odour of the yneamine had disappeared completely. The yellow solution was concentrated in a water-pump vacuum. The residue, a sirupy liquid, had the theoretically required weight and consisted of reasonably pure (about 95%) allenyl-diamide. If desired the product can be distilled (short-path distillation) in a high vacuum. It solidified upon standing at $-25^{0}C$.

*We are indebted to Prof. J. Ficini (Paris, France) for detailed instructions.

(9) *1-Allenylcyclohexanol*[142]

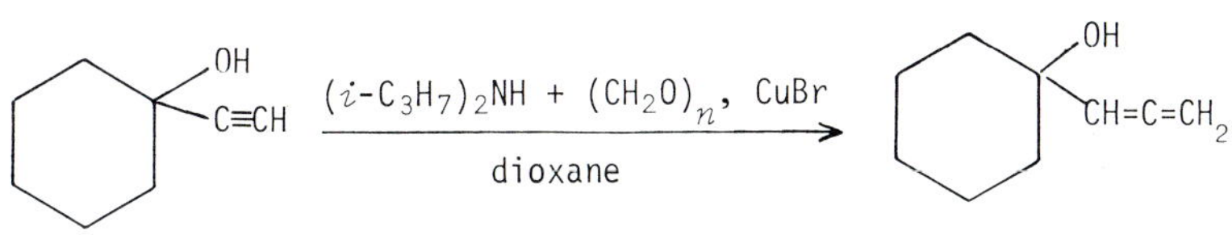

Apparatus: 500-ml round-bottomed flask and reflux condenser.

A mixture of 0.30 mol of 1-ethynylcyclohexanol (commercially available), 0.10 mol of copper(I) bromide, 0.40 mol of diisopropylamine and 15 g (corresponding to 0.50 mol of formaldehyde) paraformaldehyde and 40 ml of dioxane was heated under reflux for 3 h. After cooling to room temperature the red suspension was poured into 400 ml of 2 N hydrochloric acid, cooled at $0^{0}C$. After vigorous shaking the mixture was extracted five times with diethyl ether: a small amount of solid material that separated during this operation was discarded. The combined extracts were washed with a saturated NaCl solution and dried over magnesium sulfate. Evaporation of the solvent followed by distillation through a 20-cm Vigreux column gave 1-allenylcyclohexanol, b.p. $85^{0}C/12$ mmHg, n_D^{22} 1.5059, in 47% yield.

Chapter VIII

TRANSFORMATION OF FUNCTIONAL GROUPS IN ACETYLENES AND CUMULENES

1. UNSATURATED SULFOXIDES AND SULFONES

(1) *Allenyl methyl sulfoxide*[151]

$$H_2C{=}C{=}CHSCH_3 \xrightarrow{NaIO_4} H_2C{=}C{=}CH{-}\overset{O}{\overset{\|}{S}}{-}CH_3$$

Apparatus: 1-l flask, see Fig. 1 (without dropping funnel).

A mixture of 100 ml of water and 0.10 mol of methylthioallene (see Chapter IV, Exp. 23) was placed in the flask and 24 g of sodium periodate were added in five portions at intervals of 4 min. The mixture was agitated vigorously and the temperature was kept between 25 and 30°C by occasional cooling. After 1 h twenty extractions with 10-ml portions of chloroform were carried out. The combined extracts were dried (without previous washing) over magnesium sulfate. Removal of the chloroform by evaporation in a water-pump vacuum gave the reasonably pure (> 95%) sulfoxide, n_D^{16} 1.5456, in more than 90% yield.

(2) *Methyl 1,2-pentadienyl sulfoxide*[151]

$$C_2H_5CH{=}C{=}CH{-}SCH_3 \xrightarrow[H_2O,\ CH_3OH]{NaIO_4} C_2H_5CH{=}C{=}CH{-}\overset{O}{\overset{\|}{S}}{-}CH_3$$

Apparatus: 1-l flask, see Fig. 1.

A mixture of 0.15 mol of 1-methylthio-1,2-pentadiene (see Chapter IV, Exp. 26), 100 ml of water, 40 ml of methanol (note 1) and 0.18 mol of $NaIO_4$ was vigorously stirred. The temperature rose gradually, but was kept between 30 and 35°C by occasional cooling. After 1.5 h 500 ml of water were added to the white suspension and ten extractions with 30-ml portions of chloroform were carried out. The extracts were dried (without previous washing) over magnesium sulfate. Evaporation of the solvent in a water-pump vacuum (the last traces at 0.5-1 mmHg) gave the sulfoxide as a viscous oil in almost quantitative yield (purity *ca.* 96%). The compound solidified during storage in a refrigerator.

Note 1. This co-solvent is used in order to increase the solubility of the allenic sulfide (compare Exp. 1).

(3) *Ethyl 1-propynyl sulfoxide*[151]

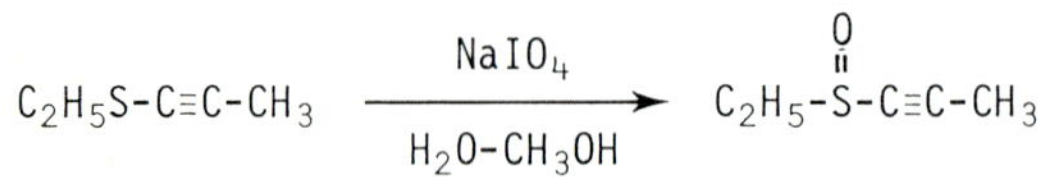

Apparatus: 500-ml flask, Fig. 1 (without dropping funnel).

A mixture of 0.05 mol of 1-ethylthio-1-propyne (see IV, Exp. 23 and Ref. 1), 15 ml of methanol, 60 ml of water and 0.07 mol of sodium periodate was stirred vigorously. The temperature rose a few degrees in 30 min. The mixture was warmed to 45°C and stirring was continued for an additional 1 h, without external warming or cooling. Water (100 ml) was added, then seven extractions with 20-ml portions of chloroform were carried out. The extracts were dried (without previous washing) over magnesium sulfate and concentrated in a water-pump vacuum. The residue, n_D^{22} 1.5012, yield *ca.* 100%, was almost pure ethyl 1-propynyl sulfoxide.

(4) *Ethyl 1-propynyl sulfone*[152]

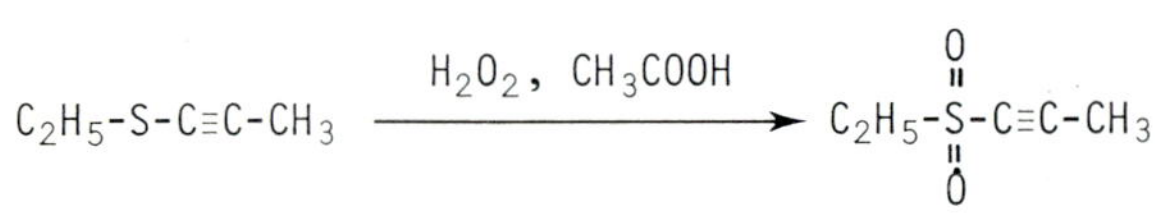

Apparatus: 500-ml flask, provided with a dropping funnel, a mechanical stirrer and a thermometer, combined with a vent.

A mixture of 0.10 mol of 1-ethylthio-1-propyne (see IV, Exp. 23 and Ref. 1) and 100 ml of glacial acetic acid was heated at 95°C, then 36 ml of 33% hydrogen peroxide were added in 15 min with occasional cooling. After this addition the mixture was kept for 1 h at 100°C, then cooled to 20°C and 200 ml of water were added. Ten extractions with 20-ml portions of chloroform were carried out and the combined extracts were dried over magnesium sulfate (without previous washing) and subsequently concentrated in a water-pump vacuum.The last traces of volatile compounds were removed in a vacuum of < 1 mmHg (bath temperature about 60°C). The desired sulfone, n_D^{20} 1.4745, was obtained in 88% yield.

In a similar way $(CH_3C{\equiv}C)_2SO_2$ (m.p. 98-100°C, after one crystallization from diethyl ether-pentane) was prepared in 78% yield from $(CH_3C{\equiv}C)_2S$ (Chapter III, Exp. 37).

(5) *Allenyl methyl sulfone*[152]

$$CH_3\text{-}\underset{\underset{O}{\|}}{S}\text{-}CH{=}C{=}CH_2 \xrightarrow{H_2O_2,\ CH_3COOH} CH_3\text{-}\overset{\overset{O}{\|}}{\underset{\underset{O}{\|}}{S}}\text{-}CH{=}C{=}CH_2$$

Apparatus: 250-ml flask with a thermometer and a reflux condenser.

A mixture of 100 ml of glacial acetic acid, 15 ml of 30% hydrogen peroxide and 6.0 g of allenyl methyl sulfoxide (see Chapter VII-1, Exp. 1) was heated for 30 min at 100°C. The colourless solution was cooled to 20°C and poured into 300 ml of ice-water. The sulfone was isolated by extracting the solution twelve times with 20-ml portions of chloroform, drying the combined extracts over $NaHCO_3$ and thoroughly removing the solvent in a water-pump vacuum. The residue, n_D^{20} 1.5079, was practically pure sulfone, yield 80%. The product solidified during storage for a few days at room temperature.

2. ACETYLENIC HALIDES

(1) *1-Iodo-2-propyne (propargyl iodide)*

$$HC{\equiv}CCH_2Br + NaI \xrightarrow[C_2H_5OH]{} HC{\equiv}CCH_2I + NaBr\downarrow$$

Apparatus: 1-l flask, provided with a mechanical stirrer and a reflux condenser.

A solution of 92 g of dry sodium iodide in 350 ml of 100% ethanol (note 1) was heated to about 70°C. Freshly distilled propargyl bromide (note 2) (0.50 mol) was added in 10 min. After heating for 20 min at 70-75°C the white suspension was cooled to room temperature and 500 ml of water were added, then the product was extracted five times with 60-ml portions of high-boiling light petroleum (b.p. $> 190^{o}C$). The combined extracts were washed with water and dried over magnesium sulfate. The propargyl iodide was isolated from the extract by heating this at 10-15 mmHg in a distillation apparatus, using a 40-cm Widmer column and a single receiver cooled at $-30^{o}C$. The distillation was stopped when the temperature in the top of the column had risen above $50^{o}C$. Redistillation of the contents of the receiver in the same apparatus gave propargyl iodide, b.p. $< 25^{o}C$/15 mmHg, (receiver cooled at $-30^{o}C$), n_D^{20} 1.5815, containing about 10% of iodoallene.

Note 1. In another experiment, performed on the same scale, 96% ethanol (instead of 100%) and sodium iodide, containing 3-4% of water, were used. 175 ml of 96% ethanol appeared to be sufficient to dissolve 90 g of this quality of NaI. The reaction with $HC{\equiv}CCH_2Br$ gave no trace of iodoallene, but instead some (10%) of $HC{\equiv}CCH_2Br$ was still present in the crude product. This amount will probably increase if more water is present since then the equilibrium cannot completely shift to the right.

Note 2. Propargyl *chloride* gives comparable amounts of propargyl iodide and iodoallene. Propargyl tosylate will probably yield a product that is free from the allene.

(2) *1,4-Dichloro-2-pentyne*

$$CH_3\text{-}CH(OH)\text{-}C{\equiv}C\text{-}CH_2Cl \xrightarrow{SOCl_2\ (pyridine)} CH_3CH(Cl)\text{-}C{\equiv}C\text{-}CH_2Cl$$

Apparatus: 500-ml round-bottomed flask with a combination of thermometer and vent, connected to a drying tube filled with lumps of anhydrous calcium chloride.

A solution of 0.25 mol of 5-chloro-3-pentyn-2-ol (note 1) and 3 ml of pyridine in 70 ml of dry diethyl ether was cooled to $-20^{o}C$. Freshly distilled thionyl chloride (0.30 mol) was added in five equal portions with intervals of five min, keeping the temperature between -20 and $-10^{o}C$. After the addition the temperature was allowed to rise to $0^{o}C$ in 2 h. The flask was then placed in an ice-bath and after 12 h the reaction mixture was cautiously poured into a solution of 20 g of $NaHCO_3$ in 300 ml of water. After successive swirling and vigorous shaking the layers were separated and the aqueous layer was extracted three times with diethyl ether. The combined solutionswere dried over magnesium sulfate and concentrated in a water-pump vacuum. Distillation of the remaining liquid through a 30-cm Vigreux column gave the acetylenic dichloride, b.p. $60^{o}C$/12 mmHg, n_D^{20} 1.4919 in 80% yield.

Note 1. Prepared by lithiating $HC{\equiv}CCH_2Cl$ at $-80^{o}C$ with BuLi in hexane-ether (1:1) and subsequently adding freshly distilled acetaldehyde (compare Ref. 1).

(*3*) *3-Chloro-1-butyne*

$$HC{\equiv}C\text{-}CH(OH)CH_3 \xrightarrow[\text{KOH, }Et_2O]{\text{tosyl chloride}} HC{\equiv}C\text{-}CH(OTs)CH_3 \xrightarrow[\text{DMSO}]{\text{LiCl}} HC{\equiv}C\text{-}CH(CH_3)Cl$$

Apparatus: see Fig. 5, volume of distillation flask 1 l.

To a mixture of 0.55 mol of anhydrous lithium chloride and 100 ml of DMSO was added a solution of 0.40 mol of the acetylenic tosylate (for a general procedure concerning the preparation of acetylenic tosylates, see Chapter VIII-3, Exp. 3) in 150 ml of DMSO. The flask was equipped for vacuum distillation (see Fig. 5). Between the receiver, which was cooled at $-75^{o}C$, and the water-pump was placed a tube filled with KOH pellets. The apparatus was evacuated (10-20 mmHg) and the flask gradually heated until DMSO began to reflux in the column. The contents of the receiver were distilled at normal pressure through a 40-cm Vigreux column, giving the acetylenic chloride b.p. $74^{o}C/760$ mmHg, n_D^{20} 1.4250, in 86% yield.

Note 1. For less volatile acetylenic chlorides (b.p. $30^{o}C/15$ mmHg) the thionyl chloride method is preferred over the method described above (see Chapter VIII-2, Exp. 2).

(*4*) *3-Bromo-1-nonyne*

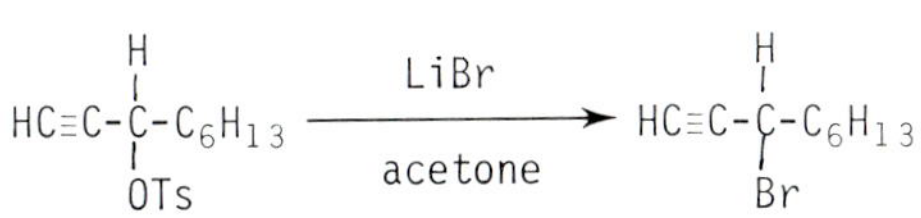

Apparatus: 1-l two-necked, round-bottomed flask with a mechanical stirrer and a reflux condenser.

A mixture of 0.30 mol of the acetylenic tosylate (prepared according to the general method described in VIII-3, Exp. 3 from the corresponding acetylenic alcohol, which was prepared from $HC{\equiv}CMgBr$ and $C_6H_{13}CH{=}O$ in THF, as described in Ref. 1) and 100 ml of dry acetone was added to a solution of 0.40 mol of anhydrous lithium bromide in 150 ml of acetone. After refluxing for 1 h 600 ml of ice-water were added and the acetylenic bromide was extracted eight times with small portions of pentane (b.p. 36-40°C). The combined extracts were washed with water and dried over magnesium sulfate. After removal of the pentane by evaporation in a water--pump vacuum the residue was distilled to give the acetylenic bromide, b.p. 82°C//15 mmHg, n_D^{16} 1.4742, in 92% yield.

(5) *3-Chloro-3-methyl-1-butyne*[143]

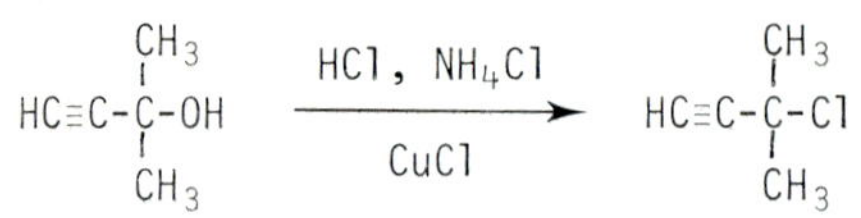

Apparatus: 1-l conical flask with a thermometer.

To 200 ml of concentrated (36%) hydrochloric acid were added 50 g of powdered NH_4Cl, 10 g of powdered copper(I) chloride (technical grade) and 2 g of copper bronze. The mixture was cooled to -15°C (internal temperature), then 0.50 mol of the acetylenic carbinol (prepared from an alkali metal acetylide and acetone[1] or commercially available) was added. Gaseous hydrogen chloride (40 g, weight increase) was introduced, in 10 min, with manual swirling of the flask, keeping the temperature of the mixture between -15 and -10°C (note 1). The flask was placed for 1 h in a cooling bath at -10°C with occasional swirling, then 300 ml of ice--water were added. The upper layer was separated off sharply. The crude chloride weighed 45-48 g and had n_D^{20} 1.421-1.423 (note 1). It was shaken with a small amount of anhydrous K_2CO_3 in a 250-ml round-bottomed flask, which was connected to a short Vigreux column, condenser and receiver, cooled at -75°C. A tube filled with $CaCl_2$ was placed between the receiver and the water-pump. The system was evacuated (10-20 mmHg) while the flask with the crude chloride and K_2CO_3 was heated at 30-40°C. The contents of the receiver were carefully redistilled through a 40-cm Widmer column. The fraction, passing over between 75 and 79°C, was almost pure 3-chloro-3-methyl-1-butyne, n_D^{20} 1.4168-1.4182, yield 70-74%.

Note 1. At higher temperatures the tertiary chloride undergoes a rearrangement, catalysed by CuCl, giving $H_2C{=}C(Cl)CH{=}CH_2$. In these cases the refractive index of the crude product is much higher.

(6) *3-Bromo-3-methyl-1-butyne*

$$HC{\equiv}C{-}C(CH_3)_2OH + 1/3\ PBr_3 \longrightarrow HC{\equiv}C{-}C(CH_3)_2Br + P(OH)_3$$

Apparatus: 500-ml three-necked, round-bottomed flask, provided with a dropping funnel, a mechanical stirrer and a thermometer, combined with a vent.

To 0.30 mol of the acetylenic carbinol (commercially available, see also Ref. 1) was added between 15 and 20°C (note 1) 0.11 mol of phosphorous tribromide. During the addition, which was carried out in 25 min, the solid crust (carbinol) on the glass wall gradually disappeared. After an additional 15 min 100 ml of water were added with vigorous stirring. The organic layer was separated off and dried over a small amount of magnesium sulfate. The liquid was transferred into a 500-ml round-bottomed flask, which was connected to a 40-cm Vigreux column, condenser and receiver, cooled in a bath of dry-ice and acetone. The apparatus (see Fig. 5) was evacuated (10-20 mmHg) and the distillation flask warmed in a bath at 35-40°C. In the receiver was collected almost pure 3-bromo-3-methyl-1-butyne, n_D^{20} 1.4668, in 50% yield. The residue consisted of the HBr adduct of the acetylenic bromide.

Note 1. A cooling bath of dry-ice and acetone is advised.

(7) *1-Chloro-1-cyclohexylacetylene*[143]

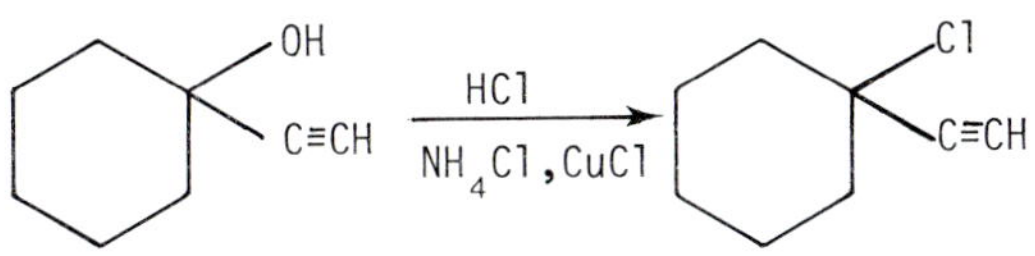

Apparatus: 2-l three-necked, round-bottomed flask with a stirrer, a gas inlet tube and a thermometer, combined with a gas outlet.

The flask was charged with 150 g of powdered NH_4Cl, 20 g of powdered copper(I) chloride (technical quality), 3 g of copper bronze and 660 ml of 36% hydrochloric acid. About 130 g of gaseous hydrogen chloride were introduced with vigorous stirring and cooling at -10 to -15°C. Ethynylcyclohexanol (1.50 mol, commercially available) was then added in 2 min and stirring was continued for 90 min, while maintaining the temperature of the mixture between -5 and 0°C. After this period ice-water (just enough to effect dissolution of the salts) and a small amount of pentane were added at 0°C. The upper layer was separated off and the aqueous layer was extracted with 100 ml of pentane. The combined solutions were washed with water and dried over magnesium sulfate. After removal of the pentane in a water--pump vacuum, the remaining liquid was carefully distilled through a 40-cm Widmer column to afford the teriairy chloride, b.p. 55°C/15 mmHg, n_D^{22} 1.4804, in 78-86% yield.

(8) 1,4-Dichloro-2-butyne[173]

$$HOCH_2C{\equiv}CCH_2OH + 2\ SOCl_2 \xrightarrow{\text{fast}} ClOSOCH_2C{\equiv}CCH_2OSOCl \xrightarrow{\text{slow}}$$

$$\longrightarrow ClCH_2C{\equiv}CCH_2Cl + 2\ SO_2\uparrow$$

Apparatus: 3-l three-necked flask with dropping funnel, thermometer and vent.

In the flask were placed 2 mol of powdered butynediol (recrystallized from a 3:10 mixture of THF and diethyl ether) and 15 ml of pyridine. Thionyl chloride (4.3 mol), precooled at -40°C, was added in 30 min with cooling in a bath of -30°C. No difficulties were encountered in keeping the reaction under control: the net temperature effect was about zero because of the abundant evolution of HCl from the reaction mixture, so that the temperature could easily be kept below -10°C. During the addition the mixture was occasionally swirled. After the addition the temperature was allowed to rise over 2 h to about 0°C. The flask was then placed in a bath of ice and ice-water (10-15 l). The vent was connected to a $CaCl_2$-tube. After about 12 h, the gasses were sucked off by evacuating the flask by means of the water pump. The bath temperature was increased gradually over 2 h from 20 to 40°C. Sucking off at this bath temperature was continued until the pressure had dropped to 40 mm. The liquid was then carefully distilled through a 40-cm Widmer column. $ClCH_2C{\equiv}CCH_2Cl$, b.p. 57°C/12 mmHg, n_D^{20} 1.5054, was obtained in 90-97% yield.

Note 1. Traces of $SOCl_2$ may be present in the product after distillation. These can be removed (if desired) by shaking the product with a $NaHCO_3$ solution. Technical butynediol can also be used: yields are somewhat lower (80-85%).

(9) 1,6-Dichloro-2,4-hexadiyne[174]

$$HOCH_2(C{\equiv}C)_2CH_2OH + 2\ SOCl_2 \xrightarrow{\text{fast}} ClOSOCH_2(C{\equiv}C)_2CH_2OSOCl \xrightarrow{\text{slow}}$$
$$(+\ 2\ HCl\uparrow)$$

$$\longrightarrow ClCH_2(C{\equiv}C)_2CH_2Cl + 2\ SO_2\uparrow$$

Apparatus: 500-ml flask, see Exp. 8.

Warning: Use rubber gloves, the product is a skin irritant.

To a mixture of 0.3 mol of the diol (see III, Exp. 60) and 5 ml of pyridine were added with swirling and cooling in a bath of -30°C 85 g of precooled (-30°C) thionyl chloride. The addition was carried out in 10-g portions over a period of 20 min. The cooling bath was then removed and the temperature of the mixture was allowed to rise to 25 or 30°C in 4 h (occasional cooling may be necessary). Some boiling stones were then added and the mixture was warmed at 40°C for 45 min in a water-pump vacuum. Subsequently, the mixture was vigorously shaken with 200 ml of ice water and 20 ml of pyridine. Three extractions with diethyl ether were carried out. The extracts were successively washed with 3 N HCl and water and then dried over magnesium sulfate. The diethyl ether was removed in a water-pump vacuum. The residue was almost pure $ClCH_2(C{\equiv}C)_2CH_2Cl$, n_D^{20} 1.5750. Distillation at about 0.2 mm (b.p. 45-50°C) gave a product with the same n_D^{20}. There was no residue. Yields were at least 85%.

(10) 1-Bromo-2-alkynes[1]

$$3\ RC{\equiv}CCH_2OH + PBr_3 \longrightarrow 3\ RC{\equiv}CCH_2Br + P(OH)_3$$

$$(R = CH_3,\ C_2H_5,\ C_4H_9)$$

Apparatus: Chapter I, Fig. 1; size of the flask, 1 l.

To a mixture of 0.9 mol of the alcohol, 250 ml of diethyl ether (note 1) and 5 ml of pyridine was added with cooling at -30 to -35°C 0.32 mol of PBr_3. During the addition which was carried out in 45 min a slow stream of nitrogen was introduced. The temperature was maintained for 2 h at -30°C; subsequently the temperature was allowed to rise gradually to 20°C over a period of 3 h. A heavy underlayer, probably $P(OH)_3$, was formed. Finally the mixture was heated at 40°C for 30 min. It was then poured into 500 ml of a saturated solution of NaCl. After shaking the upper layer was separated. The aqueous layer was extracted with a small portion of diethyl ether. The ethereal solutions were dried over magnesium sulfate. The greater part of the diethyl ether was distilled off at normal pressure through a 40-cm Widmer column. The residue was carefully distilled in a water-pump vacuum. $CH_3C{\equiv}CCH_2Br$, b.p. about 60°C/80 mmHg, n_D^{20} 1.5062, was obtained in 80% yield. $C_2H_5C{\equiv}CCH_2Br$, b.p. 38°C/10 mmHg, n_D^{20} 1.4980, was obtained in 81% yield. $C_4H_9C{\equiv}CCH_2Br$, b.p. 72°C/12 mmHg, n_D^{20} 1.4936, was obtained in 83% yield.

Note 1. When smaller amounts of diethyl ether are used, yields are lower and considerable amounts of HBr-adducts $RC(Br){=}CHCH_2Br$ are formed.

3. ACETYLENIC ESTERS AND DITHIOESTERS

(1) *Acetates*[144]

$$HC{\equiv}C\text{-}C(R^1)(R^2)OH + CH_3C({=}O)Cl \xrightarrow{PhNEt_2} HC{\equiv}C\text{-}C(R^1)(R^2)OC({=}O)CH_3$$

Apparatus: 500-ml round-bottomed flask and a thermometer.

A mixture of 0.30 mol of the tertiairy acetylenic alcohol, 0.35 mol of acetyl chloride (freshly distilled) and 0.35 mol of *NN*-diethylaniline was gradually heated with manual swirling. At 40-50°C an exothermic reaction started and the temperature rose in a few minutes to 120°C. It was kept at that level by occasional cooling. After the exothermic reaction had subsided, the mixture was heated for an additional 10 min at 125-130°C, during which the mixture was swirled by hand so that the salt that had been deposited on the glass wall was redissolved. After cooling to below 50°C a mixture of 5 ml of 36% HCl and 200 ml of ice-water was added and the obtained solution was extracted with small portions of diethyl ether. The ethereal solutions were washed with water and subsequently dried over magnesium sulfate. The solvent was removed by evaporation in a water-pump vacuum and the remaining liquid distilled through a 30-cm Vigreux column. The following acetates were prepared: $R^1 = R^2 = CH_3$, b.p. 45°C/27 mmHg, n_D^{20} 1.4193, in 78% yield, and R^1R^2C = cyclohexyl, b.p.87°C/15 mmHg, n_D^{20} 1.4669, in 88% yield.

(2) *Sulfinates*[108]

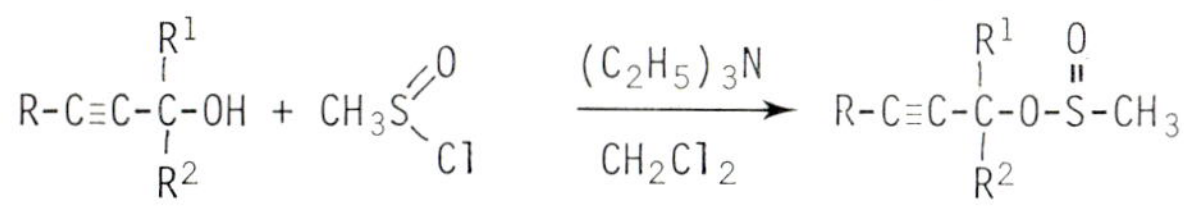

Apparatus: 1-l flask, see Fig. 1.

A mixture of 0.10 mol of the acetylenic alcohol, 0.12 mol of triethylamine and 200 ml of dichloromethane (note 1) was cooled to $-50^{o}C$. Methanesulfinyl chloride (0.12 mol) (for its preparation from CH_3SSCH_3, $(CH_3CO)_2O$ and chlorine, see Ref. 73) was added in 10 min at -40 to $-50^{o}C$. A white precipitate was formed immediately. After the addition the cooling bath was removed and the temperature was allowed to rise to $-20^{o}C$, then the mixture was vigorously shaken or stirred with 100 ml of water. The lower layer was separated off and the aqueous layer was extracted twice with 10-ml portions of CH_2Cl_2. The combined solutions were dried over magnesium sulfate and concentrated in a water-pump vacuum (note 2). The yields of the products, which are pure enough (usually > 96%) for further conversions, are normally almost quantitative.

Note 1. All reagents and solvents must be dry

Note 2. Most sulfinates are reasonably thermostable.

(3) *Tosylates*[1]

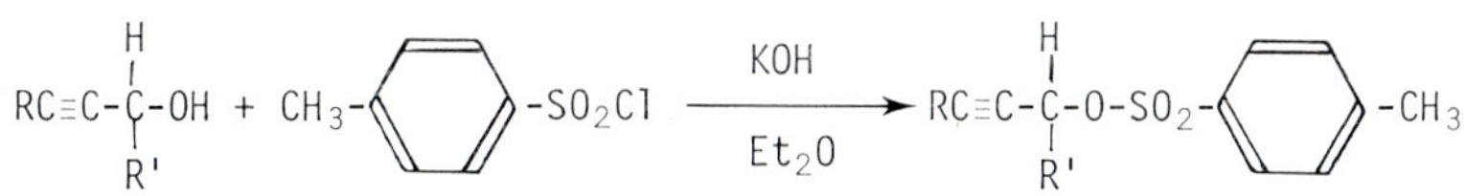

Apparatus: 1-l round-bottomed, three-necked flask, provided with a mechanical stirrer, a thermometer and a funnel for the addition of the KOH.

In the flask were placed 0.20 mol of the acetylenic alcohol, 0.24 mol of tosyl chloride and 350 ml of diethyl ether. The mixture was stirred at room temperature until the solid had passed into solution and then cooled at -5 to $-10^{o}C$ in a bath of dry-ice and acetone. Machine-powdered KOH (130 g) was added with vigorous stirring, initially in relatively small portions (*ca.* 5 g), at intervals of 2 min. The reaction was strongly exothermic at first, and efficient cooling was necessary in order to maintain the temperature of the reaction mixture between -5 and $0^{o}C$

in the case of primary alcohols, or between 0 and +5°C for secondary alcohols. After about 40 g of KOH had been added in about 20 min, the remainder was added in 10 min. Stirring at the indicated temperature was maintained for an additional 20 min in the case of primary alcohols and 1 h in the case of secondary alcohols. Working up was carried out by pouring the reaction mixture with swirling into 500 ml of ice-water. After separation of the layers the aqueous layer was extracted twice with small amounts of diethyl ether. Drying of the combined ethereal solutions over magnesium sulfate, followed by evaporation of the diethyl ether in a water-pump vacuum, gave the almost pure tosylates in yields, that were generally higher than 95%, starting from primary alcohols. The yields of the tosylates obtained from the secondary alcohols may be lower (around 90%).

(*4*) *Propargyl dithiopropionate*[145]

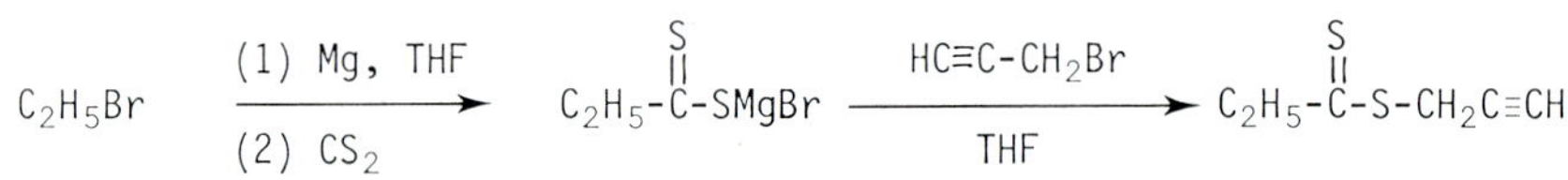

Apparatus: 2-l flask, see Fig. 1.

To a solution (partly suspension upon cooling) of ethylmagnesium bromide in THF, prepared according to Chapter II, Exp. 6, from 1.1 mol of ethyl bromide, was added 1.0 mol of carbon disulfide with cooling at 0°C (dry-ice-acetone bath). The addition took about 30 min. After an additional 1 h at 10-15°C 50 ml of dry HMPT were added and subsequently 1.2 mol of propargyl bromide. The reaction was moderately exothermic and the temperature rose slowly. After warming the mixture for 2 h at 50°C it was allowed to stand overnight at room temperature. Water (500 ml) was then added with swirling (or stirring if possible). After dissolution of the salt the mixture was shaken vigorously, the reddish upper layer was separated off and the aqueous layer was extracted twice with pentane. The combined solutions were dried over magnesium sulfate and then concentrated in a water-pump vacuum. Distillation of the residue at < 1 mmHg (note 1) afforded the dithioester, b.p. *ca.* 50°C/0.5 mmHg, n_D^{20} 1.5823, in 72% yield.

Note 1. Distillation at 10-20 mmHg may give rise to 3,3-sigmatropic rearrangement of the product and decomposition of the allenic dithioester formed.

4. ACETYLENIC AND ALLENIC NITRILES AND ACETYLENIC THIOCYANATES

(1) *3-Pentynenitrile*[8]

$$CH_3C{\equiv}C{-}CH_2Br \xrightarrow[\text{THF}]{\text{CuCN + LiBr}} CH_3C{\equiv}C{-}CH_2C{\equiv}N$$

Apparatus: 500-ml round-bottomed flask and thermometer; manual swirling.

To a solution of 9 g of anhydrous lithium bromide in 100 ml of dry THF were added successively 0.47 mol of CuCN and 0.40 mol of 1-bromo-2-butyne (for the preparation see VIII-2, Exp. 10). The mixture was warmed to about 40°C. An exothermic reaction started and the temperature rose in a few minutes to 70-80°C. The suspension disappeared and a brown solution was formed. After heating for an additional 30 min at 75°C the mixture was poured into a vigorously stirred solution of 50 g of NaCN and 50 g of NH_4Cl in 300 ml of water. The temperature was kept below 20°C by efficient cooling. The product was isolated by performing five extractions with diethyl ether, drying the extracts (without washing) over magnesium sulfate and distilling the diethyl ether at normal pressure through a 40-cm Widmer column. The residue was carefully fractionated through the same column at water-pump pressure. After a first fraction (about 4 g), consisting mainly of $CH_3C(C{\equiv}N){=}C{=}CH_2$ had been obtained, the desired product distilled at 47°C/15 mmHg, n_D^{18} 1.4432. The yield was 62%.

(2) *3-Octynenitrile*

$$n\text{-}C_4H_9C{\equiv}C{-}CH_2Br \xrightarrow[\text{THF, LiBr}]{\text{CuCN}} n\text{-}C_4H_9C{\equiv}C{-}CH_2C{\equiv}N$$

Apparatus: 250-ml round-bottomed, three-necked flask, provided with a thermometer, a mechanical stirrer and a reflux condenser.

In the flask were placed 0.40 mol of dry, powdered copper(I) cyanide, 9 g of anhydrous lithium bromide (note 1), 50 ml of dry THF and 0.30 mol of 1-bromo-2--heptyne (prepared from the corresponding alcohol and PBr_3 in diethyl ether, see VIII-2, Exp. 10). The mixture was heated until an exothermic reaction started, which occurred at about 80°C. The solid dissolved gradually. The mixture was kept for 15 min at 80°C (occasional cooling) and subsequently for 30 min at 100°C. After cooling to about 40°C (note 2) the viscous brown liquid was poured into a vigorously stirred solution of 50 g of ammonium chloride in 250 ml of 4 N HCl, which was kept at 0-5°C. The flask was also rinsed with this solution. The product was extracted 5-7 times with a 1:1 mixture of diethyl ether and pentane. The combined extracts were washed with saturated NH_4Cl solution and subsequently dried over magnesium sulfate. The residue remaining after removal of the solvents in a water-pump vacuum, was carefully distilled through a 30-cm Widmer column. The desired nitrile, b.p. 84°C/15 mmHg, n_D^{17} 1.4487, was obtained in 72% yield. The first fraction (about 5 g) consisted mainly of the 1,3-substitution product *n*-$C_4H_9C(C{\equiv}N){=}C{=}CH_2$.

Note 1. The commercial product was heated for 1 h at 160°C and 10-20 mmHg.
Note 2. Further cooling may result in partial solidification.

(3) *4-Methyl-2,3-pentadienenitrile*[8]

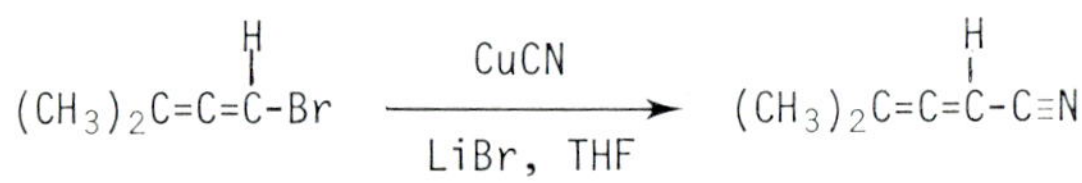

Apparatus: 500-ml round-bottomed flask and thermometer; swirling by hand.

In the flask were placed a solution of 7 g of anhydrous LiBr in 50 ml of dry THF, 0.40 mol of the allenic bromide (see Chapter VI, Exp. 31) and 0.50 mol of finely powdered copper(I) cyanide. The mixture was swirled by hand and the temperature rose in about 15 min to 60°C. It was kept between 55 and 60°C by occasional cooling in a water-bath. When the exothermic reaction had subsided, the flask was warmed for an additional 30 min at 55-60°C and the brown solution was then poured into a vigorously stirred solution of 30 g of NaCN and 100 g of NH_4Cl in 300 ml of water, to which 150 ml of diethyl ether had been added. During this operation the temperature was kept below 20°C. The reaction flask was subsequently rinsed with the NaCN solution. After separation of the layers the aqueous layer was extracted with ether. The extracts were dried over magnesium sulfate and then concentrated in a water-pump vacuum. Subsequent distillation through a 40-cm Vigreux column gave the nitrile, b.p. 50°C/15 mmHg, n_D^{20} 1.4739, in 83% yield.

(4) *1-Cyano-2-phenylacetylene*[8]

$$C_6H_5C{\equiv}C\text{-}Br \xrightarrow[\text{THF, LiBr}]{\text{CuCN}} C_6H_5C{\equiv}C\text{-}C{\equiv}N$$

Apparatus: 250-ml round-bottomed flask and thermometer, manual swirling.

A mixture of 0.12 mol of dry, powdered copper(I) cyanide, 5 g of anhydrous lithium bromide (note 1), 30 ml of dry THF and 0.10 mol of 1-bromo-2-phenylacetylene (prepared from phenylacetylene and sodium hypobromite[1]) was heated at 50°C. An exothermic reaction started and the temperature rose in about 15 min to 80-90°C and the solid gradually dissolved. External heating in a bath of 80°C was applied for an additional 10 min after the reaction had subsided. The mixture was cooled to 40°C and 150 ml of a cold (-5°C) aqueous solution of 20 g of NaCN and 25 g of NH_4Cl were added. After vigorous shaking three extractions with diethyl ether were carried out. The ethereal solutions were washed with a concentrated aqueous solution of NH_4Cl and subsequently dried over magnesium sulfate. The residue remaining after concentration of the extract in a water-pump vacuum was distilled at 0.5-1 mmHg. The nitrile, b.p. 95°C/15 mmHg, was obtained in 77% yield. The compound solidified at room temperature.

Note 1. In the absence of lithium bromide no reaction occurred.

(5) *Propargyl thiocyanate*

$$HC{\equiv}C\text{-}CH_2Br + KSC{\equiv}N \xrightarrow[H_2O]{} HC{\equiv}C\text{-}CH_2SC{\equiv}N$$

Apparatus: 500-ml three-necked, round-bottomed flask, provided with a thermometer, a mechanical stirrer and a vent.

A mixture of 0.40 mol of potassium thiocyanate, 45 ml of water and 0.30 mol of propargyl bromide was vigorously stirred for 30 min at 80°C. After cooling, the product was extracted with diethyl ether. Drying of the extract of magnesium sulfate, followed by distillation in a water-pump vacuum, gave propargyl thiocyanate, b.p. 65°C/15 mmHg, n_D^{20} 1.5152, in 96% yield.

5. ACETYLENIC AMINES

(1) *NN-Dialkylpropargylamines*[166,175]

$$HC{\equiv}CCH_2Br + 2\ R_2NH \longrightarrow HC{\equiv}CCH_2\text{-}NR_2 + R_2NH.HBr\downarrow$$

Apparatus: Same as in Exp. 4.

To a vigorously stirred solution of 0.65 mol of the secondary amine in 300 ml of dry diethyl ether was added in 15 min 0.30 mol of propargyl bromide. In all cases separation of the dialkylamine salt started immediately and the temperature rose to 30-35°C during the addition. In the case of pyrrolidine the salt deposited as a viscous oil, which became crystalline during or after work-up. After the addition of propargyl bromide stirring at 30-35°C was continued for 1 h. The reaction mixtures were then cooled to 0-10°C, the salt was filtered off on a sintered--glass funnel and rinsed well with dry diethyl ether. In the case of the reaction with pyrrolidine the solution was poured off from the lower layer, which was subsequently rinsed three times with redistilled pentane (40 ml portions). The ethereal solutions were concentrated in a water-pump vacuum. To the residue were added 150 ml of dry pentane and the small amount of salt was filtered off and rinsed with pentane. Subsequent concentration in a water-pump vacuum, followed by careful distillation through a 40-cm Vigreux column gave the propargylic amines. For the volatile *NN*-diethylpropargylamine the first filtrate was concentrated by distillation (40-cm Widmer column) at normal pressure, keeping the bath temperature below 90°C. The residue was distilled off from the small amount of salt at 10-20 mmHg, condensing the vapours in a receiver cooled below -40°C. Redistillation at normal pressure gave the pure amine. The following amines were prepared:

$HC{\equiv}CCH_2N(C_2H_5)_2$, b.p. 120°C/760 mmHg, n_D^{20} 1.4320, yield 81%;

$HC{\equiv}CCH_2\text{-}N$ 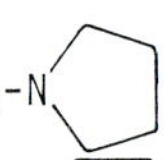 , b.p. 42°C/10 mmHg, n_D^{20} 1.4668, yield 74%;

$HC{\equiv}CCH_2\text{-}N$ 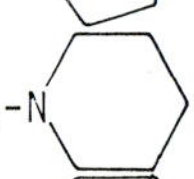 , b.p. 62°C/10 mmHg, n_D^{20} 1.4732, yield 89%;

$HC{\equiv}CCH_2\text{-}N$ 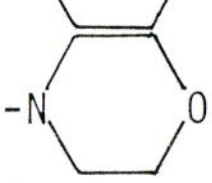 , b.p. 75°C/10 mmHg, n_D^{20} 1.4764, yield 91%;

(2) *3-Amino-3-methyl-1-butyne*[146]

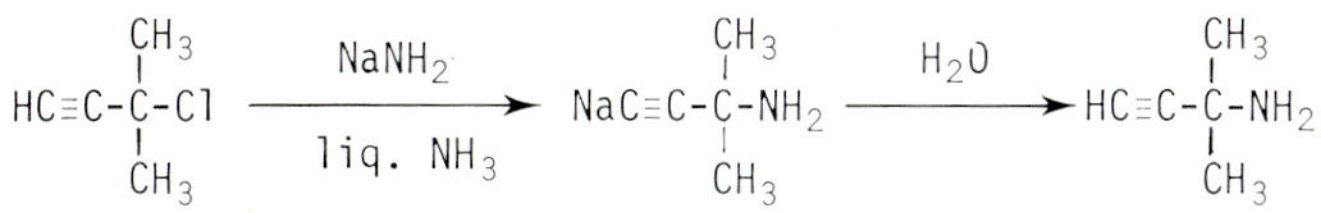

Apparatus: 1-l flask, see Fig. 2.

Sodium amide was prepared (see Chapter II, Exp. 11) from 12 g of sodium in 400 ml of liquid ammonia. To this suspension was added in 10 min 0.20 mol of the tertiary chloride (see Chapter VIII-2, Exp. 5). A thick brown suspension was formed. Ten minutes after the addition the stirrer was replaced with a stopper and the flask was placed in a water bath at 50°C. When the stream of ammonia, escaping through the gas outlet, had become very faint, the outlet was connected to the water pump and the remaining ammonia was removed by evacuation during about 30 min, while maintaining heating at about 40°C. Nitrogen was then admitted to the flask and the outlet was connected, via a plastic tube, to a trap cooled at -70°C. From the dropping funnel were added 120 ml of water in 15 min, during which the flask was swirled occasionally in order to effect complete hydrolysis of the solid mass. The contents of the cold trap were returned to the flask. The dropping funnel and gas outlet were replaced with stoppers and a 40-cm Vigreux column was placed on the middle neck of the flask. The column was connected to a condenser for downward distillation. The receiver was immersed in a bath at -30°C. The flask was gradually heated, so that a mixture of the tertiary amine and water passed over (together with some gaseous ammonia, note 1). The distillation was stopped when the temperature in the head of the column was 100°C. Finely machine-powdered KOH was added (with swirling) to the distillate, with cooling in an ice-bath, until dissolution no longer occurred. The upper layer was separated off in a very small separating funnel and transferred into a 200-ml distillation flask, containing 15 g of powdered KOH. After vigorous shaking the flask was connected to a 40-cm Vigreux column, condenser and single receiver. Between the receiver, cooled at -70°C, and the water pump was placed a tube filled with KOH pellets. The amine was "distilled" under water-pump pressure and condensed in the receiver. During this operation the distillation flask was warmed in a water bath at 40°C. The contents of the receiver consisted of pure 3-amino-3-methyl-1-butyne: n_D^{16} 1.4268, yield 68%. Distillation at normal pressure gave a b.p. of 83°C/760 mmHg.

Note 1. The distillation should therefore be conducted in a fume hood.

(3) *1-Amino-1-ethynylcyclohexane*[146]

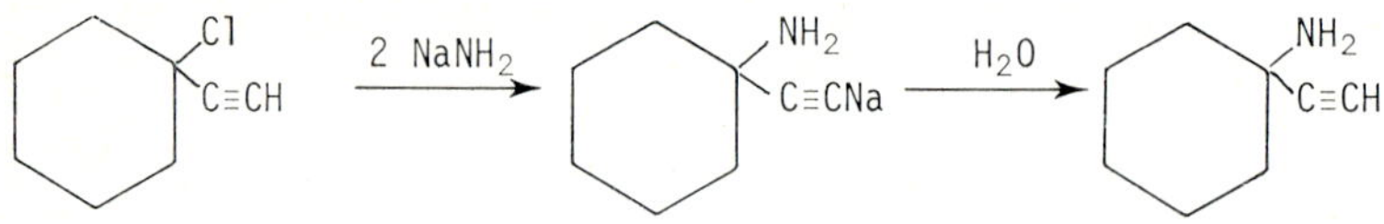

Apparatus: 3-l round-bottomed flask with a wide neck.

A suspension of sodium amide in 500 ml of anhydrous liquid ammonia was prepared from 18 g of sodium (see Chapter II, Exp. 11). To the suspension was added in 10 min with swirling a mixture of 0.30 mol of 1-chloro-1-ethynylcyclohexane (see VIII-2, Exp. 27) and 50 ml of diethyl ether. The reaction was very vigorous and a thick suspension was formed. The greater part of the ammonia was evaporated by placing the flask in a water bath at 50°C. After addition of 500 ml of ice-water the product was extracted three times with diethyl ether. The ethereal extracts were dried over anhydrous K_2CO_3 and subsequently concentrated in a water-pump vacuum. Distillation of the residue afforded the amine, b.p. 54°C/15 mmHg, n_D^{17} 1.4845, in 87% yield.

(4) *NN-Dimethylpropargylamine*[166]

$$HC{\equiv}CCH_2Br + 2\ (CH_3)_2NH \longrightarrow HC{\equiv}CCH_2N(CH_3)_2 + (CH_3)_2NH.HBr\downarrow$$

Apparatus: 1-l three-necked, round-bottomed flask, provided with a dropping funnel, a mechanical stirrer and a thermometer, combined with a vent.

Dibutyl ether (450 ml) was cooled to 0°C and mixed with 2.50 mol of liquid dimethylamine. Freshly distilled (at 100 mmHg) propargyl bromide was added dropwise with vigorous stirring. During this addition which took about 45 min the temperature of the mixture was maintained between -5 and 10°C. The dimethylamine.HBr salt separated immediately. After the propargyl bromide had been added the mixture was warmed at 20-25°C for an additional 1 h. The dropping funnel and thermometer were then replaced with stoppers and some boiling stones were introduced. The flask was connected to a distillation set, consisting of a 40-cm Widmer column, a condenser and a receiver, cooled at -75°C. Between the water pump and the receiver was placed a tube filled with KOH pellets. The system was evacuated (10-20 mmHg) and the three-necked flask warmed in a water-bath. A mixture of dimethylamine, dimethylpropargylamine and dibutyl ether condensed in the receiver. The distillation was stopped when about 300 ml of liquid had been collected. The distillate was carefully distilled at normal pressure through a 40-cm Widmer column. The greater part of the dimethylamine was condensed in a cold trap connected to the receiver. Redistillation of the fraction with b.p. 30-120^{0}C gave pure dimethylpropargylamine, b.p. 82^{0}C/760 mmHg, n_D^{20} 1.4197 in 87% yield.

6. ACETYLENIC ETHERS, SELENIDES AND KETONES

(1) *Ethers, derived from tertiary acetylenic alcohols*[149]

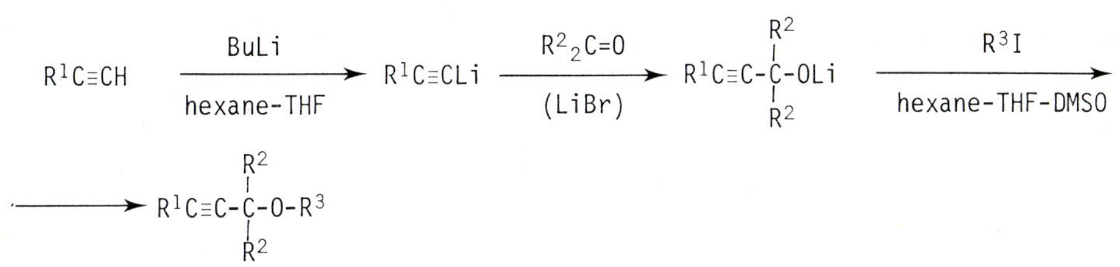

Apparatus: 1-l flask, see Chapter I, Fig. 1.

A solution of 0.20 mol of butyllithium in about 140 ml of hexane was cooled below -30^{0}C. Dry tetrahydrofuran (100 ml) was added with efficient cooling. Subsequently 0.22 mol of the alkyne (propyne and butyne were dissolved in 30 ml of THF, cooled below -20^{0}C) was introduced in 2 min and with vigorous stirring. During this addition the temperature was kept below -10^{0}C. Subsequently the ketone (0.20 mol) was added in 10 min with cooling between -30 and -50^{0}C. In the cases of the reactions of $CH_3C{\equiv}CLi$ and $CH_3OCH_2C{\equiv}CLi$ with cyclopentanone and cyclohexanone, however, a solution of 0.10 mol of anhydrous lithium bromide (freed from traces of water by heating the commercial salt at 150-170^{0}C during 30 min in a water-pump vacuum) in 20 ml of THF was added first (note 1). After the addition of the ketones the cooling bath was removed and the temperature of the mixture was allowed to rise to -5^{0}C. Methyl iodide or ethyl iodide (0.28 mol) and dry DMSO (note 2) (160 ml) were added successively. The temperature of the mixture rose to 30-35^{0}C and salt separated from the solution. The reaction was terminated by warming the mixture for 1.5 h at 45-50^{0}C. Two clear layers were formed when stirring was stopped. Ice-water (500 ml) was added and after separation of the layers four extractions with redistilled pentane were carried out. The combined solutions were washed three times with brine and were subsequently dried over magnesium sulfate. After filtration and removing the solvents in a water-pump vacuum the residue was carefully distilled through a 30-cm-Widmer column. The following compounds were prepared (yields at least 80%):

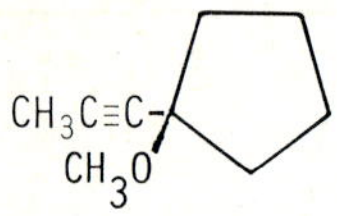

, b.p. 66°C/12 mmHg, n_D^{20} 1.4624;

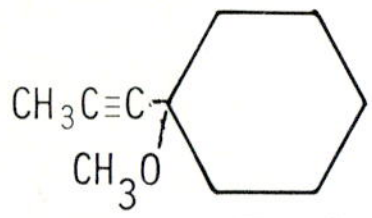

, b.p. 80°C/12 mmHg, n_D^{20} 1.4701;

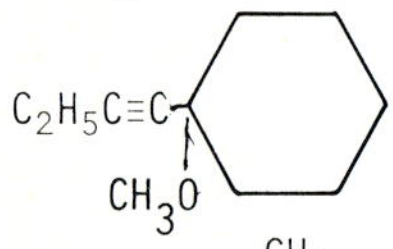

, b.p. 80°C/10 mmHg, n_D^{20} 1.4658;

$CH_3OCH_2C{\equiv}C\text{-}C(CH_3)_2\text{-}OCH_3$, b.p. 52°C/15 mmHg, n_D^{20} 1.4300;

$C_2H_5OCH_2C{\equiv}C\text{-}C(CH_3)_2\text{-}OC_2H_5$, b.p. 76°C/19 mmHg, n_D^{20} 1.4306;

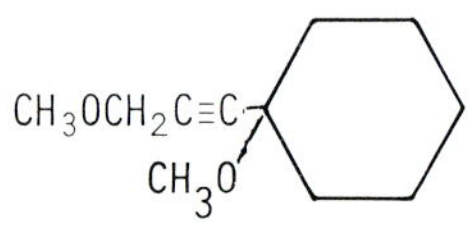

, b.p. 114°C/12 mmHg, n_D^{20} 1.4690.

Note 1. When the addition of LiBr was omitted, the methyl ethers derived from propyne and cyclopentanone or cyclohexanone were obtained in lower than 30% yields.

Note 2. Dried by stirring the commercial product during some hours with calcium hydride at 40-50°C and subsequently distilling in a water-pump vacuum. HMPT as a cosolvent gives similar results, but the reaction times are much longer.

(2) *1,4-Di-tert.-butoxy-2-butyne*[76]

$$HOCH_2C{\equiv}C\text{-}CH_2OH \xrightarrow[p\text{-toluene sulfonic acid, } CH_2Cl_2]{\text{isobutene}} (CH_3)_3C\text{-}O\text{-}CH_2\text{-}C{\equiv}C\text{-}CH_2\text{-}O\text{-}C(CH_3)_3$$

Apparatus: 2-l three-necked, round-bottomed flask, provided with a gas inlet tube, a gas-tight mechanical stirrer and a gas outlet; two washing bottles filled with paraffin oil were connected with the gas inlet and outlet.

To a suspension of 2.0 mol of finely powdered 2-butyne-1,4-diol (note 1) in 600 ml of dry dichloromethane were added 50 g of anhydrous *p*-toluenesulfonic acid (note 2). Isobutene (6 mol) was introduced with vigorous stirring. The flow was adjusted in such a way that only a small amount escaped from the solution (note 3). The reaction was slightly exothermic, so that no external cooling was applied. When all of the isobutene had been introduced, stirring was stopped (note 4) and the flask was allowed to stand overnight. The reaction mixture was then washed five times with ice-water. The organic layer was dried over K_2CO_3 and after filtration the solvent was removed in a water-pump vacuum. Distillation of the remaining liquid through a 20-cm Vigreux column afforded the bis-ether, b.p. $70^{o}C/0.5$ mmHg, n_D^{20} 1.4460, in 85% yield.

Note 1. The technical product (brown) was purified by crystallization from a 2:1 mixture of diethyl ether and THF.

Note 2. Prepared by azeotropic removal of water from the commercial, water-containing product by means of benzene. This solvent was removed in a water--pump vacuum and the remaining mass was powdered. The water-containing acid appeared to work much less efficiently as catalyst.

Note 3. This required continuous control, especially during the first 2-3 h.

Note 4. Gas-liquid chromatography showed complete conversion into the bis-ether. If this is not the case, additional isobutene must be introduced and the mixture be allowed to stand for a longer period.

(3) *1-Methoxy-1-ethynylcyclohexane*[149]

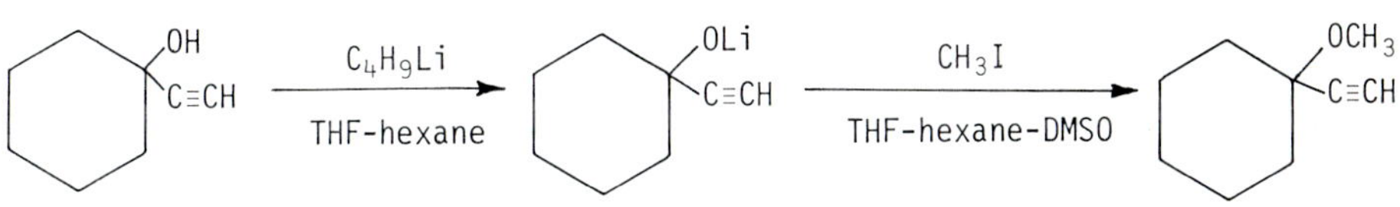

Apparatus: 1-l flask, see Fig. 1.

To a solution of 0.32 mol of butyllithium in about 220 ml of hexane were added 220 ml of THF below $-10^{o}C$. A solution of 0.30 mol of 1-ethynylcyclohexanol[1] (also commercially available) in 30 ml of THF was added in 15 min at -25 to $-20^{o}C$. Subsequently 100 ml of dry, pure DMSO were added at the same temperature and 5 min later 0.45 mol of methyl iodide were run in during 1-2 min. The cooling bath was then removed and the reaction proceeded rather slowly. After stirring for 1 h at $5-15^{o}C$ the mixture was warmed to $45^{o}C$ and held at this temperature for an additional hour. It was then poured into 400 ml of a saturated solution of sodium chloride.

After vigorous shaking the layers were separated and three extractions of the aqueous layer with diethyl ether were carried out. The combined solutions were washed twice with saturated NaCl solution and then dried over magnesium sulfate. The greater part of the solvents was distilled off at normal pressure through a 40-cm Vigreux column. Careful distillation of the remaining liquid through a 40-cm Widmer column gave the desired methyl ether, b.p. 56°C/18 mmHg, n_D^{20} 1.4620, in 86% yield.

In a similar way $(CH_3)_2C(OCH_3)C{\equiv}C\text{-}CH_2OCH_3$, b.p. 52°C/15 mmHg, n_D^{20} 1.4300, was prepared in 90% yield from $(CH_3)_2C(OH)\text{-}C{\equiv}C\text{-}CH_2\text{-}OCH_3$.

(4) *1-Trimethylsilyloxy-1-ethynylcyclohexane*[147]

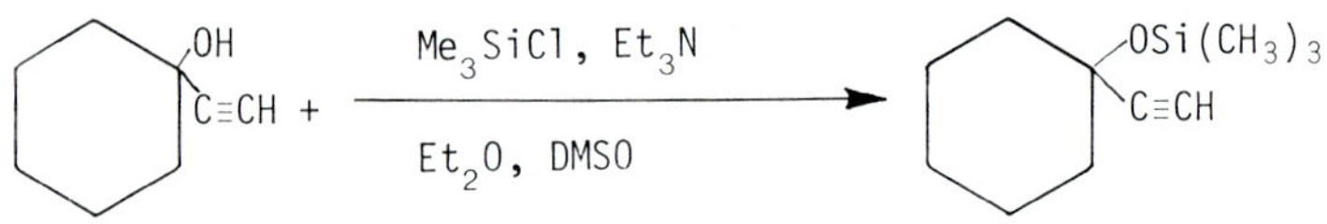

Apparatus: 1-l flask, see Chapter I, Fig. 1, the thermometer-gas outlet combination was replaced with a reflux condenser.

In the flask were placed 0.30 mol of 1-ethynylcyclohexanol (commercially available), 350 ml of dry diethyl ether, 0.35 mol of dry triethylamine and 5 ml of DMSO (or 3 ml of commercially available 1,5-diazabicyclo[5.4.0]undec-5-ene). Trimethylchlorosilane (0.35 mol) was added in 20 min with vigorous stirring. A thick precipitate was formed and refluxing of the diethyl ether started after a few minutes. When the exothermic reaction had subsided, the mixture was heated under reflux for 1 h. Ice-water (400 ml) was then added with vigorous stirring. The upper layer was separated off, washed with water and dried over magnesium sulfate. After evaporation of the diethyl ether in a water-pump vacuum the residue was distilled through a 40-cm Vigreux column. The trimethylsilyl ether, b.p. 76°C/17 mmHg, n_D^{28} 1.4453, was obtained in 91% yield.

Note 1. Other tertiary alcohols (including saturated ones) can be trimethylsilylated in a similar way.

(5) *Ethyl propargyl selenide*

$$2\ Na + 2\ Se \xrightarrow{\text{liq. } NH_3} Na_2Se_2 \qquad (1)$$

$$Na_2Se_2 + 2\ C_2H_5Br \xrightarrow{\text{liq. } NH_3} C_2H_5SeSeC_2H_5 \qquad (2)$$

$$C_2H_5SeSeC_2H_5 \xrightarrow[\text{liq. } NH_3]{2\ Li} 2\ C_2H_5SeLi \qquad (3)$$

$$C_2H_5SeLi + HC{\equiv}CCH_2Cl \xrightarrow[C_2H_5OH\ +\ H_2O]{} HC{\equiv}CCH_2SeC_2H_5 \qquad (4)$$

Apparatus: 5-l flask for reaction (1) and (2), see Fig. 4; 3-l flask for reaction (3) and (4), see Fig. 2.

To a vigorously stirred solution of 0.40 mol of sodium in 1.2 l of anhydrous liquid ammonia was added in 1 g portions 0.42 mol of powdered selenium. Thirty minutes after this addition 0.6 mol of ethyl bromide was introduced in 5-g portions at intervals of a few seconds. The ammonia was allowed to evaporate overnight and to the remaining mass were added 500 ml of water. After dissolution of the salt the diselenide was extracted with three 100-ml portions of diethyl ether. The ethereal solutions were dried over magnesium sulfate and subsequently concentrated in a water-pump vacuum, giving diethyl diselenide in more than 95% yield. The product was dissolved in 100 ml of diethyl ether and this solution was added to 1 l of liquid ammonia in the second reaction flask. Small pieces of 0.05 g (freshly cut) of lithium were introduced through the tube in 30 min with vigorous stirring. The required amount of lithium to obtain a uniformly blue solution closely approached the stoichiometric amount (less than 0.4 mol). The greater part of the ammonia was evaporated by warming the flask in a water-bath at 40°C. When the volume of the mixture had decreased to about 150 ml a mixture of 150 ml of ethanol and 40 ml of water was added cautiously. Propargyl chloride (0.40 mol) was added in 10 min without external cooling, then stirring was continued for another 1 h at about 30°C. Ice-water (500 ml) was added, then the selenide was extracted six times with small amounts of redistilled pentane. The combined solutions were washed with water and dried over magnesium sulfate. The greater part of the pentane was distilled off at normal pressure through a 40-cm Vigreux column. Ethyl propargyl selenide, b.p. 45°C/15 mmHg, n_D^{20} 1.5263, was obtained in 88% yield by subsequent distillation in a water-pump vacuum.

(6) *Ethyl propargyl ketone*[148]

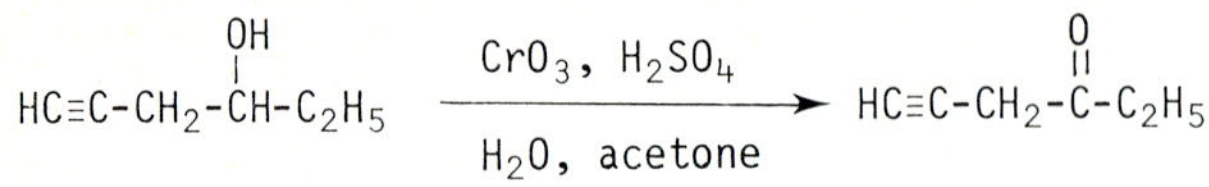

Apparatus: 1-l three-necked flask with a stirrer, thermometer and stopper.

In the flask were placed 90 ml of acetone, 0.30 mol of the propargyl carbinol[1, 176] and 100 ml of pentane (note 1). To this mixture was added with vigorous stirring a solution of 25 g of CrO_3 in 23 ml of 96% H_2SO_4 and 60 ml of water. During this addition, which was carried out in ten equal portions at intervals of about 1 min, the temperature of the reaction mixture was maintained between -5 and 0^oC by cooling with a dry-ice-acetone bath. After the addition the temperature of the mixture was allowed to rise in 35 min to 5^oC, then it was cooled to -10^oC and ten extractions with 50-ml portions of a 2:1 mixture of pentane and diethyl ether were carried out. These extractions were carried out without any delay, keeping the temperature of the aqueous layer below 0^oC. The combined extracts were washed twice with 50-ml portions of 1 N H_2SO_4 saturated with NH_4Cl. After drying the extracts over magnesium sulfate the solvents were removed in a water-pump vacuum (rotary evaporator). The remaining liquid was distilled through a short (20-30 cm) Vigreux column. The distillate was collected in a receiver, cooled at 0^oC. After a small first fraction (water-containing ketone) the main fraction was collected between 36 and 40^oC/15 mmHg, n_D^{21} 1.4350, in 70% yield. The purity (GLC) was only about 80%. In spite of several repetitions no better result could be obtained. The nature of the impurity is not known.

General note. The propargyl ketone is extremely base-sensitive. Traces of alkali on the glass walls cause partial conversion into the allenic isomer. Therefore, all glassware used in this procedure must be rinsed with 1 N H_2SO_4 and subsequently with acetone and dried with a flow of air.

Note 1. Commercial pentane may contain traces of higher boiling hydrocarbons and should therefore be distilled.

(7) *Methyl propargyl ether*[177]

$$2\ HC{\equiv}CCH_2OH + (CH_3)_2SO_4 + 2\ NaOH \longrightarrow 2\ HC{\equiv}CCH_2OCH_3 + Na_2SO_4 + 2\ H_2O$$

Apparatus: 3-l three-necked flask, fitted with a dropping funnel, gas-tight stirrer, efficient reflux condenser and a thermometer, dipping into the liquid.

To a mixture of 7.5 mol of propargyl alcohol (technical grade) and 100 ml of water was added with cooling in ice water a solution of 410 g of *sodium* hydroxide (note 1) in 650 ml of water. Subsequently, 515 g of dimethyl sulfate (technical grade) were added at such a rate (in about 2 h) that the temperature was maintained between 50 and 55°C. After the addition the mixture was heated under reflux during 2 h. The reflux condenser was then replaced by a 40-cm Vigreux column, which was connected to a descending condenser and a receiver, cooled at 0°C. The product was distilled off as quickly as possible, stirring being continued. The distillation was stopped when the thermometer in the top of the column indicated 95°C. The contents of the receiver were washed two times with a saturated NH_4Cl solution in order to remove some methanol, and were dried over magnesium sulfate. Pure $HC{\equiv}CCH_2OCH_3$, n_D^{20} 1.3969, was obtained in 79-83% yield. A small sample was distilled: b.p. 61°C (note 2). The product should be stored under nitrogen in the refrigerator.

In a similar way was prepared: $HC{\equiv}CCH_2OC_2H_5$, n_D^{20} 1.4039, yield 74-78%. During the addition of diethyl sulfate the temperature of the mixture was kept between 60 and 70°C. After 2 h of refluxing, the product was distilled off from the reaction mixture through a 20-cm Vigreux column. The distillate was washed with aqueous NH_4Cl. A small part was redistilled: b.p. 80°C/760 mmHg.

Note 1. Potassium hydroxide should not be used, since potassium sulfate is much less soluble in water than sodium sulfate. The reaction mixture would become very thick.

Note 2. A very vigorous explosion has occurred in our laboratory in the course of a distillation of a large amount of the propargyl ether, which had been stored for 3 weeks at room temperature. Traces of peroxide (the product had not been kept under nitrogen) might have been responsible for the explosion. In any case it is not advisable to distil large quantities of propargyl ethers at normal pressure.

(8) *1,4-Dimethoxy-2-butyne*[178]

$$HOCH_2C{\equiv}CCH_2OH + (CH_3)_2SO_4 + 2\ NaOH \longrightarrow CH_3OCH_2C{\equiv}CCH_2OCH_3 + Na_2SO_4 + H_2O$$

Apparatus: 2-l three-necked flask, equipped with a dropping funnel, stirrer and thermometer, combined with a vent.

To a solution of 2 mol of pure butynediol (recrystallizedfrom a 1:5 mixture of THF and diethyl ether) in 320 ml of water were added in turn during 1.5 h 20 equal portions of NaOH pellets (total amount 200 g) and 20 equal portions of dimethyl sulfate (total amount 630 g). The temperature of the mixture was kept (by occasional cooling with ice-water) between 35 and 40°C. After the addition the mixture was heated in a bath of 90°C for 3 h. Ice-water (300 ml) was then added and after cooling to 30°C five extractions with diethyl ether were carried out. The (unwashed) extracts were dried over magnesium sulfate. After the greater part of the diethyl ether had been distilled off at 760 mmHg through a 40-cm Vigreux column, the residue was distilled in a water-pump vacuum. $CH_3OCH_2C{\equiv}CCH_2OCH_3$, b.p. 54°C/ /12 mmHg, n_D^{20} 1.4392, was obtained in 88-92% yield.

In a similar way was prepared (using the same *molar* quantities): $C_2H_5OCH_2C{\equiv}CCH_2OC_2H_5$, b.p. 76°C/12 mmHg, n_D^{20} 1.4338, in 84% yield.

TYPE INDEX

$C\equiv C-(C)_4Cl$	48
$C\equiv C-(C)_3Br$	125
C≡C-C=C-C-Cl(Br)	102,193

-Allenic-

Type C=C=C-Hal	44,176-184
$C=C=C-(C)_3-Cl$	30

Hydrocarbons

-Acetylenes-

Type HC≡C-C	48,54,83,103, 119,185,186
HC≡C-C=C	186
HC≡C-C≡C	146
C-C≡C-C	119
HC≡C-C≡C-C≡CH	136
C≡C-C-C≡C	70,71
HC≡C-C=C=C	90

-Allenes-

Type $H_2C=C=C$	28-30,138,143, 157-164,190,191
-CH=C=CH-	164
>C=C=C<	141
$H_2C=C=C-C=C$	89,91,165,192
$H_2C=C=CH-C\equiv CH$	90
C=C=C-C=C=C	168

-Cumulenes-

Type C=C=C=C	141,144,145

Ketones

-Acetylenic-

Type C≡C-C(=O)	206
C≡C-C-C(=O)	193,236

-Allenic-

Type C=C=C-C(=O)	84,101
C=C=C-C-C(=O)	200

Nitriles

-Acetylenic-

Type C≡C-C≡N	62,227
C≡C-C-C≡N	225
C≡C-C=C-C-C≡N	102

-Allenic-

Type C=C=C-C≡N	173-175,226
C=C=C-C-C≡N	171
C=C=C-C=C-C≡N	102

FORMULA INDEX

$C_4H_6S_2$	$CH_3C{\equiv}C\text{-}SS\text{-}CH_3$	69	III, Exp. 48
C_5H_4	$HC{\equiv}C\text{-}CH_2C{\equiv}CH$	70	III, Exp. 49
	$HC{\equiv}C\text{-}CH{=}C{=}CH_2$	90	IV, Exp. 2
$C_5H_4O_4$	$HOOC\text{-}CH{=}C{=}CH\text{-}COOH$	50	III, Exp. 25
C_5H_5Br	$HC{\equiv}C\text{-}CH{=}CH\text{-}CH_2Br$	102	IV, Exp. 17
C_5H_5N	$CH_3C{\equiv}C\text{-}CH_2\text{-}C{\equiv}N$	225	VIII-4, Exp. 1
	$C_2H_5C{\equiv}C\text{-}C{\equiv}N$	62	III, Exp. 39
	$CH_3CH{=}C{=}CH\text{-}C{\equiv}N$	174	VI, Exp. 24
C_5H_6	$H_2C{=}C{=}CH\text{-}CH{=}CH_2$	89	IV, Exp. 1
	$H_2C{=}C{=}C{=}CH\text{-}CH_3$	144	V, Exp. 25
$C_5H_6Cl_2$	$CH_3\text{-}CH(Cl)\text{-}C{\equiv}C\text{-}CH_2Cl$	216	VIII-2, Exp. 2
C_5H_6O	$HC{\equiv}C\text{-}C(OCH_3){=}CH_2$	131	V, Exp. 13
	$HC{\equiv}C\text{-}CH{=}CH\text{-}CH_2OH$	78	III, Exp. 57
	$HC{\equiv}C\text{-}CH(OH)CH{=}CH_2$	78	III, Exp. 58
	$HC{\equiv}C\text{-}CH_2\text{-}C({=}O)\text{-}CH_3$	193	VI, Exp. 47
	$CH_3C{\equiv}C\text{-}C({=}O)\text{-}CH_3$	206	VII-3, Exp. 1
$C_5H_6O_2$	$CH_3C{\equiv}C\text{-}CH_2\text{-}COOH$	105	IV, Exp. 22
	$C_2H_5C{\equiv}C\text{-}COOH$	105	IV, Exp. 22
	$CH_3CH{=}C{=}CH\text{-}COOH$	33	III, Exp. 6
	$H_2C{=}C{=}C(CH_3)\text{-}COOH$	35	III, Exp. 7
C_5H_6S	$H_2C{=}CH\text{-}C{\equiv}C\text{-}SCH_3$	135	V, Exp. 16
C_5H_7Br	$HC{\equiv}C\text{-}C(CH_3)_2Br$	219	VIII-2, Exp. 6
	$C_2H_5C{\equiv}C\text{-}CH_2Br$	221	VIII-2, Exp. 10
	$(CH_3)_2C{=}C{=}CHBr$	181	VI, Exp. 31
C_5H_7Cl	$HC{\equiv}C\text{-}C(CH_3)_2Cl$	218	VIII-2, Exp. 5
C_5H_7ClS	$C_2H_5C{\equiv}C\text{-}S\text{-}CH_2Cl$	73	III, Exp. 52
C_5H_7ClSe	$C_2H_5C{\equiv}C\text{-}Se\text{-}CH_2Cl$	73	III, Exp. 52
C_5H_8	$(CH_3)_2C{=}C{=}CH_2$	190	VI, Exp. 44
	$C_3H_7C{\equiv}CH$	83	III, Exp. 62
	$CH_3CH{=}C{=}CH\text{-}CH_3$	160	VI, Exp. 5
C_5H_8O	$H_2C{=}C{=}CH\text{-}CH_2CH_2OH$	210	VII-3, Exp. 7
	$HC{\equiv}CCH_2\text{-}OC_2H_5$	237	VIII-6, Exp. 7
	$H_2C{=}C{=}CH\text{-}CH(CH_3)OH$	188	VI, Exp. 42
C_5H_8OS	$H_2C{=}C{=}C(SCH_3)\text{-}OCH_3$	42	III, Exp. 16
	$C_2H_5\text{-}S({=}O)\text{-}C{\equiv}C\text{-}CH_3$	214	VIII-1, Exp. 3
$C_5H_8O_2S$	$CH_3C{\equiv}C\text{-}SO_2\text{-}C_2H_5$	214	VIII-1, Exp. 4
C_5H_8S	$H_2C{=}C{=}CH\text{-}SC_2H_5$	106	IV, Exp. 23
	$CH_3C{\equiv}C\text{-}SC_2H_5$	106	IV, Exp. 23
C_5H_8Se	$HC{\equiv}C\text{-}CH_2\text{-}SeC_2H_5$	235	VIII-6, Exp. 5
	$H_2C{=}C{=}CH\text{-}SeC_2H_5$	100	IV, Exp. 15

C_5H_9N	$HC{\equiv}C\text{-}C(CH_3)_2NH_2$	229	VIII-5, Exp. 2
	$HC{\equiv}CCH_2N(CH_3)_2$	230	VIII-5, Exp. 4
	$H_2C{=}C{=}CH\text{-}N(CH_3)_2$	98	IV, Exp. 12
$C_5H_{10}Si$	$HC{\equiv}C\text{-}Si(CH_3)_3$	55	III, Exp. 31
C_6H_2	$HC{\equiv}C\text{-}C{\equiv}C\text{-}C{\equiv}CH$	136	V, Exp. 17
$C_6H_4Cl_2$	$ClCH_2C{\equiv}CC{\equiv}CCH_2Cl$	221	VIII-2, Exp. 9
C_6H_4Se	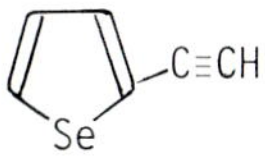	207	VII-3, Exp. 3
C_6H_5N	$H_2C{=}C{=}CH\text{-}CH{=}CH\text{-}C{\equiv}N$	102	IV, Exp. 17
	$HC{\equiv}C\text{-}CH{=}CH\text{-}CH_2\text{-}C{\equiv}N$	102	IV, Exp. 17
C_6H_6	$HC{\equiv}C\text{-}CH_2C{\equiv}C\text{-}CH_3$	71	III, Exp. 50
C_6H_6O	$H_2C{=}C{=}CH\text{-}C{\equiv}C\text{-}CH_2OH$	194	VI, Exp. 49
	$HC{\equiv}C\text{-}C{\equiv}C\text{-}CH(OH)CH_3$	83	III, Exp. 61
C_6H_6OS	$CH_3C{\equiv}C\text{-}S({=}O)\text{-}C{\equiv}C\text{-}CH_3$	61	III, Exp. 38
$C_6H_6O_2$	$HOCH_2C{\equiv}CC{\equiv}CCH_2OH$	81	III, Exp. 60
$C_6H_6O_2S$	$CH_3C{\equiv}C\text{-}SO_2\text{-}C{\equiv}C\text{-}CH_3$	214	VIII-1, Exp. 4
$C_6H_6O_4$	$HOOC\text{-}C(CH_3){=}C{=}CH\text{-}COOH$	51	III, Exp. 26
C_6H_6S	$CH_3C{\equiv}C\text{-}S\text{-}C{\equiv}CCH_3$	60	III, Exp. 37
C_6H_7Br	$HC{\equiv}CCH{=}CH\text{-}CH(Br)CH_3$	193	VI, Exp. 48
C_6H_7Cl	$HC{\equiv}CCH{=}CH\text{-}CH(Cl)CH_3$	193	VI, Exp. 48
C_6H_7N	$C_3H_7C{\equiv}C\text{-}C{\equiv}N$	62	III, Exp. 39
	$CH_3CH{=}C{=}CH\text{-}CH_2\text{-}C{\equiv}N$	171	VI, Exp. 20
	$(CH_3)_2C{=}C{=}CH\text{-}C{\equiv}N$	226	VIII-4, Exp. 3
C_6H_8	$H_2C{=}C{=}CH\text{-}CH{=}CH\text{-}CH_3$	192	VI, Exp. 46
C_6H_8O	$HC{\equiv}C\text{-}CH_2C({=}O)\text{-}C_2H_5$	236	VIII-6, Exp. 6
	$HC{\equiv}C\text{-}CH(OH)CH{=}CH\text{-}CH_3$	78	III, Exp. 58
	$H_2C{=}CH\text{-}C{\equiv}C\text{-}OC_2H_5$	131	V, Exp. 12
	$H_2C{=}C{=}CH\text{-}C({=}O)\text{-}C_2H_5$	101	IV, Exp. 16
	$CH_2{=}CH\text{-}C{\equiv}C\text{-}CH_2CH_2OH$	64	III, Exp. 41
	$HC{\equiv}C\text{-}C(OCH_3){=}CHCH_3$	131	V, Exp. 13
	$H_2C{=}C{=}C{=}CH\text{-}OC_2H_5$	126	V, Exp. 7
$C_6H_8S_2$	$C_2H_5\text{-}C({=}S)\text{-}SCH_2C{\equiv}CH$	224	VIII-3, Exp. 4
C_6H_9Cl	$H_2C{=}C{=}CH(CH_2)_3Cl$	30	III, Exp. 2
	$HC{\equiv}C\text{-}(CH_2)_4Cl$	48	III, Exp. 23
C_6H_9N	$HC{\equiv}C\text{-}CH{=}CH\text{-}N(CH_3)_2$	104	IV, Exp. 20
	$H_2C{=}CH\text{-}C{\equiv}C\text{-}N(CH_3)_2$	132	V, Exp. 14
C_6H_9NO	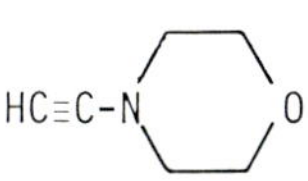	122	V, Exp. 4

C_6H_{10}	$(CH_3)_3C-C\equiv CH$	119	V, Exp. 1
	$C_4H_9-C\equiv CH$	83	III, Exp. 62
$C_6H_{10}O$	$(CH_3)_2C=C=CH-CH_2OH$	189	VI, Exp. 43
$C_6H_{10}OS$	$H_2C=C=C(SCH_3)CH_2CH_2OH$	46	III, Exp. 21
	$C_2H_5CH=C=CH-S(=O)-CH_3$	213	VIII-1, Exp. 2
$C_6H_{10}O_2$	$CH_3OCH=C=CHCH_2OCH_3$	94	IV, Exp. 7
	$CH_3OCH_2C\equiv CCH_2OCH_3$	238	VIII-6, Exp. 8
	$H_2C=C=C(OCH_3)CH_2CH_2OH$	39	III, Exp. 12
	$H_2C=C=C(OCH_3)CH(CH_3)OH$	43	III, Exp. 17
$C_6H_{10}O_2S$	$(CH_3)_2C=C=CH-SO_2-CH_3$	198	VII-1, Exp. 3
$C_6H_{10}S$	$C_2H_5CH=C=CH-SCH_3$	111	IV, Exp. 26
$C_6H_{11}ClSSi$	$(CH_3)_3Si-C\equiv C-S-CH_2Cl$	73	III, Exp. 52
$C_6H_{11}ClSi$	$(CH_3)_3Si-C\equiv C-CH_2Cl$	66	III, Exp. 44
$C_6H_{12}OSi$	$(CH_3)_3Si-C\equiv C-CH_2OH$	58	III, Exp. 34
$C_6H_{12}SSi$	$CH_3C\equiv C-S-Si(CH_3)_3$	68	III, Exp. 47
C_7H_9N	$C_4H_9C\equiv C-C\equiv N$	62	III, Exp. 39
C_7H_9NS	$(CH_3)_3C-C\equiv C-S-C\equiv N$	67	III, Exp. 46
$C_7H_{10}O$	$H_2C=C(CH_3)C\equiv C-CH_2OCH_3$	97	IV, Exp. 11
	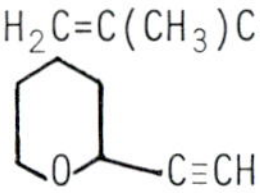	172	VI, Exp. 22
	$HC\equiv C-C(CH_3)=CHOC_2H_5$	188	VI, Exp. 41
	$HC\equiv C-CH=C(CH_3)-OC_2H_5$	125	V, Exp. 6
	$H_2C=C(CH_3)CH=C=CH-OCH_3$	97	IV, Exp. 11
	$H_2C=C=C=C(CH_3)-OC_2H_5$	45	III, Exp. 20
	$(CH_3)_2C=C=C=CH-OCH_3$	127,130	V, Exp. 8 and 11
	$H_2C=C=CHC(CH_3)_2-C(=O)H$	204	VII-2, Exp. 6
$C_7H_{10}O_2$	$C_3H_7C\equiv C-CH_2COOH$	105	IV, Exp. 22
	$C_4H_9C\equiv C-COOH$	105	IV, Exp. 22
	$CH_3-C(=O)OC(CH_3)_2C\equiv CH$	212	VIII-3, Exp. 1
	$H_2C=C=CH-CH_2COOC_2H_5$	203	VII-2, Exp. 4
	$(CH_3)_2C=C=CH-O-C(=O)-CH_3$	205	VII-2, Exp. 7
$C_7H_{10}S_2$	$H_2C=C=CH-CH_2-C(=S)-SC_2H_5$	201	VII-2, Exp. 3
$C_7H_{11}Br$	$C_4H_9C\equiv C-CH_2Br$	221	VIII-2, Exp. 10
$C_7H_{11}BrO$	$BrCH_2CH(OC_2H_5)CH_2C\equiv CH$	125	V, Exp. 6
$C_7H_{11}ClS$	$(CH_3)_3C-C\equiv C-SCH_2Cl$	73	III, Exp. 52
$C_7H_{11}NO$	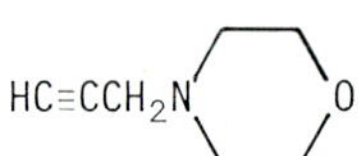	228	VIII-5, Exp. 1
	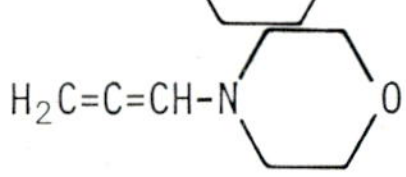	99	IV, Exp. 13

C_7H_{12}	$(CH_3)_3C\text{-}CH_2C{\equiv}CH$	103	IV, Exp. 18
	$(CH_3)_2C{=}C{=}C(CH_3)_2$	141	V, Exp. 22
	$C_4H_9CH{=}C{=}CH_2$	157	VI, Exp. 1
	t.-$C_4H_9CH{=}C{=}CH_2$	157	VI, Exp. 2
$C_7H_{12}O$	$HC{\equiv}C(CH_2)_5OH$	104	IV, Exp. 21
	$C_4H_9C{\equiv}C\text{-}CH_2OH$	77	III, Exp. 56
	$H_2C{=}C{=}CH(CH_2)_4OH$	30	III, Exp. 3
	$(CH_3)_2C{=}C{=}CHCH(CH_3)OH$	189	VI, Exp. 43
	$H_2C{=}C{=}CH\text{-}OC(CH_3)_3$	92	IV, Exp. 5
$C_7H_{12}O_2$	$H_2C{=}C{=}C(OCH_3)CH(C_2H_5)OH$	43	III, Exp. 17
	$H_2C{=}C{=}C(OCH_3)C(CH_3)_2OH$	43	III, Exp. 17
	$H_2C{=}C{=}C(OCH_3)CH_2OC_2H_5$	40	III, Exp. 13
	$HC{\equiv}C\text{-}CH(OC_2H_5)_2$	148	V. Exp. 28
$C_7H_{12}S$	$CH_3C{\equiv}C\text{-}S\text{-}C(CH_3)_3$	109	IV, Exp. 24
	$C_3H_7CH{=}C{=}CH\text{-}SCH_3$	209	VII-3, Exp. 6
	$H_2C{=}C{=}CH\text{-}SC(CH_3)_3$	109	IV, Exp. 24
	$H_2C{=}C{=}C(C_2H_5)\text{-}SC_2H_5$	46	III, Exp. 21
$C_7H_{12}S_2$	$(CH_3)_2C{=}C{=}C(SCH_3)_2$	74	III, Exp. 53
$C_7H_{13}N$	$HC{\equiv}C\text{-}CH_2N(C_2H_5)_2$	228	VIII-5, Exp. 1
$C_7H_{13}NO$	$CH_3O\text{-}CH_2C{\equiv}CCH_2N(CH_3)_2$	132	V, Exp. 14
	$CH_3O\text{-}CH{=}C{=}CH\text{-}CH_2N(CH_3)_2$	113	IV, Exp. 28
	$CH_3O\text{-}CH_2CH{=}C{=}CH\text{-}N(CH_3)_2$	112	IV, Exp. 27
$C_9H_{14}O$	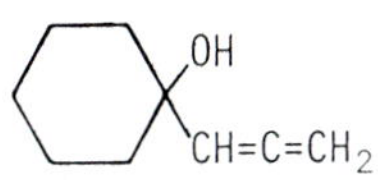	211	VII-3, Exp. 9
$C_7H_{14}OSi$	$(CH_3)_3Si\text{-}C{\equiv}C\text{-}OC_2H_5$	57	III, Exp. 33
	$H_2C{=}C{=}C(OCH_3)Si(CH_3)_3$	41	III, Exp. 15
$C_7H_{14}SSi$	$HC{\equiv}C\text{-}CH(SCH_3)\text{-}Si(CH_3)_3$	54	III, Exp. 30
C_8H_5Cl	$C_6H_5C{\equiv}C\text{-}Cl$	67	III, Exp. 45
C_8H_{10}	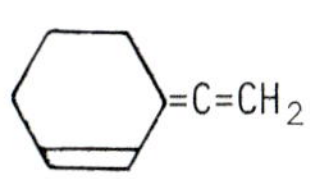	91	IV, Exp. 3
$C_7H_{10}O$	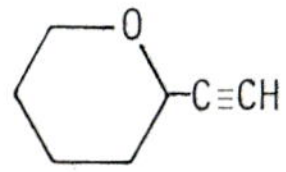	172	VI, Exp. 22
$C_8H_{10}O$	$CH_3\text{-}C{\equiv}C\text{-}CH_2\text{-}C{\equiv}C\text{-}OC_2H_5$	73	III, Exp. 51
	$H_2C{=}CH\text{-}C{\equiv}C\text{-}CH{=}CH\text{-}OC_2H_5$	137	V, Exp. 18
$C_8H_{10}OS$	$C_2H_5C{\equiv}C\text{-}S({=}O)\text{-}C{\equiv}C\text{-}C_2H_5$	61	III, Exp. 38
$C_8H_{10}S$	$C_2H_5C{\equiv}C\text{-}S\text{-}C{\equiv}C\text{-}C_2H_5$	60	III, Exp. 37

$C_8H_{11}Cl$	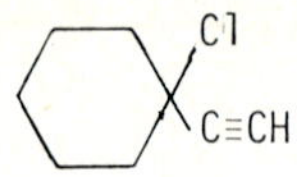	219	VIII-2, Exp. 7
$C_8H_{11}N$	$C_4H_9C{\equiv}C\text{-}CH_2C{\equiv}N$	225	VIII-4, Exp. 2
C_8H_{12}	cyclooctyne	119	V, Exp. 2
	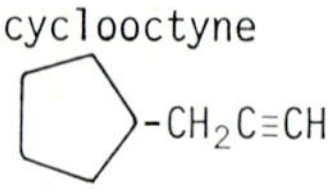	185	VI, Exp. 37
	$H_2C{=}C(C_4H_9)\text{-}C{\equiv}CH$	186	VI, Exp. 40
	$(CH_3)_2C{=}C(C_2H_5)\text{-}C{\equiv}CH$	186	VI, Exp. 40
	$H_2C{=}C(CH_3)\text{-}C(C_2H_5){=}C{=}CH_2$	165	VI, Exp. 12
	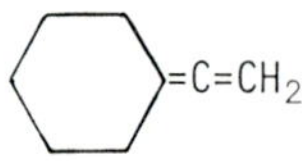	191	VI, Exp. 45
	$t.\text{-}C_4H_9CH{=}C{=}C{=}CH_2$	145	V, Exp. 26
	$(CH_3)_2C{=}C{=}C{=}C(CH_3)_2$	141	V. Exp. 22
$C_8H_{12}O$	$(CH_3)_2C{=}C{=}CH\text{-}CH_2\text{-}C({=}O)\text{-}CH_3$	200	VII-2, Exp. 2
	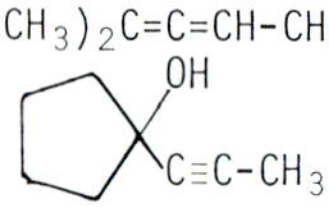	75	III, Exp. 54
$C_8H_{12}O_2$	$CH_3C{\equiv}C\text{-}CH_2CH_2COOC_2H_5$	207	VII-3, Exp. 2
	$CH_3CH{=}C{=}CH\text{-}CH_2COOC_2H_5$	204	VII-2, Exp. 5
	$C_4H_9CH{=}C{=}CHCOOH$	33	III, Exp. 6
	$t.\text{-}C_4H_9CH{=}C{=}CHCOOH$	33	III, Exp. 6
	$C_2H_5O\text{-}CH{=}C{=}C{=}CH\text{-}OC_2H_5$	128	V, Exp. 9
$C_8H_{12}Si$	$HC{\equiv}C\text{-}CH_2C{\equiv}C\text{-}Si(CH_3)_3$	71	III, Exp. 50
$C_8H_{13}ClO$	$(CH_3)_3C\text{-}CH(OH)\text{-}C{\equiv}C\text{-}CH_2Cl$	145	V, Exp. 26
$C_8H_{13}N$	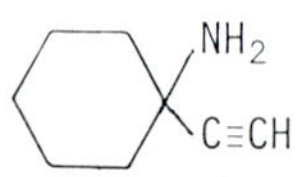	230	VIII-5, Exp. 3
	$HC{\equiv}C\text{-}CH{=}CH\text{-}N(C_2H_5)_2$	104	IV, Exp. 20
	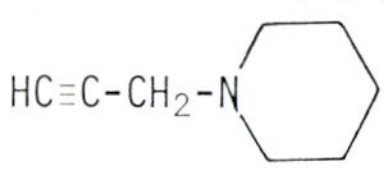	228	VIII-5, Exp. 1
	$H_2C{=}CH\text{-}C{\equiv}C\text{-}N(C_2H_5)_2$	134	V, Exp. 15
$C_8H_{14}O$	$HC{\equiv}C\text{-}CH(C_3H_7)CH_2CH_2OH$	53	III, Exp. 28
	$H_2C{=}C{=}C(C_4H_9)\text{-}OCH_3$	37	III, Exp. 10
	$C_2H_5C(CH_3){=}C{=}CH\text{-}OC_2H_5$	167	VI, Exp. 14
	$C_3H_7CH{=}C{=}CH\text{-}CH_2\text{-}CH_2OH$	208	VII-3, Exp. 4
$C_8H_{14}O_2$	$CH_3O\text{-}CH_2C{\equiv}C\text{-}C(CH_3)_2\text{-}OCH_3$	231	VIII-6, Exp. 1
	$C_2H_5O\text{-}CH(CH_3)\text{-}C{\equiv}C\text{-}CH_2OCH_3$	96	IV, Exp. 9
	$H_2C{=}C{=}C(OCH_3)\text{-}CH(CH_3)\text{-}OC_2H_5$	40	III, Exp. 14
	$C_2H_5O\text{-}CH(CH_3)\text{-}CH{=}C{=}CH\text{-}OCH_3$	96	IV, Exp. 9
	$C_2H_5O\text{-}CH{=}C{=}CH\text{-}CH_2\text{-}OC_2H_5$	95	IV, Exp. 8

$C_8H_{14}S$	$CH_3CH{=}C{=}C(C_2H_5){-}SC_2H_5$	46	III, Exp. 21
	$H_2C{=}C{=}C(\textit{i}.{-}C_3H_7){-}SC_2H_5$	46	III, Exp. 21
$C_8H_{18}Si_2$	$(CH_3)_3Si{-}C{\equiv}C{-}Si(CH_3)_3$	56	III, Exp. 32
C_9H_5N	$C_6H_5C{\equiv}C{-}C{\equiv}N$	227	VIII-4, Exp. 4
C_9H_7Br	$C_6H_5CH{=}C{=}CHBr$	183	VI, Exp. 33
C_9H_7Cl	$HC{\equiv}C{-}CH(Cl){-}C_6H_5$	177	VI, Exp. 27
	$C_6H_5CH{=}C{=}CHCl$	177	VI, Exp. 27
C_9H_7ClS	$C_6H_5C{\equiv}C{-}SCH_2Cl$	73	III, Exp. 52
C_9H_7I	$C_6H_5CH{=}C{=}CHI$	184	VI, Exp. 35
C_9H_8	$C_6H_5CH_2C{\equiv}CH$	186	VI, Exp. 38
	$C_6H_5CH{=}C{=}CH_2$	159	VI, Exp. 4
C_9H_8O	$C_6H_5O{-}CH{=}C{=}CH_2$	94	IV, Exp. 6
	$C_6H_5O{-}CH_2C{\equiv}CH$	94	IV, Exp. 6
	$C_6H_5CH(OH)C{\equiv}CH$	177,80	VI, Exp. 27, III, Exp. 59
C_9H_8OS	$H_2C{=}C{=}CH{-}S({=}O){-}C_6H_5$	197	VII-1, Exp. 2
C_9H_8S	$C_6H_5S{-}CH_2C{\equiv}CH$	100	IV, Exp. 14
	$C_6H_5S{-}C{\equiv}C{-}CH_3$	100	IV, Exp. 14
	$C_6H_5S{-}CH{=}C{=}CH_2$	110	IV, Exp. 25
C_9H_{12}	$HC{\equiv}C{-}CH_2C{\equiv}C{-}C_4H_9$	71	III, Exp. 50
$C_9H_{12}O$	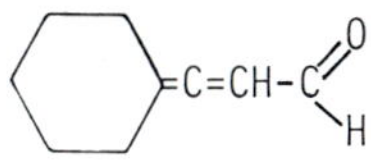	84	III, Exp. 63
C_9H_{14}	1,2-Cyclononadiene	139	V, Exp. 20
	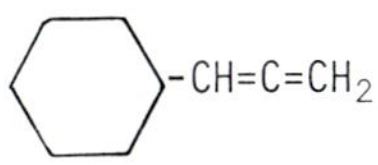	158	VI, Exp. 3
$C_9H_{14}O$	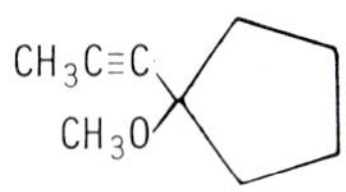	231	VIII-6, Exp. 1
	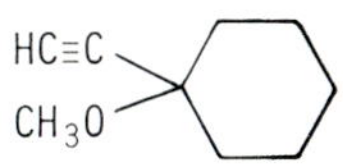	233	VIII-6, Exp. 3
	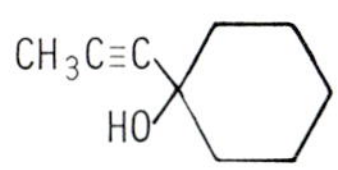	75	III, Exp. 54
	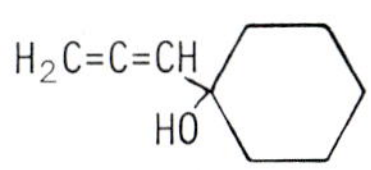	211	VII-3, Exp. 9
	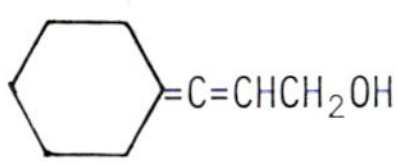	189	VI, Exp. 43

	$CH_3CH=C=C=CH-O-t.-C_4H_9$	128	V, Exp. 10
	$C_4H_9CH=C=CH-C(=O)-CH_3$	84	III, Exp. 63
$C_9H_{15}Br$	$HC\equiv C-CH(C_6H_{13})-Br$	217	VIII-2, Exp. 4
	$C_6H_{13}CH=C=CHBr$	182	VI, Exp. 32
C_9H_{16}	$H_2C=C=CH-C_6H_{13}$	28	III, Exp. 1
$C_9H_{16}O$	$C_4H_9CH=C=CH-OC_2H_5$	166	VI, Exp. 13
	$t.-C_4H_9CH=C=CH-OC_2H_5$	167	VI, Exp. 15
	$C_2H_5CH=C=CH(CH_2)_4OH$	172	VI, Exp. 22
	$t.-C_4H_9CH=C=CHCH_2CH_2OH$	36	III, Exp. 8
$C_9H_{16}OS$	$C_3H_7CH=C=C(SCH_3)-CH_2CH_2OH$	46	III, Exp. 21
$C_9H_{16}O_2$	$H_2C=C=C(OCH_3)-C(C_2H_5)_2OH$	43	III, Exp. 17
	$t.-C_4H_9O-CH_2C\equiv C-CH(CH_3)-OH$	128	V, Exp. 10
$C_9H_{17}N$	$H_2C=C=C(CH_3)-CH_2N(C_2H_5)_2$	169	VI, Exp. 18
$C_9H_{17}NO$	$CH_3O-CH_2C\equiv C-CH_2N(C_2H_5)_2$	134	V, Exp. 15
$C_{10}H_{10}$	$C_6H_5CH(CH_3)-C\equiv CH$	54	III, Exp. 29
	$C_6H_5CH_2CH_2C\equiv CH$	48,103	III, Exp. 22; IV, Exp. 19
	$C_6H_5C(CH_3)=C=CH_2$	164	VI, Exp. 10
	$C_6H_5CH_2CH=C=CH_2$	138	V, Exp. 19
$C_{10}H_{10}O$	$H_2C=C=CH-CH(C_6H_5)-OH$	188	VI, Exp. 42
$C_{10}H_{10}OS$	$CH_3CH=C=CH-S(=O)-C_6H_5$	197	VII-1, Exp. 2
$C_{10}H_{14}$	$H_2C=C(c.-hexyl)C\equiv CH$	186	VI, Exp. 40
	$H_2C=C=CH(CH_2)_4CH=C=CH_2$	28	III, Exp. 1
	1,2,3-Cyclodecatriene	140	V, Exp. 21
$C_{10}H_{14}O_2$	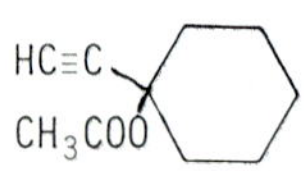	221	VIII-3, Exp. 1
$C_{10}H_{15}N$	$n.-C_6H_{13}CH=C=CH-C\equiv N$	175	VI, Exp. 25
$C_{10}H_{16}$	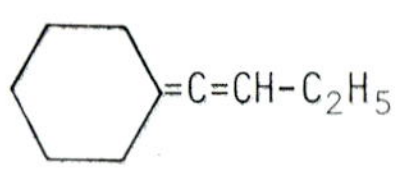	163	VI, Exp. 8 and 9
$C_{10}H_{16}O$	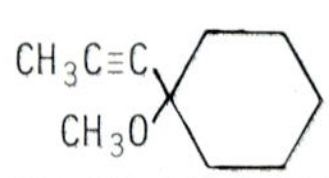	231	VIII-6, Exp. 1
$C_{10}H_{16}O_2$	$CH_3CH=CHC\equiv CCH(OC_2H_5)_2$	137	V, Exp. 18
	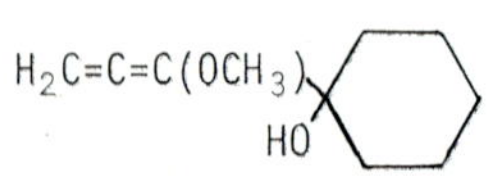	43	III, Exp. 17
	$CH_3CH=C=CH-CH_2COOt.-C_4H_9$	170	VI, Exp. 19
$C_{10}H_{17}NO$	$H_2C=C=CH-CH(CH_3)-C(=O)-N(C_2H_5)_2$	200	VII-2, Exp.

$C_{10}H_{18}O$	$H_2C{=}C{=}C(OCH_3){-}C_6H_{13}$	38	III, Exp. 11
	$C_2H_5C(CH_3){=}C{=}CH{-}O\textit{t}.{-}C_4H_9$	44	III, Exp. 19
$C_{10}H_{18}O_3$	$C_2H_5O{-}CH_2C{\equiv}C{-}CH(OC_2H_5)_2$	63	III, Exp. 40
	$C_2H_5O{-}CH{=}C{=}CH{-}CH(OC_2H_5)_2$	96	IV, Exp. 10
$C_{10}H_{18}S$	$CH_3CH{=}C{=}C(SC_2H_5){-}C_4H_9$	46	III, Exp. 21
$C_{10}H_{18}SSi_2$	$(CH_3)_3SiC{\equiv}C{-}S{-}C{\equiv}C{-}Si(CH_3)_3$	60	III, Exp. 37
$C_{10}H_{20}Si$	$HC{\equiv}C{-}CH(C_4H_9){-}Si(CH_3)_3$	52	III, Exp. 27
	$H_2C{=}C{=}C(C_4H_9){-}Si(CH_3)_3$	169	VI, Exp. 17
	$\textit{t}.{-}C_4H_9CH{=}C{=}CH{-}Si(CH_3)_3$	36	III, Exp. 9
$C_{11}H_{12}$	$C_6H_5{-}C(C_2H_5){=}C{=}CH_2$	161	VI, Exp. 6
$C_{11}H_{12}S$	$(CH_3)_2C{=}C{=}CH{-}S{-}C_6H_5$	195	VI, Exp. 50
$C_{11}H_{14}Si$	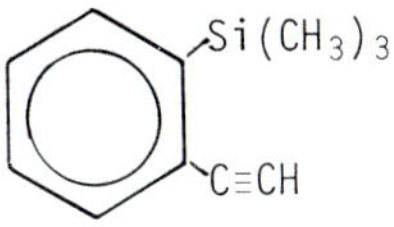	84	III, Exp. 64
$C_{11}H_{18}O_2$	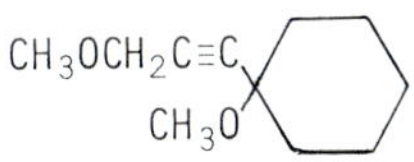	231	VIII-6, Exp. 1
$C_{11}H_{20}$	$(C_4H_9)_2CH{-}C{\equiv}CH$	186	VI, Exp. 39
	$(\textit{t}.{-}C_4H_9)_2C{=}C{=}CH_2$	162	VI, Exp. 7
$C_{11}H_{20}O_2$	$H_2C{=}C{=}CH(CH_2)_4OCH(CH_3){-}OC_2H_5$	30	III, Exp. 3
$C_{11}H_{20}OSi$	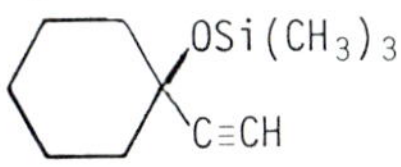	231	VIII-6, Exp. 4
$C_{11}H_{21}N$	$C_2H_5CH_2CH{=}C{=}CH{-}CH_2N(C_2H_5)_2$	209	VII-3, Exp. 5
$C_{12}H_{14}O$	$C_6H_5C(CH_3){=}C{=}C(CH_3){-}CH_2OH$	171	VI, Exp. 21
$C_{12}H_{15}NS$	$(C_2H_5)_2N{-}C{\equiv}C{-}S{-}C_6H_5$	121	V, Exp. 3
$C_{12}H_{16}Si$	$HC{\equiv}C{-}CH(C_6H_5){-}Si(CH_3)_3$	54	III, Exp. 29
$C_{12}H_{18}OS$	$(CH_3)_3C{-}C{\equiv}C{-}S({=}O){-}C{\equiv}C{-}C(CH_3)_3$	61	III, Exp. 38
$C_{12}H_{18}S$	$(CH_3)_3C{-}C{\equiv}C{-}S{-}C{\equiv}C{-}C(CH_3)_3$	60	III, Exp. 37
$C_{12}H_{22}O_2$	$\textit{t}.{-}C_4H_9O{-}CH{-}C{\equiv}C{-}CH{-}O\textit{t}.{-}C_4H_9$	232	VIII-6, Exp. 2
$C_{12}H_{24}OSi$	$(CH_3)_3Si{-}C(C_2H_5){=}C{=}CH{-}O\textit{t}.{-}C_4H_9$	44	III, Exp. 19
$C_{12}H_{24}O_2Si$	$(CH_3)_3SiO{-}CH(CH_3){-}C{\equiv}C{-}CH_2{-}O\textit{t}.{-}C_4H_9$	128	V, Exp. 10
$C_{14}H_{22}$	$H_2C{=}C{=}C(\textit{t}.{-}C_4H_9){-}C(\textit{t}.{-}C_4H_9){=}C{=}CH_2$	168	VI, Exp. 16
$C_{15}H_{12}$	$C_6H_5CH{=}C{=}CH{-}C_6H_5$	164	VI, Exp. 11
$C_{15}H_{13}OP$	$H_2C{=}C{=}CH{-}P({=}O)(C_6H_5)_2$	199	VII-1, Exp. 4
$C_{15}H_{26}N_2O_2$	$(C_2H_5)_2N{-}C({=}O){-}C(CH_3){=}C{=}C(CH_3){-}C({=}O){-}N(C_2H_5)_2$	210	VII-3, Exp. 8
$C_{15}H_{30}Sn$	$H_2C{=}C{=}CH{-}Sn(C_4H_9)_3$	32	III, Exp. 4
$C_{16}H_{32}OSn$	$H_2C{=}C{=}C(OCH_3)Sn(C_4H_9)_3$	41	III, Exp. 15

NMR parameters of the allene moiety*

R.H.A.M. Janssen, R.J.J. Ch. Lousberg and M.J.A. de Bie

For many years the synthesis and chemical properties of allenes has been a major area of interest in this laboratory. In our NMR research group we studied the ^{1}H and ^{13}C behaviour of these compounds.

The proton chemical shifts of the protons directly attached to the basic three carbon skeleton are found between 5.0 and 6.8 ppm. The $^{4}J(H,H)$ between these protons is about -6 Hz. The shift region is similar to the region for similarly substituted alkenes, although the spread in shifts is smaller and the allene proton resonances are slightly upfield from the alkene resonances. We could not establish a reliable additivity rule for the allene proton shifts as we could for the ^{13}C shifts (*vide infra*) and therefore we found the proton shifts much less valuable for the structural analysis of the allene moiety than the ^{13}C NMR data on the basic three-carbon system.

A remarkable feature of the allene ^{13}C NMR spectra is the lowfield position of the central carbon-atom, *viz.*, 185-210 ppm relative to TMS, whereas the exterior carbon-atom resonances are generally found between 60 and 130 ppm.
Over the past decade it has been established that for various substituents the ^{13}C chemical shift increment is a constitutive property. This applies to many systems e.g. benzenes, alkanes and alkenes. The availability of over 200 allenes, randomly substituted with groups of different nature, enabled us to prove that in the case of allenes the ^{13}C chemical shift increment is a constitutive property too, thus establishing a convenient method for estimating δ (^{13}C) values for allenes.

Using a multiple linear regression computer program, a set of substituent parameters was calculated for a number of the most commonly occurring groups. The calculated substituent effects allow a prediction of the chemical shifts of the exterior and central carbon atoms of the allene with standard deviations of 1.6 and 2.3 ppm, respectively Although most compounds were measured as neat liquids, for a number of compounds duplicate measurements were obtained in various solvents. The solvent effects are generally less than 1 ppm, which are well within the error bounds of the standard deviations of the calculated shifts. It is possible, however, that under extreme conditions larger deviations may be observed.

*The chapter is an abridged and adapted version of a manuscript originally published in Recl. Trav. Chim. Pays-Bas, 100 (1981), 85.
The reader is referred to this publication for an extended discussion of the material presented and literature references on the subject.

In describing the chemical shifts of the exterior and central allene carbons, the substituent effect symbols depicted in Fig. 1 were used.

Fig. 1. Substituent and steric contributions

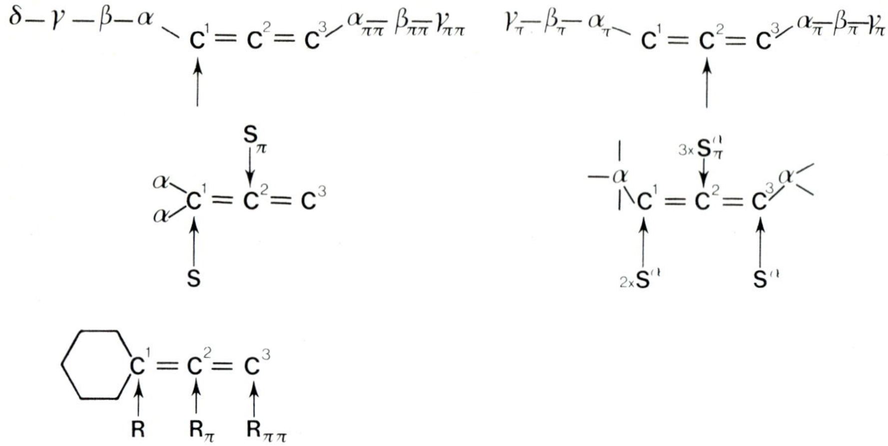

The arrows indicate the carbon to be calculated.

Exterior carbons of the allene

The chemical shift can be calculated according the following equation:

$$\delta C^1 = B + \sum_{i=\alpha}^{\delta} X_i + \sum_{i=\alpha}^{\gamma} X_{i\pi\pi} + S^{\alpha} + S + R + R_{\pi\pi} + C^{\gamma} \qquad (1)$$

where:

- B = a basic value;
- X_i = increment of substituents on C to be calculated;
- $X_{i\pi\pi}$ = increment of substituent on the other side of the allene;
- S^{α} = steric correction term;
- S = disubstitution correction term;
- $R;R_{\pi\pi}$ = ring correction terms;
- C^{γ} = γ-s-*trans*-correction term.

The steric correction term S^{α}, correcting for crowding at the β position in the case of N or C as α-substituent, is applied once for each non hydrogen β-substituent attached to the α-substituent above the first one. E.g., isopropyl and $N(CH_3)_2$ require the inclusion of S^{α} only once, whereas *tert.*-butyl groups require this correction twice. From the analysis S^{α} was found to be -4.2 ppm. The parameter S accounts for double substitution of the allene carbon to be calculated; the value of this correction term was computed to be -4.6 ppm. In compounds in which the C^1 of the allene is part of a five- or a six-membered ring, the descriptor S has to be included because of the 1,1-disubstitution of C^1. In the calculation of the shifts each ring carbon is counted only once, taking as its contribution the increment corresponding to the minimum number of bonds to the allene chromophore. In addition, one has to correct for the fact that the substituents are not freely rotating and are part of a ring. This is accounted for by the ring correction term R and $R_{\pi\pi}$ (-4.7 and -0.4 ppm, respectively, in the case of a six-membered ring and -6.1 and 0.8 ppm, respectively, in the case of a five-membered ring). For substituents not part of a ring which includes an allene carbon but in an otherwise fixed so-called γ-s-*trans* (antiperiplanar) orientation, the γ-s-*trans* correction of 2.5 ppm has to be applied.

Central carbon of the allene chromophore

The substituent effects on the chemical shift of the central carbon C^2 are given in Fig. 1 and are described by eq. 2:

$$\delta C^2 = B_{\pi} + \sum_{i=\alpha}^{\gamma} x_{i\pi} + S_{\pi} + S_{\pi}^{\alpha} + R_{\pi} \qquad (2)$$

where:

B_{π} = a basic value;

$x_{i\pi}$ = increment of substituent;

S_{π} = disubstitution correction term;

S_{π}^{α} = steric correction term;

R_{π} = ring correction term.

S_{π}^{α} and S_{π} are correction terms for steric crowding, similar to those for the exterior carbons. S_{π} is applied twice if the allene is 1,1,3,3-tetra substituted. Table I summarizes the compounds from which the substituent effects were derived and Table II lists the substituent effects.

It should be noted that if a complete group is given as a substituent then the increment also includes the $S_{(\pi)}^{\alpha}$ effect.

In Table III a comparison is made between calculated and observed chemical shift values of a variety of allenes that were not included in the data set of Table I, from which the substituent effects were calculated.

Note: for highly substituted allenes containing silicon, germanium or tin we observed deviations of the calculated chemical shifts from the measured values, and one should therefore treat these particular cases with special care.

Examples of calculation

A) 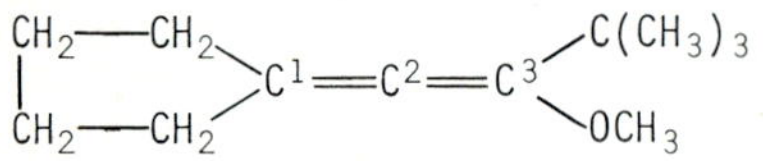

B) 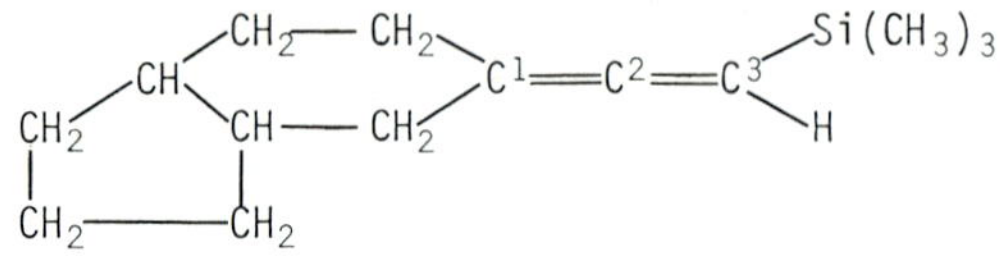

A)

	C^1			C^3			C^2	
B		74.5	B		74.5	B		214.1
2x	αC	23.2	1x	αC	11.6	3x	$\alpha_{\pi} C$	-13.2
2x	βC	15.2	1x	αO	40.3	1x	$\alpha_{\pi} O$	-12.5
1x	$\alpha_{\pi\pi} C$	-0.2	4x	βC	30.4	6x	$\beta_{\pi} C$	-9.0
1x	$\alpha_{\pi\pi} O$	14.9	2x	$\alpha_{\pi\pi} C$	-0.4			
4x	$\beta_{\pi\pi} C$	1.2	2x	$\beta_{\pi\pi} C$	0.6			
		128.8			157.0			179.4
corrections								
1x	R^5	-6.1	1x	$R^5_{\pi\pi}$	-0.4	1x	R^5_{π}	-1.7
1x	S	-4.6	1x	S	-4.6	1x	S_{π}	4.2
			2x	S^{α}	-8.4	2x	S^{α}_{π}	2.4
		118.1			143.6			184.3

B)

	C^1			C^3			C^2	
B		74.5	B		74.5	B		214.1
2x	αC	23.2	1x	αSiR_3	8.9	2x	$\alpha_{\pi} C$	-8.8
2x	βC	15.2	2x	$\alpha_{\pi\pi} C$	-0.4	1x	$\alpha_{\pi} SiR_3$	2.3
2x	γC	-3.4	2x	$\beta_{\pi\pi} C$	0.6	2x	$\beta_{\pi} C$	-3.0
2x	δC	1.8	2x	$\gamma_{\pi\pi} C$	-2.0	2x	$\gamma_{\pi} C$	1.4
1x	$\alpha_{\pi\pi} SiR_3$	-7.8						
		103.5			81.6			206.0
corrections								
1x	R^6	-4.7	1x	$R^6_{\pi\pi}$	-0.4	1x	R^6_{π}	-1.5
1x	S	-4.6				1x	S_{π}	2.1
1x	C^{γ}	2.5						
		96.7			81.2			206.6

Table Ia

Data on allenes used in substituent-effect computations[a]

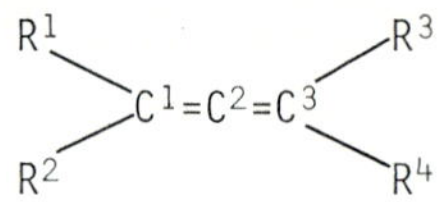

Molecular formula	R^1	R^2	R^3	R^4	δC^1_{obs}	δC^2_{obs}	δC^3_{obs}
C_3H_4[b]	H	H	H	H	74.8	213.5	74.8
C_4H_6[c]	CH_3	H	H	H	84.4	210.4	74.1
C_5H_8[c]	C_2H_5	H	H	H	91.7	208.9	75.3
C_5H_8[d]	CH_3	CH_3	H	H	93.4	207.3	72.1
C_5H_8[c]	CH_3	H	CH_3	H	85.4	207.1	85.4
C_6H_{10}[e]	C_2H_5	CH_3	H	H	100.1	206.3	74.3
C_7H_{12}[d]	CH_3	CH_3	CH_3	CH_3	92.6	200.2	92.6
C_7H_{12}[d]	C_2H_5	H	C_2H_5	H	93.2	202.3	93.2
C_7H_{12}[b]	*t.*-C_4H_9	H	H	H	97.7	202.4	72.1
C_8H_{14}[b]	C_5H_{11}	H	H	H	90.1	209.3	74.4
C_8H_{14}[b]	*t.*-C_4H_9	CH_3	H	H	106.8	205.8	74.3
C_9H_{14}[b]	*c.*-C_6H_{11}	H	H	H	96.1	208.0	75.4
C_9H_{16}[b]	*i.*-C_4H_9	H	CH_3	CH_3	87.3	202.7	93.8
$C_{10}H_{18}$[b]	*neo*-C_5H_{11}	H	C_2H_5	H	88.1	205.0	91.1
$C_{11}H_{20}$[b]	*t.*-C_4H_9	*t.*-C_4H_9	H	H	119.6	207.0	75.8
$C_{15}H_{28}$[b]	*t.*-C_4H_9	*t.*-C_4H_9	*t.*-C_4H_9	H	122.1	198.8	103.1
C_4H_6O[f]	OCH_3	H	H	H	123.1	202.0	90.3
C_4H_6O[b]	CH_2OH	H	H	H	91.4	208.0	75.8
C_5H_8O[f]	OC_2H_5	H	H	H	121.7	202.3	89.6
C_5H_8O[f]	OCH_3	H	CH_3	H	122.4	194.9	102.0
C_5H_8O[b]	CH_2OCH_3	H	H	H	88.3	209.8	75.2
C_5H_8O[b]	$CHOHCH_3$	H	H	H	96.1	207.3	76.7
$C_6H_{10}O$[f]	OC_3H_7	H	H	H	122.1	202.2	90.3

Table Ia.2
cont.

$C_6H_{10}O$[b]	$OCH(CH_3)_2$	H	H	H	120.4	202.7	87.7
$C_7H_{12}O$[b]	OC_4H_9	H	H	H	121.8	201.9	89.8
$C_7H_{12}O$[f]	O*t*.-C_4H_9	H	H	H	115.3	204.2	86.2
$C_7H_{12}O$[b]	$CHOHCH_3$	H	CH_3	CH_3	95.4	200.3	96.7
$C_8H_{14}O$[b]	OCH_3	C_4H_9	H	H	132.6	197.0	86.3
$C_8H_{14}O$[b]	OC_2H_5	H	CH_3	C_2H_5	121.1	189.1	116.6
$C_8H_{14}O$[b]	OC_2H_5	H	*i*.-C_3H_7	H	122.3	192.0	113.4
$C_9H_{16}O$[b]	OC_2H_5	H	C_4H_9	H	121.6	194.2	106.4
$C_9H_{16}O$[b]	C_4H_8OH	H	C_2H_5	H	92.8	204.0	91.6
$C_9H_{16}O$[b]	OC_2H_5	H	*t*.-C_4H_9	H	122.4	189.9	117.3
$C_{10}H_{18}O$[b]	O*t*.-C_4H_9	H	C_2H_5	CH_3	114.0	192.3	111.7
$C_{10}H_{18}O$[b]	CH_2OH	CH_3	*t*.-C_4H_9	CH_3	99.2	197.4	108.9
$C_{11}H_{18}O$[b]	OC_2H_5	H	*c*.-C_6H_{11}	H	121.8	192.2	111.7
$C_8H_{14}O_2$[b]	OCH_3	$CHCH_3OC_2H_5$	H	H	135.7	198.9	90.6
$C_8H_{14}O_2$[b]	OCH_3	H	$CHCH_3OC_2H_5$	H	124.1	193.9	110.9
C_4H_6S[d]	SCH_3	H	H	H	90.0	206.1	81.3
C_5H_8S[d]	SC_2H_5	H	H	H	88.5	206.9	80.9
C_5H_8S[d]	CH_2SCH_3	H	H	H	88.6	201.9	76.1
C_5H_8S[d]	SCH_3	CH_3	H	H	99.9	203.6	80.1
C_5H_8S[d]	SCH_3	H	CH_3	H	93.1	202.0	90.3
$C_6H_{10}S$[d]	SCH_3	C_2H_5	H	H	107.2	203.0	82.2
$C_6H_{10}S$[b]	SCH_3	H	C_2H_5	H	91.0	199.9	100.0
$C_6H_{10}S$[d]	SC_2H_5	CH_3	H	H	97.9	204.0	79.0
$C_6H_{10}S$[f]	$SCH(CH_3)_2$	H	H	H	86.9	206.7	79.4
$C_6H_{10}S$[b]	$CHCH_3SCH_3$	H	H	H	93.6	208.1	76.4
$C_7H_{12}S$[f]	SC_2H_5	H	C_2H_5	H	89.3	200.6	99.4
$C_7H_{12}S$[f]	S-*t*.-C_4H_9	H	H	H	83.6	210.6	76.8

Table Ia.3
cont.

$C_7H_{12}S$[d]	SCH_3	*i.*-C_3H_7	H	H	112.2	201.9	81.8
$C_8H_{14}S$[d]	SC_2H_5	*i.*-C_3H_7	H	H	110.0	201.9	80.8
$C_9H_{16}S$[b]	SCH_3	H	*i.*-C_4H_9	CH_3	88.0	198.5	106.3
C_6H_9S[f]	S-cyclo-C_3H_6	H	H	H	88.8	206.2	79.8
$C_5H_8S_2$[f]	SCH_3	H	SCH_3	H	99.0	195.5	99.0
C_5H_8OS[b]	OCH_3	H	SCH_3	H	128.9	187.5	105.6
C_5H_8OS[b]	OCH_3	SCH_3	H	H	133.0	193.6	90.7
C_9H_8[b]	Ø	H	H	H	94.4	210.0	78.8
$C_{10}H_{10}$[b]	Ø	CH_3	H	H	99.5	208.7	76.5
$C_{10}H_{10}$[b]	Ø*p*-CH_3	H	H	H	93.9	209.7	78.5
$C_{10}H_{10}$[b]	CH_2-Ø	H	H	H	89.8	209.2	75.0
$C_{11}H_{12}$[b]	Ø	C_2H_5	H	H	109.7	211.3	81.5
$C_{27}H_{20}$[d]	Ø	Ø	Ø	Ø	113.6	209.5	113.6
C_9H_8O[b]	O-Ø	H	H	H	117.9	203.1	88.9
$C_{12}H_{14}O$[b]	Ø	CH_3	CH_2OH	CH_3	102.1	201.0	101.0
$C_{15}H_{20}O$[b]	Ø	*t.*-C_4H_9	CH_2OH	CH_3	117.3	198.3	100.4
$C_{17}H_{16}O$[b]	Ø	Ø	CH_2OH	CH_3	111.5	201.5	103.4
C_9H_8S[b]	S-Ø	H	H	H	86.2	209.1	78.6
$C_{10}H_{10}S$[f]	Ø	H	SCH_3	H	101.8	200.6	94.8

[a] Chemical shift relative to TMS in ppm

[b] Synthesized in our laboratory

[c] R.A. Friedel and H.L. Retcofsky, J. Amer. Chem. Soc. *85*, (1963), 1300.

[d] R. Steur, J.P.C.M. van Dongen, M.J.A. de Bie, W. Drenth, J.W. de Haan and L.J.M. van de Ven, Tetrahedron Lett. *36*, (1971), 3307.

[e] ^{13}C Data bank, Vol. 1, Bruker, Karlsruhe, 1976.

[f] J.P.C.M. van Dongen, Ph.D. Thesis, State University Utrecht, 1974.

Table Ib

Data on allenes used in substituent-effect calculations[a]

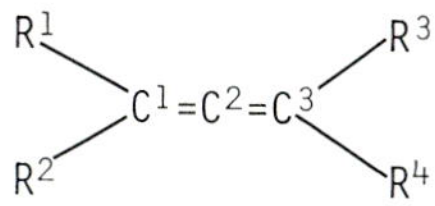

Molecular formula	R^1	R^2	R^3	R^4	δC^1_{obs}	δC^2_{obs}	δC^3_{obs}
C_5H_8O[b]	C_2H_4OH	H	H	H	86.8	209.5	74.8
$C_8H_{14}O$[b]	C_2H_4OH	H	C_3H_7	H	87.6	205.1	90.7
$C_9H_{16}O$[b]	C_2H_4OH	CH_3	C_3H_7	H	95.6	201.0	90.3
$C_{10}H_{18}O$[b]	C_2H_4OH	C_2H_5	C_3H_7	H	102.4	200.1	92.8
C_3H_3F[b]	F	H	H	H	129.8	200.2	93.9
C_3H_3Cl[b]	Cl	H	H	H	88.8	207.9	84.5
C_4H_5Cl[b]	Cl	H	CH_3	H	87.9	203.5	96.9
C_5H_7Cl[b]	Cl	H	CH_3	CH_3	86.0	199.9	107.5
C_6H_9Cl[e]	Cl	H	C_2H_5	CH_3	87.9	199.1	113.9
$C_8H_{13}Cl$[b]	Cl	*i.*-C_3H_7	C_2H_5	H	112.9	198.1	101.9
C_9H_7Cl[b]	Cl	H	Ø	H	91.2	203.3	103.7
C_3H_3Br[b]	Br	H	H	H	72.7	207.6	83.8
C_4H_5Br[b]	Br	H	CH_3	H	71.5	203.0	95.5
C_5H_7Br[b]	Br	H	CH_3	CH_3	70.0	199.9	106.6
$C_8H_{13}Br$[b]	Br	H	*i.*-C_4H_9	CH_3	70.6	199.6	109.9
$C_9H_{15}Br$[b]	Br	H	*neo*-C_5H_{11}	CH_3	70.6	201.0	108.5
C_3H_3I[b]	I	H	H	H	35.3	208.0	78.3
$C_7H_{11}I$[b]	I	H	*t.*-C_4H_9	H	36.1	203.4	106.9
$C_8H_{13}I$[b]	I	H	*i.*-C_4H_9	CH_3	35.0	203.4	105.6
$C_9H_{15}I$[b]	I	H	*neo*-C_5H_{11}	CH_3	35.2	204.6	104.2
$C_4H_4O_2$[b]	COOH	H	H	H	88.1	217.7	80.0
$C_5H_6O_2$[b]	$COOCH_3$	H	H	H	86.7	215.2	78.7

Table Ib.2 cont.

$C_5H_6O_2$[b]	COOH	CH_3	H	H	95.0	214.9	78.1
$C_7H_{10}O_2$[e]	$COOCH_3$	C_2H_5	H	H	101.9	213.8	79.3
$C_8H_{12}O_2$[e]	$COOCH_3$	C_2H_5	CH_3	H	101.8	210.5	90.2
$C_9H_{14}O_2$[b]	COOH	H	*i.*-C_4H_9	CH_3	86.1	212.0	103.3
$C_{10}H_{16}O_2$[b]	COOH	H	*neo*-C_5H_{11}	CH_3	85.8	213.2	101.8
$C_{10}H_{16}O_2$[b]	CH_2COO*t*C_4H_9	H	CH_3	H	86.3	206.0	84.6
C_4H_3N[b]	-C≡N	H	H	H	67.4	218.7	80.7
C_5H_5N[b]	-C≡N	H	CH_3	H	67.0	215.9	92.1
$C_9H_{13}N$[b]	-C≡N	H	*i.*-C_4H_9	CH_3	66.1	213.5	104.9
$C_{10}H_{15}N$[b]	-C≡N	H	C_6H_{13}	H	67.6	215.1	96.8
$C_{10}H_{15}N$[b]	-C≡N	H	C_5H_{11}	CH_3	66.8	213.1	106.3
$C_{10}H_{15}N$[b]	-C≡N	H	*neo*-C_5H_{11}	CH_3	65.8	214.5	103.3
C_6H_7N[b]	$-CH_2C≡N$	H	CH_3	H	82.2	205.8	89.7
C_4H_3N[b]	-N≡C	H	H	H	86.1	210.1	86.1
C_5H_6[b]	CH_2=CH	H	H	H	93.9	211.4	75.1
C_8H_{12}[b]	$CH_2=C(CH_3)$	C_2H_5	H	H	109.0	210.0	78.0
$C_{10}H_{16}$[b]	$CH_2=C(CH_3)$	H	C_4H_9	H	91.9	206.7	94.4
$C_{13}H_{22}$[b]	*t.*-C_4H_9CH=CH	H	*t.*-C_4H_9	H	96.4	204.2	103.8
$C_7H_{10}O$[b]	$CH_2=C(CH_3)$	H	OCH_3	H	116.4	196.7	125.2
$C_{11}H_{22}Si$[b]	$Si(CH_3)_3$	H	*i.*-C_4H_9	CH_3	85.0	210.3	93.6
$C_{12}H_{24}Si$[b]	$Si(CH_3)_3$	H	C_6H_{13}	H	82.4	210.8	83.5
$C_{12}H_{24}Si$[b]	$Si(CH_3)_3$	H	C_5H_{11}	CH_3	86.0	208.9	96.2
$C_{12}H_{24}Si$[b]	$Si(CH_3)_3$	H	*neo*-C_5H_{11}	CH_3	84.0	210.3	91.9

Table Ib.3
cont.

$C_7H_{14}OSi$[b]	$Si(CH_3)_3$	OCH_3	H	H	132.2	203.2	85.8
$C_{10}H_{20}OSi$[b]	$Si(CH_3)_3$	O*t*.C_4H_9	H	H	122.6	204.0	81.5
$C_{13}H_{28}OSi_2$[b]	$Si(CH_3)_3$	$Si(CH_3)_3$	O*t*.-C_4H_9	H	107.2	212.3	107.4
$C_{13}H_{30}OSi_3$[b]	$Si(CH_3)_3$	$Si(CH_3)_3$	$Si(CH_3)_3$	OCH_3	106.5	210.8	123.9
$C_8H_{12}Si$[b]	-C≡C$Si(CH_3)_3$	H	H	H	75.5	217.0	77.0
$C_{10}H_{16}Si_2$[b]	-C≡C$Si(CH_3)_3$	H	CH_3	CH_3	73.4	210.1	97.4
$C_{10}H_{18}OSi_3$[b]	-CO$Si(CH_3)_3$	H	C_2H_5	CH_3	101.9	212.2	105.1
C_5H_4[b]	-C≡CH	H	H	H	74.8	217.7	77.3
C_8H_{10}[b]	-C≡CH	C_2H_5	CH_3	H	88.9	208.3	88.9
C_8H_{10}[b]	-C≡CH	CH_3	CH_3	CH_3	81.6	207.5	96.8
C_9H_{12}[b]	-C≡CH	C_2H_5	CH_3	CH_3	88.1	206.5	98.1
$C_{11}H_{16}$[b]	-C≡C-*t*.-C_4H_9	H	CH_3	CH_3	73.5	209.3	96.4
$C_{12}H_{18}$[b]	-C≡CCH_3	*t*.-C_4H_9	CH_3	CH_3	97.2	204.1	98.4
$C_{10}H_{14}O$[b]	-C≡CCH_2OH	H	C_4H_9	H	75.5	212.6	93.3
$C_{11}H_8$[b]	-C≡CØ	H	H	H	75.6	216.6	77.5
$C_{13}H_{12}$[b]	-C≡CØ	H	CH_3	CH_3	73.5	210.2	97.5
$C_{15}H_{16}$[b]	-C≡CØ	H	*t*.-C_4H_9	H	77.3	209.9	104.7
$C_9H_{14}Si$[b]	-C≡C$Si(CH_3)_3$	H	CH_3	H	75.1	213.3	87.6
$C_{14}H_{24}Si$[b]	-C≡C$Si(CH_3)_3$	*t*.-C_4H_9	CH_3	CH_3	98.0	204.8	100.8
$C_{10}H_{20}Sn$[b]	$Sn(CH_3)_3$	H	*i*.-C_4H_9	H	76.8	208.9	79.5
$C_{11}H_{22}Sn$[b]	$Sn(CH_3)_3$	H	*neo*-C_5H_{11}	H	75.6	209.0	78.5
$C_{11}H_{22}Sn$[b]	$Sn(CH_3)_3$	H	*i*.-C_3H_9	CH_3	76.2	207.9	87.9
$C_{12}H_{24}Sn$[b]	$Sn(CH_3)_3$	H	C_5H_{11}	CH_3	76.9	207.2	89.2
$C_{12}H_{24}Sn$[b]	$Sn(CH_3)_3$	H	*neo*-C_5H_{11}	CH_3	75.5	209.3	86.6

Table Ib.4 cont.

$C_{11}H_{22}Ge$[b]	$Ge(CH_3)_3$	H	*i.*-C_4H_9	CH_3	81.3	206.8	90.8
$C_{12}H_{24}Ge$[b]	$Ge(CH_3)_3$	H	C_5H_{11}	CH_3	82.0	206.3	92.1
$C_{12}H_{24}Ge$[b]	$Ge(CH_3)_3$	H	*neo*-C_5H_{11}	CH_3	80.9	208.7	89.9
C_5H_9N[b]	$N(CH_3)_2$	H	H	H	113.1	204.2	85.5
$C_{11}H_{21}N$[b]	$-CH_2N(C_2H_5)_2$	H	C_3H_7	H	87.9	205.3	90.2
$C_{10}H_{14}$[b]	$C=C=C(CH_3)_2$	H	CH_3	CH_3	88.8	203.5	96.7
$C_{12}H_{18}$[b]	$(CH_3)C=C=C(CH_3)_2$	CH_3	CH_3	CH_3	98.5	201.5	96.3
$C_{14}H_{22}$[b]	(*t.*-C_4H_9)$C=C=CH_2$	*t.*-C_4H_9	H	H	109.3	206.0	79.3
$C_{16}H_{26}$[b]	(*t.*-C_4H_9)$C=C=CHCH_3$	*t.*-C_4H_9	H	CH_3	108.6	203.1	86.7
$C_{18}H_{30}$[b]	(*t.*-C_4H_9)$C=C=C(CH_3)_2$	*t.*-C_4H_9	CH_3	CH_3	107.4	200.8	94.8
C_5H_8Se[b]	SeC_2H_5	H	H	H	77.0	205.8	78.3
C_7H_{10}[b]	$(CH_2)_4$		H	H	102.2	202.7	75.0
C_8H_{12}[b]	$(CH_2)_5$		H	H	100.8	203.9	72.6
$C_{10}H_{16}$[b]	$(CH_2)_5$		H	C_2H_5	102.9	198.6	90.7
$C_9H_{14}O$[b]	$(CH_2)_5$		CH_2OH	H	103.3	198.3	90.1
$C_{10}H_{16}O$[b]	$(CH_2)_5$		$CH(OH)CH_3$	H	104.0	196.9	95.1
$C_9H_{13}Cl$[b]	$(CH_2)_5$		Cl	CH_3	111.3	193.6	98.2
C_6H_8O[b]	$-COC_2H_5$	H	H	H	95.3	215.7	78.5
C_6H_8O[b]	-CHO	H	CH_3	CH_3	96.7	216.9	101.0
$C_7H_{10}O$[b]	-CHO	H	CH_3	C_2H_5	97.9	215.9	106.8
$C_7H_{10}O$[b]	$C(CH_2)_2CHO$	H	H	H	93.3	207.7	78.5
$C_9H_{12}O$[b]	-CHO	H	$(CH_2)_5$		96.3	212.2	109.8

Table Ib.5 cont.

$C_{11}H_{10}O$[b]	-CHO	H	Ø	CH_3	100.0	220.3	106.5
$C_{10}H_8O$[b]	-COØ	H	H	H	92.9	216.7	78.8
$C_7H_{10}O_2$[b]	$-OCOH_3$	H	H	H	107.5	189.5	110.4

[a–e] See Table Ia

Table II

Substituent	Substituent effects in ppm									
	α	β	γ	δ	$\alpha_{\pi\pi}$	$\beta_{\pi\pi}$	$\gamma_{\pi\pi}$	α_{π}	β_{π}	γ_{π}
C	11.6	7.6	-1.7	0.9	-0.2	0.3	-1.0	-4.4	-1.5	0.7
O	40.3	5.2	-5.3	3.3	14.9	2.2	-0.5	-12.5	-0.8	1.2
S	9.1	4.2			6.1	2.9		-6.8	0.2	
Ph	19.4	3.0	1.4		2.9	-0.6	0.0	-1.3	1.1	0.0
N	27.6	4.3			10.7	0.4		-10.0	-0.6	
F	55.3				19.4			-13.9		
Cl	15.5				9.6			-7.1		
Br	-2.9				10.7			-6.9		
I	-39.2				3.0			-5.2		
$Si(CH_3)_3$	8.9	9.3	0.0		-7.8	-0.1	0.0	2.3	3.8	0.0
$Ge(CH_3)_3$	9.0				-6.8			0.0		
$Sn(CH_3)_3$	4.0				-10.0			-0.1		
Se	-3.4				4.5			-7.5		
Li	10.1				-24.8			-18.0		
C=C	19.8				0.6			-3.2		
C≡C	-0.3				3.0			2.6		
C=C=C	11.7				4.0			-3.5		
C≡N	-6.5	-3.7			6.3	3.8		6.1	0.5	
N≡C	11.6				11.6			-4.0		
COO...	13.8	-0.3			5.4	-1.3		3.6	0.7	
CO...(keto)	18.2				4.5			2.5		
CHO	22.5	0.4			6.3	3.6		9.9	-1.4	
OCO...	35.1				18.3			-17.9		

$B = 74.5$

$S = -4.6$

$S^{\alpha} = -4.2$

$C^{\gamma} = 2.5$

$B_{\pi} = 214.1$

$S_{\pi} = 2.1$

$S_{\pi}^{\alpha} = 1.2$

Ring corrections	R	$R_{\pi\pi}$	R_{π}
$(CH_2)_4C$=C=C	-6.1	0.8	-1.7
$(CH_2)_5C$=C=C	-4.7	-0.4	-1.5

Table III

Deviations of calculated shifts from observed shifts[a]

$\Delta\delta C = \delta C_{obs} - \delta C_{calc}$

Mol.form	R^1	R^2	R^3	R^4	δC^1_{obs}	$\Delta\delta C^1$	δC^2_{obs}	$\Delta\delta C^2$	δC^3_{obs}	$\Delta\delta C^3$
C_6H_{10}[b]	C_3H_7	H	H	H	89.9	-2.1	209.6	+0.7	74.4	-0.2
C_7H_{12}[b]	*i.*-C_4H_9	H	H	H	88.6	-1.7	209.9	+0.3	73.8	+1.2
C_8H_{14}[b]	*neo*-C_5H_{11}	H	H	H	86.6	-2.0	210.3	0.0	73.2	+1.6
C_9H_{14}[b]	H	$(CH_2)_6$		H	92.3	+0.1	206.3	-2.6	92.3	+0.1
C_9H_{16}[b]	*neo*-C_5H_{11}	H	CH_3	H	87.0	-1.4	206.4	+0.5	83.8	+0.6
$C_{10}H_{14}$[b]	*t.*-C_4H_9	H	$-C\equiv CCH_3$	H	104.1	+1.6	209.3	-1.6	77.1	+3.7
$C_{10}H_{18}$[b]	*neo*-C_5H_{11}	H	CH_3	CH_3	85.6	-2.6	203.9	+0.3	92.8	+2.6
$C_{11}H_{20}$[b]	*t.*-C_4H_9	H	*t.*-C_4H_9	H	101.7	+0.5	202.9	+1.8	101.7	+0.5
$C_{12}H_{22}$[b]	*neo*-C_5H_{11}	H	*t.*-C_4H_9	H	89.4	+0.1	203.0	-0.8	101.7	+4.1
$C_6H_{10}O$[b]	$-CH(CH_3)OCH_3$	H	H	H	93.1	+0.1	208.8	+0.7	75.5	-0.3
$C_6H_{10}O_2$[b]	OCH_3	H	CH_2OCH_3	H	124.0	+0.6	195.4	-0.2	105.3	+0.5
$C_8H_{14}O_2$[b]	OC_2H_5	H	$CH_2OC_2H_5$	H	122.4	+0.7	196.3	+0.1	104.7	+0.1
$C_8H_{14}O_2$[b]	OCH_3	H	$C(CH_3)_2OCH_3$	H	124.0	0.0	192.3	-2.7	113.0	+1.4
$C_9H_{16}O_2$[b]	OC_2H_5	H	$CHCH_3OC_2H_5$	H	121.4	-0.6	193.2	-0.2	108.6	+0.5

Table III.2
cont.

$C_{10}H_{18}O_2$[b]	$CH(OC_2H_5)_2$	H	CH_3	CH_3	89.8	-0.5	202.5	-1.5	96.8	+1.5
$C_{10}H_{18}O_2$[b]	OC_2H_5	H	$C(CH_3)_2OC_2H_5$	H	122.3	-0.1	195.6	+3.3	111.5	-1.2
$C_7H_{12}S$[b]	SCH_3	H	C_3H_7	H	89.2	-1.0	199.9	0.0	97.8	-0.6
$C_8H_{14}S$[b]	SCH_3	H	*i.*-C_4H_9	H	89.2	-0.1	201.0	-0.6	96.9	+0.2
$C_8H_{14}S$[b]	SCH_3	H	C_3H_7	CH_3	88.2	-2.1	197.5	-0.8	107.4	+2.0
$C_9H_{16}S$[b]	SCH_3	H	*neo*-C_5H_{11}	H	88.1	-0.2	201.4	-0.6	94.9	-0.1
$C_9H_{16}S$[b]	SCH_3	H	C_5H_{11}	H	89.3	-1.0	200.0	-0.6	98.3	-1.0
$C_{10}H_{18}S$[b]	SCH_3	H	C_5H_{11}	CH_3	88.3	-1.8	197.4	+1.2	107.4	+1.1
$C_{10}H_{18}S$[b]	SCH_3	H	*neo*-C_5H_{11}	CH_3	87.4	-0.7	199.8	+0.1	104.3	+2.3
$C_{11}H_{20}OS$[b]	-$CH(OC_2H_5)S$ *neo*-C_5H_{11}	H	H	H	91.7	+2.0	208.7	-3.0	77.5	-3.1
$C_8H_{13}Br$[b]	Br	H	*neo*-C_5H_{11}	H	71.1	+2.1	203.0	-0.4	97.8	-1.5
$C_9H_{15}Br$[b]	Br	H	C_5H_{11}	CH_3	71.3	+0.8	199.6	-0.1	110.9	+0.3
$C_9H_{13}N$[b]	-C≡N	H	C_4H_9	CH_3	66.7	-0.2	213.2	+0.5	106.1	-0.1
$C_{11}H_{17}N$[b]	-C≡N	H	*t.*-C_4H_9	C_3H_7	68.5	+0.7	212.0	+1.4	119.7	-0.4
$C_{13}H_{21}N$[b]	-C≡N	H	*t.*-C_4H_9	*neo*-C_5H_{11}	68.7	+2.9	213.0	+1.0	115.8	-0.5

Table III.3
cont.

$C_7H_{10}O_2$ [e]	$-COOCH_3$	CH_3	CH_3	H	95.0	+1.6	211.0	-0.7	88.7	-1.6
$C_9H_{14}O_2$ [e]	$-COOCH_3$	C_2H_5	C_2H_5	H	103.1	+2.1	209.6	+0.9	97.3	-0.9
$C_7H_8O_4$ [e]	$-COOCH_3$	H	$-COOCH_3$	H	92.2	+1.2	219.7	-3.0	92.2	+1.2
$C_{15}H_{20}O_8$ [e]	$-COOC_2H_5$	$-COOC_2H_5$	$-COOC_2H_5$	$-COOC_2H_5$	102.5	-2.2	227.4	-8.1	102.5	-2.2
$C_{10}H_8O_2$ [e]	-COOH	H	Ø	H	91.4	+0.2	216.3	-0.1	99.0	-0.3
$C_{13}H_{14}O_2$ [e]	$-COOCH_3$	H	Ø	C_2H_5	91.5	+1.9	213.7	+0.4	112.5	-0.4
					$\overline{\lvert\Delta\delta C^1\rvert}$ = 1.5		$\overline{\lvert\Delta\delta C^2\rvert}$ = 1.9		$\overline{\lvert\Delta\delta C^3\rvert}$ = 1.5	

REFERENCES

1. L. Brandsma, "Preparative Acetylenic Chemistry", Elsevier Publishing Company, Amsterdam, London, New York, (1971).
2. R.A. Raphael, "Acetylenic Compounds in Organic Synthesis", Butterworths Scientific Publications, London, (1955).
3a. H.G. Viehe, "Chemistry of Acetylenes", Marcel Dekker, New York, (1969).
3b. V. Jäger and H.G. Viehe, "Alkine, Di- und Polyine, Allene, Kumulene", in "Houben-Weyl, Methoden der Organischen Chemie Band 5/2a", Georg Thieme Verlag, Stuttgart, (1977).
4. T.F. Rutledge, "Acetylenes and Allenes", Reinhold Book Corporation, New York, Amsterdam, London, (1969).
5. W. Ziegenbein, "Einführung der Äthinyl- und Alkinyl-Gruppe in Organische Verbindungen", Verlag Chemie, Weinheim, (1963).
6. R. Baudouy, F. Delbecq and J. Goré, Tetrahedron Lett., (1979), 937.
7. J.C. Clinet and G. Linstrumelle, Tetrahedron Lett., (1978), 1137.
8. L. Brandsma, J. Meijer, A.J. de Jong, unpublished results.
9. Y. Leroux and R. Mantione, J. Organometal. Chem. 30, (1971), 295 and R. Mantione and Y. Leroux, J. Organometal. Chem. 31, (1971), 5 and references cited therein.
10a. D.R. Taylor, "The Chemistry of Allenes", in Chem. Revs. 67, (1967), 317.
10b. M. Murray, "Methoden zur Herstellung und Umwandlung von Allenen bzw. Kumulenen", in "Houben-Weyl, Methoden der Organischen Chemie, Band 5/2a", Georg Thieme Verlag, Stuttgart, (1977).
10c. J.-L. Moreau, "Organometallic Derivatives of Allenes and Ketenes", Chapter 10 in "The Chemistry of Ketenes, Allenes and Related Compounds", S. Patai (ed.), John Wiley and Sons, Chichester, New York, Brisbane, Toronto, (1980).
10d. W.D. Huntsman, "Rearrangements involving Allenes", Chapter 15 in "The Chemistry of Ketenes, Allenes and Related Compounds", S. Patai (ed.), John Wiley and Sons, Chichester, New York, Brisbane, Toronto, (1980).
10e. H. Hopf, "The Preparation of Allenes and Cumulenes", Chapter 20 in "The Chemistry of Ketenes, Allenes and Related Compounds", S. Patai (ed.), John Wiley and Sons, Chichester, New York, Brisbane, Toronto, (1980).
11. J.C. Craig and M. Moyle, J. Chem. Soc., (1963), 4402 and 5356.
12. S. Bhanu, E.A. Khan and F. Scheinmann, J. Chem. Soc. (Perkin I), (1976), 1609.
13. J. Klein and S. Brenner, J. Amer. Chem. Soc., 91, (1969), 3094.
14. W. Priester and R. West, J. Amer. Chem. Soc., 98, (1976), 8421 and 8426.
15. L. Brandsma and E. Mugge, Recl. Trav. Chim. Pays-Bas, 92, (1973), 628.
16. H. Hommes, H.D. Verkruijsse and L. Brandsma, Recl. Trav. Chim. Pays-Bas, 99, (1980), 113.
17. J.-C. Clinet and S. Julia, J. Chem. Research (S), (1978), 125; J. Chem. Research (M), (1978), 1714.
18. F. Jaffé, J. Organometal. Chem., 23, (1970), 53.
19. R. Baudouy, F. Delbecq and J. Goré, J. Organometal. Chem., 177, (1979), 39.
20. L.F. Fieser and M. Fieser, "Reagents for Organic Synthesis" (Vol. I, p. 911) John Wiley and Sons, New York, London, Sydney (1967).
21. R. Mantione, A. Alves, P.P. Montijn, G.A. Wildschut, H.J.T. Bos and L. Brandsma, Recl. Trav. Chim. Pays-Bas, 89, (1970), 97.
22. G. Balme, A. Douteau, J. Goré and M. Malacria, Synthesis, (1979), 508.
23. L. Brandsma, H.E. Wijers and J.F. Arens, Recl. Trav. Chim. Pays-Bas, 82, (1963), 1040.
24. G.R. Khan, K.A. Pover and F. Scheinmann, J. Chem. Soc. (Chem. Comm.), (1979), 215.
25. X. Creary, J. Amer. Chem. Soc., 99, (1977), 7632.
26. A. Schaap, L. Brandsma and J.F. Arens, Recl. Trav. Chim. Pays-Bas, 86, (1967), 393.
27. L. Brandsma, H.E. Wijers and C. Jonker, Recl. Trav. Chim. Pays-Bas, 83, (1964), 208.

28. G. Linstrumelle and D. Michelot, J. Chem. Soc. (Chem. Comm.), (1975), 561. J.C. Clinet, G. Linstrumelle, Nouv. J. Chimie, 1, (1977), 373.
29. S. Hoff, L. Brandsma and J.F. Arens, Recl. Trav. Chim. Pays-Bas, 87, (1968), 916.
30. S. Hoff, L. Brandsma and J.F. Arens, Recl. Trav. Chim. Pays-Bas, 87, (1968), 1179.
31. J.A. Rompes, S. Hoff, P.P. Montijn, L. Brandsma and J.F. Arens, Recl. Trav. Chim. Pays-Bas, 88, (1969), 1445.
32. S. Hoff and L. Brandsma, Recl. Trav. Chim. Pays-Bas, 88, (1969), 845.
33. S. Hoff, B.H. Steenstra, L. Brandsma and J.F. Arens, Recl. Trav. Chim. Pays-Bas, 88, (1969), 1284.
34. P.H.M. Schreurs, J. Meijer, P. Vermeer and L. Brandsma, Tetrahedron Lett., (1976), 2387.
35. H.A. Selling, J.A. Rompes, P.P. Montijn, S. Hoff, J.H. van Boom, L. Brandsma and J.F. Arens, Recl. Trav. Chim. Pays-Bas, 88, (1969), 119.
36. Y. Leroux and C. Roman, Tetrahedron Lett., (1973), 2585.
37. R. Mantione and Y. Leroux, Tetrahedron Lett., (1971), 593.
38. G.A. Wildschut, J.H. van Boom, L. Brandsma, J.F. Arens, Recl. Trav. Chim. Pays-Bas, 87, (1968), 1447.
39. E.J. Corey and P.L. Fuchs, Tetrahedron Lett., (1972), 3769.
40. F. Bernadou, D. Mesnard and L. Miginiac, J. Chem. Research (S), (1978), 106; J. Chem. Research (M), (1978), 1501.
41. W.L. Collier and R.S. Macomber, J. Org. Chem., 38, (1973), 1367.
42. J. Villieras, P. Perriot and J.F. Normant, Synthesis, (1979), 502.
43. D. Reisdorf and H. Normant, Organometal. Chem. Syn., 1, (1972), 393.
44. L. Brandsma and H.D. Verkruijsse, Synthesis, (1978), 290.
45. E.V. Dehmlow, M. Lissel, Justus Liebigs Ann. Chem., (1980), 1.
46. P.P. Montijn, J.H. van Boom, L. Brandsma and J.F. Arens, Recl. Trav. Chim. Pays-Bas, 86, (1967), 115.
47. W. Verboom, R.H. Everhardus, H.J.T. Bos, L. Brandsma, Recl. Trav. Chim. Pays-Bas, 98, (1979), 508.
48. R.H. Everhardus and L. Brandsma, Synthesis, (1978), 359.
49. J.B. Armitage, C.L. Cook, E.R.H. Jones and M.C. Whiting, J. Chem. Soc., (1952), 2010.
50. M. Makosza and M. Wawrzyniewicz, Tetrahedron Lett., (1969), 4659.
51. W. von E. Doering and P.M. LaFlamme, Tetrahedron, 2, (1958), 75;
W.R. Moore and H.R. Ward, J. Org. Chem., 25, (1960), 2073;
W.R. Moore and H.R. Ward, J. Org. Chem., 27, (1962), 4179;
W.R. Moore and H.R. Ward, J. Amer. Chem. Soc., 85, (1963), 86.
52. L. Skattebøl, Tetrahedron Lett., (1961), 167;
L. Skattebøl, Acta Chem. Scand., 17, (1963), 1683;
L. Skattebøl, J. Org. Chem., 31, (1966), 1554.
53. H.N. Cripps and E.F. Kiefer, Org. Synthesis, 42, (1962), 12.
54. P.P. Montijn, L. Brandsma and J.F. Arens, Recl. Trav. Chim. Pays-Bas, 86, (1967), 129; compare also H. Maurer and H. Hopf, Angew. Chem., 88 (1976), 687.
55. H. Kleijn, H. Westmijze, A. Schaap, H.J.T. Bos and P. Vermeer, Recl. Trav. Chim. Pays-Bas, 98, (1979), 209.
56. M. Le Quan and P. Cadiot, Bull. Soc. Chim. France, (1965), 45; J.-C. Masson, M. Le Quan and P. Cadiot, Bull. Soc. Chim. France, (1967), 777.
57. J.M. Oostveen, H. Westmijze and P. Vermeer, J. Org. Chem., 45, (1980), 1158.
58. D.R.M. Walton and F. Waugh, J. Organometal. Chem., 37, (1972), 45.
59. J. Meijer, unpublished.
60. W. Verboom, M.Schoufs, J. Meijer, H.D. Verkruijsse and L. Brandsma, Recl. Trav. Chim. Pays-Bas, 97, (1978), 244.
61. R.A. van der Welle and L. Brandsma, Recl. Trav. Chim. Pays-Bas, 92, (1973), 667.
62. A. Schaap, L. Brandsma and J.F. Arens, Recl. Trav. Chim. Pays-Bas, 84, (1965), 1200.

63. J.R. Nooi and J.F. Arens, Recl. Trav. Chim. Pays-Bas, 80, (1961), 244.
64. L.I. Zakharkin, V.V. Gavrilenko and B.A. Paley, J. Organomet. Chem., 21, (1970), 269.
65. J. Meijer and L. Brandsma, Recl. Trav. Chim. Pays-Bas, 90, (1971), 1098.
66. R.S. Sukhai, J. Meijer and L. Brandsma, Recl. Trav. Chim. Pays-Bas, 96, (1977), 179.
S.J. Harris and D.R.M. Walton, J. Chem. Soc. (Chem. Comm.), (1976), 1008.
67. J. Meijer, H.E. Wijers and L. Brandsma, Recl. Trav. Chim. Pays-Bas, 91, (1972), 1423.
68. H.J. Backer, Recl. Trav. Chim. Pays-Bas, 67, (1948), 894.
69. H.D. Verkruijsse and M. Hasselaar, Synthesis, (1979), 292; compare also D.A. Ben-Efraim, F. Sondheimer, Tetrahedron, 25, (1969), 2823.
H. Taniguchi, I.M. Mathai and S.I. Miller, Tetrahedron, 22, (1966), 867.
D. van der Steen, H.J.J. Pabon and D.A. van Dorp, Recl. Trav. Chim. Pays-Bas. 82, (1963), 1015.
70. J. Ficini , J. Besseyre, A. Krief, Bull Soc. Chim. France, (1976), 987.
71. L. Brandsma, Recl. Trav. Chim. Pays-Bas, 82, (1963), 1119.
72. M. Bertrand and C. Rouvier, Bull. Soc. Chim. France, (1968), 220.
73. I.B. Douglass and R.V. Norton, J. Org. Chem., 33, (1968), 2104.
74. G. Brauer, "Handbuch der Präparativen Anorganischen Chemie", Band I, 3. Aufl., p. 380, Ferdinand Enke Verlag, Stuttgart, (1975).
75. Ya.I. Ginzburg, J. Gen. Chem. (USSR), 10, (1940), 513; Chem. Abstr., 34, (1940), 7843.
G.F. Hennion and J.J. Sheenan, J. Amer. Chem. Soc., 71, (1949), 1964.
M. Bertrand, Bull Soc. Chim. France, (1956), 461.
E.R.H. Jones, H.H. Lee and M.C. Whiting, J. Chem. Soc., (1960), 341.
76. R. Mantione, Bull. Soc. Chim. France, (1969), 4523.
77. J. Grimaldi and M. Bertrand, Bull. Soc. Chim. France, (1971), 947.
78. W.J. Gensler and J. Casella, J. Amer. Chem. Soc., 80, (1958), 1376.
79. G. Pourcelot and P. Cadiot, Bull Soc. Chim. France, (1966), 3016.
80. P. E. van Rijn and R.H. Everhardus, to be published.
81. J.P.C.M. van Dongen, A.J. de Jong, H.A. Selling, P.P. Montijn, J.H. van Boom and L. Brandsma, Recl. Trav. Chim. Pays-Bas, 86, (1967), 1077.
82. M. Bertrand and J. Le Gras, Compt. Rend. (C), 264, (1967), 520.
83. Ch.A. Brown and A. Yamashita, J. Chem. Soc. (Chem. Comm.), (1976), 959; J.C. Lindhoudt, G.L. van Mourik and H.J.J. Pabon, Tetrahedron Lett., (1976), 2565;
H. Hommes and L. Brandsma, Recl. Trav. Chim. Pays-Bas, 96, (1977), 160;
S.R. Macaulay, J. Org. Chem., 45, (1980), 734.
84. J.F. Normant, A. Alexakis and J. Villieras, J. Organometal. Chem., 57, (1973), C99.
A. Alexakis, A. Commerçon, J. Villieras and J.F. Normant, Tetrahedron Lett., (1978), 3935.
85. P. Vermeer, H. Westmijze, H. Kleijn and L.A. van Dijck, Recl. Trav. Chim. Pays-Bas, 97, (1978), 56.
86. G. Tadema, R.H. Everhardus, H. Westmijze and P. Vermeer, Tetrahedron Lett., (1978), 3935.
87. L.A. van Dijck, B.J. Lankwerden, J.G.C.M. Vermeer and A.J.M. Weber, Recl. Trav. Chim. Pays-Bas, 90, (1971), 801.
88. W.T. Borden and E.J. Corey, Tetrahedron Lett., (1969), 313.
89. A. Claesson, L.-I. Olsson, G.R. Sullivan and H.S. Mosher, J. Amer. Chem. Soc., 97, (1975), 2919.
90. L.-I. Olsson and A. Claesson, Acta Chem. Scand., B 31, (1977), 614.
91. J.L. Luche, E. Barreiro, J.M. Dollat and P. Crabbé, Tetrahedron Lett., (1975), 4615, see also J. Chem. Soc. (Chem.Comm.), (1977), 761.
92. D.J. Pasto, R.H. Shults, J.A. McGrath and A. Waterhouse, J. Org. Chem., 43, (1978), 1382.
93. L.-I. Olsson, Dissertation, Uppsala (1979).
94. F. Serratosa, Tetrahedron Lett., (1964), 895.

95. L. Brandsma and J.F. Arens, Recl. Trav. Chim. Pays-Bas, 86, (1967), 734.
96. Y. Pasternak and M. Délépine, Compt. Rend. (C), (1962), 3429.
97. T.L. Jacobs and R.A. Meijers, J. Amer. Chem. Soc., 86, (1966), 5244.
98. P. Vermeer, J. Meijer and L. Brandsma, Recl. Trav. Chim. Pays-Bas, 94, (1975), 112.
99. J.-L. Moreau and M. Gaudemar, J. Organometal. Chem., 108, (1976), 159.
100. G. Tadema, P. Vermeer, J. Meijer and L. Brandsma, Recl. Trav. Chim. Pays-Bas, 95, (1976), 66.
101. P. Vermeer, J. Meijer, C. de Graaf and H. Schreurs, Recl. Trav. Chim. Pays-Bas, 93, (1974), 46.
P.R. Ortiz de Montellano, J. Chem. Soc. (Chem. Comm.), (1973), 709.
102. A. Claesson and Ch. Sahlberg, Tetrahedron Lett., (1978), 1319.
103. J. Meijer and P. Vermeer, Recl. Trav. Chim. Pays-Bas, 93, (1974), 183.
104. H. Kleijn, J. Meijer, H. Westmijze and P. Vermeer, Recl. Trav. Chim. Pays-Bas, 96, (1977), 251.
105. P. Rona and P. Crabbé, J. Amer. Chem. Soc., 91, (1969), 3289.
106. L.-I. Olsson and A. Claesson, Acta Chem. Scand., B 33, (1979), 679.
W.H. Pirkle and C.W. Boeder, J. Org. Chem. 43, (1978), 1950.
107. A. Claesson and L.-I. Olsson, J. Chem. Soc. (Chem. Comm.), (1979), 524.
108. H. Westmijze and P. Vermeer, Synthesis, (1979), 390.
109. M. Montury, B. Psaume and J. Goré, Tetrahedron Lett., 21, (1980), 163.
110. H. Kleijn, C.J. Elsevier, H. Westmijze, J. Meijer and P. Vermeer, Tetrahedron Lett., (1979), 3101.
111. G. Balme, M. Malacria and J. Goré, Tetrahedron Lett., (1979), 7.
112. R.A. Amosand and J.A. Katzenellenbogen, J. Org. Chem., 43, (1978), 555.
113. Th.L. Jacobs and W.F. Brill, J. Amer. Chem. Soc., 75, (1953), 1314.
114. O.J. Muscio, Y.M. Jun and J.B. Philip, Tetrahedron Lett., (1978), 2379.
115. P.M. Greaves, M. Kalli, Ph.D. Landor and S.R. Landor, J. Chem. Soc. (C), (1971), 667 and references, cited therein; S.R. Landor, A.N. Patel, P.F. Whiter and P.M. Greaves, J. Chem. Soc. (C), (1966), 1223.
116. C.S.L. Baker, Ph.D. Landor, S.R. Landor and A.N. Patel, J. Chem. Soc., (1965), 4348.
117. P. Kurtz, H. Gold and H. Disselnkötter, Justus Liebigs Ann. Chem., 624, (1959), 1.
118. P.M. Greaves, S.R. Landor and D.R.J. Laws, J. Chem. Soc. (Chem. Comm.), (1965), 321.
Y. Pasternak and G. Pfeiffer, Compt. Cend. (C), 259, (1964), 1142.
119. W.J. Bailey and C.R. Pfeifer, J. Org. Chem., 20, (1955), 1337.
Ph.D. Landor, S.R. Landor and E.S. Pepper, J. Chem. Soc. (C), (1967), 185.
120. J.S. Cowie, Ph.D. Landor and S.R. Landor, J. Chem. Soc. (Chem. Comm.), (1969), 541.
A. Claesson, L.-I. Olsson and C. Bogentoft, Acta Chem. Scand., 27, (1973), 2941.
121. R. Ya Levina and E.A. Victorova, Uspekhi. Khim., 27, (1958), 162.
122. J.P. Dulcêre, M.L. Roumestant and J. Goré, Tetrahedron Lett., (1972), 4465.
123. J.P. Dulcêre, J. Goré and M.L. Roumestant, Bull. Soc. Chim. France, (1974), 1119.
124. M. Sipenhou Simo, A. Jean and M. Le Quan, J. Organometal. Chem., 35, (1972), C23.
125. H.G. Kuivila and J.C. Cochran, J. Amer. Chem. Soc., 89, (1967), 7152.
126. J. Oostveen and P. Vermeer, to be published.
127. T. Nakai, K. Tanaka, H. Seto and N. Ishikawa, Bull. Soc. Chim. Japan, 50, (1977), 3069.
128. Inorganic Synthesis, Vol. II, p. 3, (1946).
129. G. Smith and C.J.M. Stirling, J. Chem. Soc. (C), (1971), 1530.
130. L. Horner and V. Binder, Justus Liebigs Ann. Chem., 757, (1972), 33.
131. S. Braverman and H. Mechoulam, Tetrahedron, 30, (1974), 3883.
132. A. Sevin and W. Chodkiewicz, Tetrahedron Lett., (1967), 2975.

133. J. Ficini, N. Lumbroso-Bader and J. Pouliquen, Tetrahedron Lett., (1968), 4139.
134. G. Saucy and R. Marbet, Helv.Chim. Acta, 50, (1967), 1158.
135. P.J.W. Schuijl, H.J.T. Bos and L. Brandsma, Recl. Trav. Chim. Pays-Bas, 88, (1969), 597.
136. J. Meijer, P. Vermeer and L. Brandsma, Recl. Trav. Chim. Pays-Bas, 92, (1973), 601.
137. J.K. Crandall and G.L. Tindell, J. Chem. Soc. (Chem. Comm.), (1970), 1411.
138. L. Brandsma, H.J.T. Bos and J.F. Arens in "Chemistry of Acetylenes", H.G. Viehe (ed.), Marcel Dekker, New York, (1969); J. Ficini, Bull. Soc. Chim. France, (1954), 1367.
139. S.I. Radchenko, L.N. Cherkasov, B.S. Kupin and A.N. Krivosheya, J. Org. Chem. USSR, 7, (1971), 104; Zh. Organ. Khim., 7, (1971), 104.
L.N. Cherkasov, V.A. Kormer and Kh.V. Bal'yan, J. Gen. Chem. USSR, 35, (1965), 618; Zh. Obshchei Khim., 35, (1965), 616.
S.I. Radchenko and A.A. Petrov, J. Org. Chem. USSR, 1, (1965), 1000; Zh. Org. Khim., 1, (1965), 993; C.A., 63, (1965), 11340f.
140. E.B. Bates, E.R.H. Jones and M.C. Whiting, J. Chem. Soc., (1954), 1854.
141. J. Ficini and J. Pouliquen, J. Amer. Chem. Soc., 93, (1971), 3295.
142. P. Crabbé, D. André and H. Fillion, Tetrahedron Lett., (1979), 893.
143. G.F. Hennion and A.P. Boisselle, J. Org. Chem., 26, (1961), 725.
144. G. Saucy, R. Marbet, H. Lindlar and O. Isler, Helv. Chim. Acta, 42, (1959), 1945.
145. J. Meijer, P. Vermeer, H.J.T. Bos and L. Brandsma, Recl. Trav. Chim. Pays-Bas, 92, (1973), 1067.
146. G.F. Hennion and E.G. Teach, J. Amer. Chem. Soc., 75, (1953), 1653.
147. R.G. Visser, H.J.T. Bos and L. Brandsma, Recl. Trav. Chim. Pays-Bas, 99, (1980), 70.
148. J. le Gras, Compt. Rend. Acad. Sc. Paris (C), 263, (1966), 1460.
149. R.G. Visser, P.E. van Rijn, H.D. Verkruijsse, S. Mommers and L. Brandsma, to be published.
150. D.J. Nelson and W.J. Miller, J. Chem. Soc. (Chem. Comm.), (1973), 444.
151. N.J. Leonard and C.R. Johnson, J. Org. Chem., 27, (1962), 282.
152. V.N. Ipatieff, H. Pines and B.S. Friedman, J. Amer. Chem. Soc., 60, (1938), 2731; W.E. Truce, H.E. Hill and M.M. Boudakian, J. Amer. Chem. Soc., 78, (1956), 2760.
153. W. Verboom, R.H. Everhardus, J.W. Zwikker and L. Brandsma, Recl. Trav. Chim. Pays-Bas, 99, (1980).
154. J.C. Martin, U.S. 3, 131, 234, C.A., 61, (1964), 2969f.
155. D.W. In der Mauer, R.H. Everhardus, H. Hommes, W. Verboom, H.D. Verkruijsse and L. Brandsma, to be published.
156. R. Mantione and B. Kirschläger, Compt. Rend. Acad. Sci. Paris (C), 272, (1971), 786.
157. J.-C. Clinet and S. Julia, J. Chem. Research (S), (1978), 125; (M), (1978), 1714.
158. D.G. Oelberg and M.D. Schiavelli, J. Org. Chem., 42, (1977), 1804 and references cited therein.
159. D.K. Black and S.R. Landor, J. Chem. Soc., (1965), 6784.
160. A.I. Vogel, "Vogel's Textbook of Practical Organic Chemistry, including Qualitative Organic Analysis", 4th Edition, p. 326, Longman, London and New York, (1978).
161. L.J. Haynes, I.M. Heilbron, E.R.H. Jones and F. Sondheimer, J. Chem. Soc., (1947), 1583.
162. F. Bohlmann and H.G. Viehe, Chem. Ber., 87, (1954), 712.
163. W. Reppe, Justus Liebigs Ann. Chem., 596, (1955), 70.
164. J.-C. Clinet and G. Linstrumelle, Nouveau Journal de Chimie, 1, (1977), 373.
165. J.C. Hamlet, H.B. Henbest and E.R.H. Jones, J. Chem. Soc., (1951), 2652.
166. H.D. Verkruijsse, L.J. de Noten, H.J.T. Bos and L. Brandsma, Recl. Trav. Chim., (1981).

167. M.F. Shostakowskii and A.V. Bogdanova, "The Chemistry of Diacetylenes", John Wiley, New York, Toronto, 1974, p. 11.
168. J.P. Ward and D.A. van Dorp, Recl. Trav. Chim., 85, (1966), 117.
169. E.R.H. Jones, G. Eglington, M.C. Whiting and B.L. Shaw, Organic Synthesis, 34, (1954), 46.
170. F.Ya. Perveev, Zhur. Obshchei Khim., 19, (1949), 1304; C.A., 44, (1950), 1008a.
171. I.M. Heilbron, E.R.H. Jones, R.N. Racey, J.T. McCombie and R. Raphael, J. Chem. Soc., (1945), 77.
172. A. Sevin, W. Chodkiewicz and P. Cadiot, Tetrahedron Lett., (1965), 1953.
173. W. Reppe, Justus Liebigs Ann. Chem., 596, (1955), 78.
174. J.B. Armitage and M.C. Whiting, J. Chem. Soc., (1952), 2005.
175. J.H. Biel and F. DiPierro, J. Amer. Chem. Soc., 80, (1958), 4609.
176. C. Prevost, M. Gaudemar, L. Miginiac, F. Bardone-Gaudemar and M. Andrac, Bull. Soc. Chim., France, (1959), 679.
177. W. Reppe, Justus Liebigs Ann. Chem., 596, (1955), 74.
178. G.F. Hennion and F.P. Kupiecki, J. Org. Chem., 18, (1953), 1601.
179. B. Wojtkowiak, Ann. Chim., 9, (1964), 5.
180. A.J. Bridges, Tetrahedron Lett., (1980), 4401.